Handbook of Biosensors

Handbook of Biosensors

Edited by **Marvin Heather**

CLANRYE
INTERNATIONAL

New Jersey

Published by Clanrye International,
55 Van Reypen Street,
Jersey City, NJ 07306, USA
www.clanryeinternational.com

Handbook of Biosensors
Edited by Marvin Heather

© 2015 Clanrye International

International Standard Book Number: 978-1-63240-256-1 (Hardback)

Contents

Preface

This book has been an outcome of determined endeavour from a group of educationists in the field. The primary objective was to involve a broad spectrum of professionals from diverse cultural background involved in the field for developing new researches. The book not only targets students but also scholars pursuing higher research for further enhancement of the theoretical and practical applications of the subject.

Advanced information regarding biosensors has been elucidated in this comprehensive book. A rapidly advancing world needs effective methods and theories to enable designers to deal with the new product development landscape successfully and make a difference in an excessively interconnected world. Frontiers of further practice are being constantly pushed ahead as result of designers continuously extending the boundaries of their discipline and developing new ways in interdisciplinary domains. This book further develops the existing concepts and diversifies into new areas of practice and theory in industrial design. The aim of this book is to help readers formulate their own new design research which is relevant and matches the present challenges of this fascinating field.

It was an honour to edit such a profound book and also a challenging task to compile and examine all the relevant data for accuracy and originality. I wish to acknowledge the efforts of the contributors for submitting such brilliant and diverse chapters in the field and for endlessly working for the completion of the book. Last, but not the least; I thank my family for being a constant source of support in all my research endeavours.

Editor

Biorecognition Techniques

GFP-Based Biosensors

Donna E. Crone, Yao-Ming Huang, Derek J. Pitman,
Christian Schenkelberg, Keith Fraser,
Stephen Macari and Christopher Bystroff

Additional information is available at the end of the chapter

1. Introduction

Green fluorescent protein (GFP) is a 27 kD protein consisting of 238 amino acid residues [1]. GFP was first identified in the aquatic jellyfish *Aequorea victoria* by Osamu Shimomura *et al.* in 1961 while studying aequorin, a Ca^{2+}-activated photoprotein.Aequorin and GFP are localized in the light organs of *A. victoria* and GFP was accidentally discovered when the energy of the blue light emitted by aequorin excited GFP to emit green light.Unlike most fluorescent proteins which contain chromophores distinct from the amino acid sequence of the protein, the chromophore of GFP is internally generated by a reaction involving three amino acid residues [2]. This unique property allows GFP to be easily cloned into numerous biological systems, both prokaryotic and eukaryotic, which has paved the way for its utilisation in a variety of biological applications, most notably in biosensing.

1.1. The three dimensional structure

The molecular structure of GFP was first determined in 1996 using X-ray crystallography [1].One of the most obvious features of its tertiary structure is a beta-barrel composed of 11 mostly-antiparallel beta strands. The molecular structure of GFP is illustrated in Figure 1 along with a cartoon representation showing the organization of the secondary structure elements that compose the beta barrel.Each beta strand is 9 to 13 residues in length and hydrogen bonds with adjacent beta strands to create an enclosed structure.The bottom of the barrel contains both termini and two distorted helical crossover segments, and the top has one short crossover and one distorted helical crossover segment.The beta-barrel (sometimes referred to as a "beta can" because it contains a central alpha-helical segment) consists of three anti-parallel three-stranded beta-meander units and a two-stranded beta-hairpin (shown in blue, green, and yellow, and red in Figure 1 respectively).The very distorted central alpha helix contains three residues which participate in an auto-catalyzed cyclization/ oxidation chromophore maturation reaction which generates the p-hydroxybenzylidene-imidazolidone chromophore.In the unfolded state, the chromophore is non-fluorescent, presumably because water molecules and molecular oxygen can interact with and quench the fluorescent signal [3].Therefore, the closed beta barrel structure is essential for fluorescence by shielding the chromophore from bulk solvent.

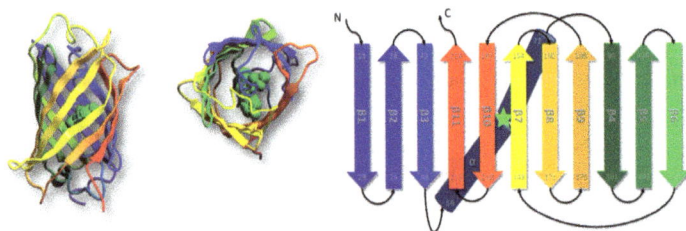

Figure 1. Tertiary structure of GFP as determined by x-ray crystallography (PDB code 2B3P).Shown on the bottom is a cartoon depicting the secondary structure elements, all anti-parallel beta strand pairings except β1 to β6, which is parallel. Numbers indicate the start and end of each secondary structure element.

The interior of the GFP beta barrel is unusually polar.There is an interior cavity filled with four water molecules on one side of the central helix, while the other side contains a cluster of hydrophobic side chains which is more typical of a protein core.Several polar side chains interact with and stabilize the GFP chromophore.Three of these, His148, Thr203, and Ser205, form hydrogen bonds with the phenolic hydroxyl group of the chromophore.Arg96 and Gln94 interact with the carbonyl group of the imidazolidone ring. Figure 2 depicts these stabilizing hydrogen bonding interactions with the chromophore.Additionally, a number of internal residues interact with and stabilize Arg96, a side chain that is known to be required for the maturation of the chromophore.Specifically, Thr62 and Gln183 form hydrogen bonds with the protonated form of Arg96 stabilizing a buried positive charge within the GFP beta barrel, which in turn stabilizes a partial negative charge on the carbonyl oxygen of the imidazolidone ring.

1.2. Thermodynamic and kinetic properties

Wild type GFP has a number of interesting characteristics that can potentially complicate its applicability to biosensing.One is its tendency to aggregate in the cell, especially when expressed in high concentrations. Aggregation is typically caused by exposed hydrophobicity, which may be due to either the presence of hydrophobic patches on the surface of the protein, or to low thermostability, or to slow folding.Surface hydrophobic-to-hydrophilic mutations decrease the aggregation tendency of GFP [4], but some biosensing applications require surface mutations that may increase aggregation. Most likely, GFP's low *in vivo* solubility is due to its extremely slow folding and unfolding kinetics.Refolding of GFP consists of at least two observable phases, depending on the variant and the method being used to measure the kinetics.Multi-phase folding kinetics indicates the existence of multiple parallel folding pathways, some fast and some slow, holding out the hope that engineered GFPs could be made to fold faster by favoring the faster folding pathway. Indeed, GFP has been engineered to eliminate the slowest phase of folding, as discussed later in this chapter. For Cycle3, a mutant whose chromophore matures correctly at 37°C, the kinetic phases range from 10 s^{-1} to 10^{-2} s^{-1} [5] (half-lives of folding ranging from 0.1 s to 100 s). Although it folds slowly, GFP unfolds *extremely* slowly, with a rate of 10^{-6} s^{-1} ($t_{1/2}$=8 days) in 3.0M GndHCl [6], such that when extrapolated to 0M GndHCl, the theoretical unfolding half-life in on the order of $t_{1/2}$= 22 years.GFP is phenomenally kinetically stable once it is folded to its native state.

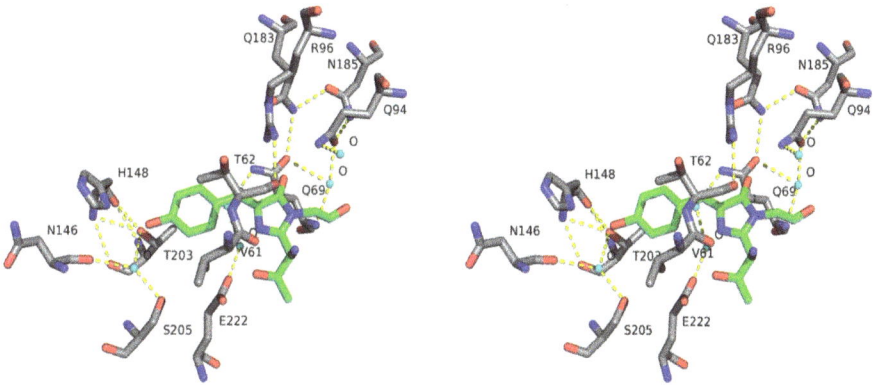

Figure 2. Stereo image of the hydrogen bonding patterns of the internal GFP residues with the chromophore (green), including four crystallographic waters (cyan). Drawn from superfolder GFP, PDB ID 2B3P.

1.3. Maturation of the chromophore

The chromophore of the native GFP structure is generated by an internal, autocatalytic reaction involving three residues on the interior alpha helix.Cyclization and oxidation of internal residues of Ser65, Tyr66, and Gly67, generate a *p*-hydroxybenzylidene-imidazolidone chromophore that maximally absorbs light at 395 nm and 475 nm [1].Excitation at either absorption peak results in emission of green light at 508 nm.Interestingly, the sidechains of the chromophore triplet65-SerTyrGly-67 can be mutated to other sidechains without loss of function. Tyrosine 66 can be mutated to any aromatic sidechain [7].This allows for the synthesis of numerous variants of GFP that alter the chromophore structure or its surrounding environment to absorb and emit light at different wavelengths, producing a wide array of fluorescent protein colors [8].

The three-step mechanism for the spontaneous generation of the chromophore consists of cyclization, oxidation, and dehydration [9]. Figure 3 illustrates the mechanism, beginning with the original triplet of amino acids. The slow step in chromophore maturation is the diffusion of molecular oxygen into the active site of the closed beta barrel (step 3). The positioning of side chains surrounding the chromophore is crucial for stabilizing the intermediates in the process of chromophore maturation,especially Arg96, which stabilizes the enolate form of intermediate 1 by forming a salt bridge with the negatively-charged oxygen atom, and Glu222, which receives protons from the water molecules to cycle between the protonated and deprotonated states.The two coplanar aromatic rings of the chromophore adopt the *cis* conformation across the Tyr66 alpha-beta carbon double bond.Photobleaching, the light-induced loss of fluorescence, is caused by short wavelength light that causes the chromophore to isomerize to the *trans* form, accompanied by distortion of its planar geometry and surrounding side chain packing [10].This type of photobleaching appears to be a slowly reversible process for GFP and other fluorescent proteins.

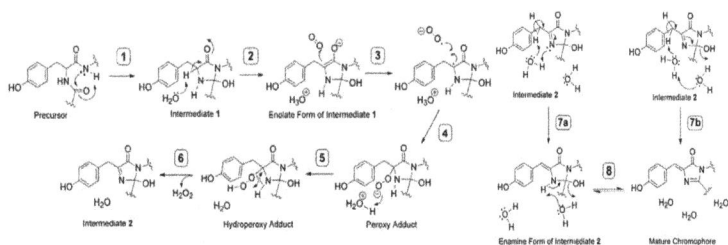

Figure 3. Mechanism of the maturation of the GFP chromophore. Steps 1-6 include the cyclization and deoxidation steps while step 7 indicates two possible pathways for the dehydration step. Used with permission from [9]

The two spectral absorbances of the GFP chromophore have been found to be highly sensitive to pH changes [11].At physiological pH, GFP exhibits maximal absorption at 395 nm while absorbing lesser amounts of light at 475 nm.However, increasing the pH to about 12.0 causes the maximal absorption of light to occur around 475 nm while diminishing the absorption at 395 nm.The two absorption maxima correspond to different protonated states of the chromophore.The pK$_a$ for the side chain hydroxyl group of Tyr66 is about 8.1 [12] and therefore, the maximal absorbance for the neutral chromophore occurs at 395 nm while maximal absorbance occurs at 470 nm for the anionic form of the chromophore.At acidic pHs lower than 6 or alkaline pHs above 12, fluorescence is diminished as GFP is denatured and the chromophore is quenched.

1.4. Wavelength variants and FRET

Starting with homologous green and red fluorescent proteins, a rainbow of different-colored fluorescent proteins have been developed. Mutating Tyr66 of the GFP chromophore to a tryptophan produces cyan fluorescence, while a histidine mutation produces blue fluorescence. Mutating a threonine on beta strand 10 to a tyrosine introduces a pi-stacking interaction which produces yellow fluorescence. See [3] for more details. At the other end of the color spectrum, the coral-derived DsRed fluorescent protein, a structural homolog of GFP, was diversified into the mFruits library, producing eight fluorescent proteins with emission maxima ranging from 537 to 610 nm [13]. Far-red fluorescent proteins, which have potential for use in deep tissue imaging due to the penetration of these wavelengths, have been discovered [14-16], while others have been developed in the lab [17] and even using computational approaches [18]. Further enhancement of these wavelength-shifted variants has improved their biophysical properties and made them available to more applications.

GFP and its derivatives have seen significant use as fluorescent pairs for Förster Resonance Energy Transfer (FRET) experiments. FRET emission arises when the emission spectrum of one chromophore overlaps with the excitation spectrum of another chromophore. If the two chromophores are physically close (on the order of a few nanometers) and in the correct orientation, then excitation of the first chromophore will excite the second chromophore through non-radiative energy transfer and produce fluorescence at the second chromophore's emission wavelength (Figure4). This phenomenon can be used to detect when two fluorescent proteins (FPs) are within a certain distance, which may be induced by a ligand-dependent conformational change in a linking domain between the two fluorescent proteins, or by binding of interacting domains fused to fluorescent proteins. The canonical pairing for FRET using fluorescent proteins is cyan fluorescent protein (CFP) and yellow fluorescent protein (YFP) [19]; but this pairing has issues concerning overlapping emission spectra, stability to photobleaching, and sensitivity to the chemical environment. The study in [20] had the goal of producing a cyan fluorescent protein more suitable for use in FRET experiments. Other pairings, such as GFP and the the DsRed-based variant mCherry red fluorescent protein, have been proposed as consistent, reliable alternatives [21]. A full review of the development and usage of fluorescent proteins as tools for FRET can be found in [22]. The genetic and physical ease of use of GFP-derived fluorescent proteins, in conjunction with their wide range of colors and spectral overlaps, makes them ideal molecules for the design of FRET-based biosensors.

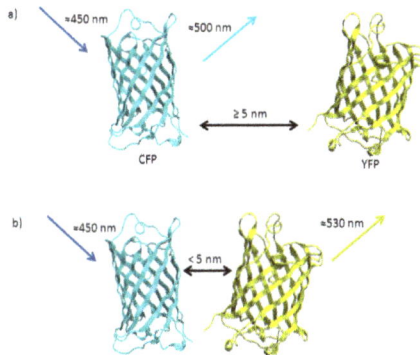

Figure 4. Illustration of the FRET phenomenon using the traditional CFP/YFP donor/acceptor pairing. a) If the two fluorescent moieties are too far apart, excitation of the donor molecule only produces observable emission from the donor. b) When in range, excitation of the donor is propagated to the acceptor molecule through non-radiative photon transfer, and emission from the acceptor is observed.

1.5. Mutants with improved features

Because of the aforementioned slow folding, low solubility and slow chromophore maturation, a significant effort has been put forth to improve these properties in GFP. These strategies range from specific, directed rational mutations based on structural and biophysical
information to fully randomized approaches such as error-prone PCR [23] and DNA shuffling [24]. By mutating the chromophore residue serine 65 to a threonine (S65T) and phenylalanine 64 to a leucine (F64L), an "enhanced" GFP (EGFP, gi:27372525) was produced with
the excitation maximum shifted from ultraviolet to blue and with better folding efficiency in
E. coli[25]. Blue excitation is favorable because it matches up with the wavelengths of laser
light used in modern cell sorting machines. Three rounds of DNA shuffling produced a mutant of GFP termed "cycle3" or GFPuv (gi:1490533) which contains three point mutations at
or near the surface of the protein (F100S, M154T, V164A). This mutant has 16- to 18-fold
brighter fluorescence than wild type GFP, attributed to a reduction of surface hydrophobicity and, subsequently, aggregation *in vivo* which prevents chromophore maturation [6].
Combining these sets of mutations produces a "folding reporter" GFP (gi: 83754214) which
is monomeric and highly fluorescent [26], but does not fold and fluoresce strongly when
fused to other poorly folded proteins. Four rounds of DNA shuffling starting with this GFP
variant produced a mutant with six additional mutations, called "superfolder" GFP (gi:
391871871), which can fold even when fused to a poorly folding protein [27]. Superfolder
GFP also showed increased resistance to chemical denaturation and faster refolding kinetics.
This GFP variant also has exceptional tolerance to circular permutation compared to the
"folding reporter" mutant of GFP (circular permutation will be discussed in Sequential rearrangements and truncations). A common theme emerges from these sets of mutations: a reduction in surface hydrophobicity leads to reduced aggregation tendency, which increases
the fraction of chromophore able to mature and, consequently, the brightness of the protein
in vivo.The hydrophobicity of the wild type GFP is hypothesized to serve as a binding site
to aequorin in jellyfish [4].

Mutating surface polar residues to increase the net charge, called "supercharging", may be
one solution to the problem of aggregation. Armed with the knowledge that the net surface
charge does not often affect protein folding or activity, [28] demonstrated that mutating the
surface residues either to majority positive or to majority negative side chains does not significantly affect fluorescence. Furthermore, these "supercharged" variants of GFP showed
increased resistance to both thermally and chemically-induced aggregation with a minimal
decrease in thermal stability. The only side effects are the unwanted binding of positively
supercharged GFP to DNA, and the formation of a fluorescent precipitate when oppositely
supercharged variants are mixed.

Disulfide bonds have been known to confer additional stability to proteins. Two externally-
placed disulfides were engineered into cycle3 GFP,one predicted to have no effect on stability, the other predicted to have a stabilizing effect [29]. The predictions, based on estimations
of local disorder, were correct. Adding a disulfide where the chain is more disordered improved stability the most.

In recent, unpublished work in our lab [30], a faster-folding GFP has been made by eliminating a conserved cis-peptide bond. The slowest phase of folding of superfolder GFP has been known to be related to cis/trans isomerization of a peptide bond preceding a proline [5]. We targeted Pro89 for mutation, since the peptide bond is cis at that position in the crystal structure, but modeling studies suggested that a simple point mutation would not have worked. Instead, we added two residues creating a longer loop, and then selected new side chains for four residues based on modeling. The new variant, called "all-trans" or AT-GFP, folds faster, lacking the slow phase. A 2.7Å crystal structure, in progress, shows clearly that the backbone is indeed composed of all trans peptide bonds in the new loop region.

All of the variants discussed so far are derived from *Aequorea* GFP, but homologous fluorescent proteins from other species have also played a role in advancing the science. Rational design of a homologous GFP from the marine arthropod *Pontellina plumata* resulted in "TurboGFP" which folds and matures much faster than EGFP with reduced *in vitro* aggregation relative to its parent protein [31]. TurboGFP and its parent protein lack *cis*-peptide bonds, known to contribute to the slow phase of GFP folding [5]. The crystal structure of TurboGFP reveals a pore to the chromophore, which mutagenesis shows to be a key component to fast maturation [31]. This makes sense, since the diffusion of molecular oxygen into the core is the rate limiting step in chromophore formation.This result represents the first successful designed improvements to a non-Cnidarian fluorescent protein. Random directed mutagenesis of beta strands 7 and 8 in the cyan fluorescent protein derivative mCerulean produced a mutant with six mutations and a T65S reversion mutation in the chromophore. This construct, termed mCerulean3, has an increased quantum yield and demonstrates minimal photobleaching and photoswitching effects, making it a better FRET donor molecule [20].

A novel fluorescent protein was developed using the consensus engineering approach, synthesizing a consensus sequence gene from 31 homologs of the monomeric Azami green protein, a distant homolog of *Aequorea* GFP. The resulting protein CGP (consensus green protein) has comparable expression to the parent protein with increased brightness and slightly decreased stability [32]. A novel directed evolution process was then carried out on CGP to stabilize it by inserting destabilizing loops into the protein, then evolving it to tolerate the insertions, then removing the destabilizing loops. After three rounds of this process, a mutant called eCGP123 demonstrated exceptional thermal stability compared to CGP and the parent Azami green protein [33]. Distantly-related fluorescent proteins have contributed much to the structural and biophysical understanding and application of the larger family.

1.6. Sequential rearrangements and truncations

Circular permutation is the repositioning of the N and C-termini of the protein to different regions of the sequence, connecting the original termini with a flexible peptide linker to produce a continuous, shuffled polypeptide. Many proteins retain their structure and function after permutation, provided the permutation site is not disruptive to secondary structural elements. This process demonstrates the tolerance of the protein's overall structure to significant rearrangements of primary sequence [34], enabling the design of biosensors based on split GFPs as discussed later.

GFP's rigid structure, extreme stability and unique post-translational chromophore formation reaction do not seem to suggest that it would tolerate circular permutation, and for the
most part, it does not. All permutations that disrupt beta strands do not form the chromophore, and about half of the permutations in loop regions cannot form the chromophore.
However, one particular permutation, starting the protein at position 145 (just before beta
strand 7) expresses and fluoresces well, although it is less stable and less bright than the
wild type GFP [34]. This circular permutation can also tolerate protein fusions to its new termini (positions 145 and 144 in wild type numbering), and position 145 in the wild type can
accept a full protein insertion, such as calmodulin or a zinc finger binding domain [35]. The
"superfolder" GFP reported in [27] was able to fluoresce after 13 of the 14 possible circular
permutations, whereas the folding-reporter GFP only tolerated 3 of those 14 permuta-
tions.Figure 5summarizes permutation and loop insertion results.

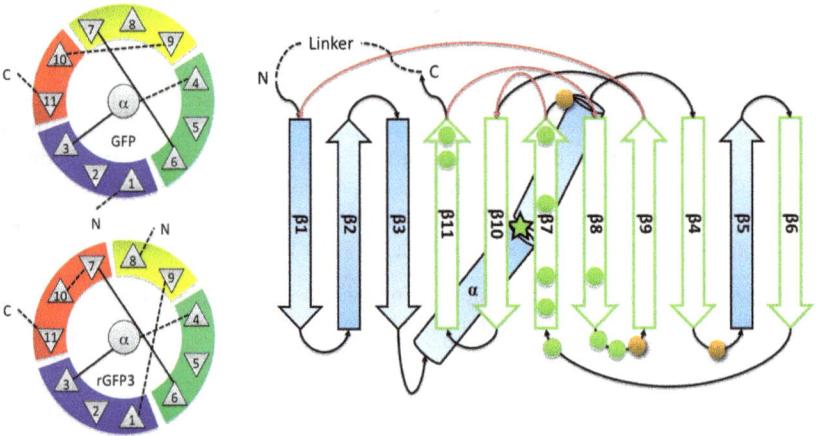

Figure 5. a) The wild type GFP, and (b) rewired GFP topology as drawn using the TOPS conventions [37]. Solid lines are
connections at the top, dashed lines at the bottom of the barrel. (c) Green dots mark locations of the termini in viable
circular permutants. Orange dots mark places where long insertions have been made [38] Green arrows mark beta
strands that can be left out and added back to reconstitute fluorescence. Red lines are connections created in rGFP3,
rewired GFP [36]. Topological changes and truncations are the least tolerated in the N-terminal 6 beta strands.

Circular rearrangements preserve the overall "ordering" of the secondary structural elements; however, non-circular rearrangement of the secondary structural elements is also
possible. Using rational computational modeling and knowledge about GFP's folding pathway, [36] designed a "rewired" GFP with identical fluorescence properties and stability as a
variant of superfolder GFP, but with the beta strands connected in a different order. These
experiments demonstrate the selective robustness of GFP's structure to large-scale rearrangements in sequence, which has implications for deciphering the GFP folding pathway,
as well as for design of split-GFP biosensors.

1.7. "Leave-One-Out" GFP

GFP can also be engineered to omit one of its secondary structural elements, either at one end or in the middle of the sequence by truncating a circular permutant. Truncation may be accomplished either at the genetic level or at the protein level, the latter by using proteolysis and gel filtration. Constructs missing one secondary structure element have been named "Leave-One-Out" or LOO, borrowing the term from a method for statistical cross-validation. When synthesized directly via the genetic approach, LOO-GFPs are non-fluorescent or weakly fluorescent. However, if co-expressed with the omitted piece, fluorescence sometimes develops *in vivo*, depending on which of the secondary structure elements was left out [39,40]. Expressing the full-length GFP and removing the beta strand by proteolysis, denaturation and gel filtration produces similar results [41]. A complete beta barrel is necessary for chromophore maturation. Once the chromophore has matured, LOO-GFP develops fluorescence rapidly upon introduction of the omitted beta strand from an external source.

That Leave-One-Out works is non-intuitive. In general, protein folding is an all-or-none process and leaving out any whole secondary structure element leads to an unfolded protein which aggregates in the cell. Yet, [40] has shown that it is possible to reconstitute LOO-GFP after truncation at several positions in the sequence. The key to understanding why LOO is sometimes possible is in the protein folding pathway. Although folding appears to be an all-or-none process by most experimental metrics, it proceeds along a loosely defined sequence of nucleation and condensation events called a folding pathway [42]. If the sequence segment that is removed is in the part of the protein that folds last, then a kinetic intermediate exists whose structure closely resembles the native state with one piece removed. This intermediate need not be the lowest energy state and may not be visible by equilibrium measurements, but its minute presence diminishes the energetic barrier of folding enough that the addition of a peptide can push the protein to the folded state. In short, Leave-One-Out uses the idea that some cyclically permuted, truncated proteins are natural sensors of the part left out.

In vivo solubility experiments performed on twelve LOO-GFPs (individually omitting each of 11 beta strands and the alpha helix) showed that there are significant differences in tolerance to the removal of particular secondary structural elements (SSE) as a function of solubility. The variability is best explained in terms of the order of folding of the SSEs. SSEs that are required for the early steps in folding leave a more completely unfolded polypeptide behind when they are left out. SSEs that fold late and not required for most of the folding pathway, leaving behind a mostly-folded protein which is more soluble. Leave-One-Out solubility analysis provides a unique insight into the folding pathway of GFP [40]. Omitting strand 7 (LOO7-GFP) appears to be the least detrimental to the overall structure of GFP, suggesting that strand 7 folds last. Binding kinetics data for LOO7-GFP to its missing beta strand as a synthetic peptide gives a K_d value of roughly 0.5 · M [11]. Surprisingly, when it is omitted by circular permutation and proteolysis, the central alpha helix can be reintroduced as a synthetic peptide to the "hollow" GFP barrel and chromophore maturation proceeds and produces fluorescence [41].However, refolding from the denatured state was required.

Some LOO-GFPs also show interesting reactions to ambient light. LOO11-GFP (beta strand 11 omitted) does not bind strand 11 when kept completely in the dark, but does bind it upon irradiation with light [43]. Raman spectroscopy showed that, in the dark, the chromophore assumes a *trans* conformation, and that light induces a switch to the native *cis* conformation. After irradiation, the chromophore relaxes back to the *trans* conformation. Following up on this result, [44] showed that using a circularly permuted LOO10-GFP construct (beta strand 10 omitted) and introducing two synthetic forms of strand 10, the wild-type strand and a strand with a mutation to cause yellow-shifted fluorescence, light irradiation increased the frequency of "peptide exchange" between the two strand 10 forms. The presence of this peptide exchange suggests that the cis/trans isomerization of the chromophore requires partial unfolding of the protein.

2. GFP-based biomarkers

The term biomarker has accumulated a variety of definitions over the years. Herein, biomarkers are defined as genetically encoded molecular indicators of state that are linked to specific genes.The utility of GFP as a biomarker was first demonstrated using GFP reporter constructs [45]. When GFP is used as a transcription reporter, a cellular promoter drives expression of the fluorescent protein resulting in fluorescent signal that temporally and locally reflects expression from the promoter *in vivo*. In the initial experiments, GFP cDNA [46] was expressed from the T7 promoter in *E. coli* or from the mec-7 (beta tubulin) promoter in *C. elegans* [45].*E. coli* cells fluoresced and the expression in *C. elegans* mirrored the pattern known from antibody detection of the native protein.Subsequently, GFP transcriptional reporters have been used in a wide variety of organisms; GFP expression has minimal effect on the cells and can be monitored noninvasively using techniques such as fluorescence microscopy and fluorescence assisted cell sorting (reviewed in [47]).

GFP fusion proteins (generated by combining the fluorescent protein coding region with the coding region of the cellular protein) are used as markers for visualization of intracellular protein tracking and interactions (reviewed in [47-49]).The GFP moiety may be N-terminal, C-terminal or even internal to the cellular protein.The availability of a color palette of fluorescent proteins allows multicolor imaging of distinct fluorescent protein fusions in the cellular environment. GFP fusion proteins are a major component of the molecular toolkit in cell biology.

2.1. Using GFP as an *in vivo* solubility marker

GFP has been used as a genetically encoded reporter for folding of expressed proteins.Expression of recombinant proteins in *E.coli* is a powerful tool for obtaining large quantities of purified protein; however, some overexpressed recombinant proteins improperly fold and aggregate. Manipulation of conditions to generate soluble protein can be a laborious process. Directed evolution can be employed to increase the solubility of the recombinant proteins, but detection of specific mutants with improved solubility is a challenge.However,

GFP biomarking can be utilized to address this challenge.Since GFP chromophore formation requires proper protein folding and GFP folds poorly when fused to misfolded proteins, fluorescence of a GFP fusion protein can serve as an internal signal of a specific soluble (not aggregated) protein [26]. When used as a folding reporter, GFP is fused C-terminally to the protein of interest using a short linker between the two protein domains.Detection of fluorescence indicates that GFP domain is properly folded and that the protein of interest therefore must be soluble.If the protein of interest misfolds and aggregates, the fused slow-folding GFP aggregates along with it and fluorescence is not detected. Therefore, this folding reporter assay can be used as a screening tool for soluble recombinant proteins in the context of directed evolution.

Split GFP may also be used to assay folding and solubility of a protein of interest in vivo by "tagging" the recombinant protein with the smaller portion of the split GFP sequence, and expressing the larger portion separately or adding it exogenously. The small size of a protein tag makes it less likely to interfere with the folding and function of the protein of interest.In the split GFP complementation assay a large fragment of GFP folding reporter (GFP1-10) is coexpressed with tagged GFP protein (GFP_{11}-protein x) [50]. As shown in Figure 6, neither GFP1-10 nor the GFP11-tagged protein fluoresce alone; however, if both components are soluble,GFP1-10 and the GFP11-tagged protein reconstitute the native structure and fluorescence.For successful implementation of the assay, directed evolution of superfolder GFP1-10 was required. This resulted in GFP1-10 OPT which has an 80% increased solubility over the corresponding superfolder GFP1-10.GFP OPT contains 7 new mutations (N39I, T105K, E111V, I128T, K166T, I167V and S205T) in addition to the superfolder mutations [50].Directed evolution of GFP11 resulted in GFP11 optima tag that had the dual properties of 1) complementation with GFP1-10 OPT and 2) minimized perturbation of the protein of interest. Note that full-length GFP OPT was subsequently found to be more tolerant of circular permutation and truncation than superfolder GFP [40].

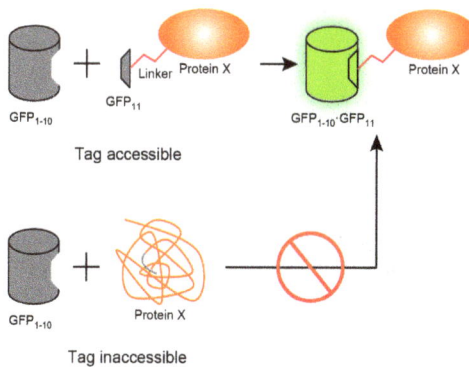

Figure 6. Mechanism of LOO11-GFP (GFP_{1-10}) as an in vivo solubility indicator for proteins tagged with strand 11 (GFP_{11}). Modified with permission from [50].

In addition to providing a less laborious method for detecting protein variants and reaction conditions for generating soluble recombinant protein, the split GFP complementation assay also serves as an assay of aggregation in living cells. For example, aggregates of the microtubule associated protein tau are found in neurofibrillary tangles but their role in the pathology of Alzheimer disease and Parkinson disease is not clear [51].The split GFP complementation assay enables monitoring of the aggregation process in living mammalian cells [52,53] and was validated using GFP_{1-10} and GFP_{11}-tau variants.Cells containing soluble tagged protein show visible fluorescence but aggregates have little or no fluorescence. Protein aggregates of GFP_{11}-tau sequestered the GFP_{11} tag, leading to decreased complementation of GFP_{1-10} and decreased fluorescence. Thus the split GFP complementation assay using tagged-GFP tau showed that it could be used as an *in vivo* model for studying factors that influence aggregation.

2.2. GFP biomarkers for single molecule imaging

It is also possible to utilize GFP biomarkers for single-molecule localization, a form of super-resolution microscopy. High affinity single chain camelid antibodies (nanobodies) to GFP can be used to deliver organic fluorophores to GFP tagged proteins that are in turn used in single molecule "nanoscopy." [54, 55]. This novel approach combines the molecular specificity of genetic tagging with the high photon yield of the organic dyes. Additionally, by varying the buffer conditions used, many organic dyes can become photoswitchable. The small size of camelid antibodies and their high affinity allow for access to regions that are generally inaccessible to conventional antibodies and targets that are expressed at very low levels [56].

One should caution that the overexpression of FRET biomarkers in transgenic animals carries some concerns that this could lead to the perturbation of endogenous signaling pathways and even retardation of animal development [57]. Additionally, in compact tissue, such as the brain tissue, cell type identification is particularly tedious due the diffused expression of the biomarkers.

3. GFP biosensors

Biosensors are distinct from biomarkers in that they are not linked to the expression of a specific gene product. Biosensors may function *in vivo* or *in vitro*. GFP variants that exhibit analyte-sensitive properties are genetically encoded biosensors, acting *in vivo*.GFP biosensors that contain amino acid substitutions that enable detection of pH changes, specific ions (Cl⁻or Ca^{2+}), reactive oxygen species, redox state, and specific peptides have been reported [39, 58-60]. In addition, modifications have been reported that enable selective activation (irreversible or reversible) of the fluorescence [61,62].Genetically encoded GFP biosensors may be single GFP domains or FRET pairs.In the following subsections we describe selected examples of GFP-based biosensors used *in vivo* or *in vitro*, with special emphasis on computationally designed biosensors.

3.1. *In vivo* pH biosensors

Within the cell, pH varies from the neutral pH of the cytosol to the acidic pH of the lyso-some lumen and protons may serve as cellular signals.Genetically encoded pH biosensors enable subcellular detection of pH and can provide insight into the regulation of cellular ac-tivities by pH. Addition of intracellular targeting tag directs the pH biosensor to particular subcellular compartments.

Many GFP variants show sensitivity to pH which results from protonation and deprotona-tion of the chromophore (see **Maturation of the GFP Chromophore**)(reviewed in [58]). The rapid and reversible response of EGFP to pH changes in the cells enabled EGFP to be used as an intracellular pH indicator [63] in place of chemical pH indicators such as fluorescein.A range of GFP based pH biosensors have been generated from modification of wtGFP and EGFP which resulted from amino acid substitutions primarily in and around the region of the chromophore.

Two classes of GFP pH indicators have been described: ratiometric and nonratiometric [58, 64].In the ratiometric pH indicators, the chromophore environment is such that the GFP bio-sensor has two sets of excitation/ emission spectra, one that varies with pH and another that does not.For these GFP variants, a calibration curve can be generated for the ratio of the spectra versus the pH.Nonratiometric GFP variants,such as EGFP [63] or ecliptic GFP [64], have pH dependent emission from the anionic chromophore (deprotonated) but almost no fluorescence of the neutral chromophore (protonated). These variants are used for reporting pH changes within cells when used as single molecule pH sensor or used in tandem with pH insensitive fluorescent partner (described below).

Ratiometric GFP pH biosensors have been generated by modification of a few key amino acids in the vicinity of the chromophore.Ratiometric pHluorin (RaGFP), the first ratiometric GFP described,contains a key S202H mutation and shows pH dependent change in excita-tion ratio between pH 5.5 and pH 7.5 [64].TheS202H mutation was shown to be important for the ratiometric property; pHlourins lacking the S202H were non ratiometric.Another class of GFP ratiometric pH sensors, deGFPs were generated from mutagenesis of the S65T GFP variant [65] resulting in substitutions H148G (deGFP1) or H148C (deGFP4) and T203C.The deGFPs are dual emission ratiometric GFPs emitting both blue and green light; blue light emission decreases with increase pH while green light emission increases with in-creased pH.

Variants pH GFP (H148D) [66] and E^2GFP (F64L/S65T/T203Y/L231H)[67] function as dual excitation ratiometric pH indicators with pH-dependent excitation at 488 nm and relatively pH-independent excitation at458 nm).In addition to its pH sensing properties, fluorescence emission from E^2GFP is affected by the concentration of certain ions, including Cl⁻. The chloride ion sensitivity of E^2GFP is a key component of the GFP–based chloride ion and pH sensor ClopHensor [68] (discussed in section **Fluorescent proteins as intrinsic ion sensors**).

In addition to single molecule based pH biosensors, ratiometric pH biosensors using tandem fluorescent protein variants have been constructed in which a pH sensitive GFP variant is linked to a less sensitive or pH insensitive GFP.GFpH and YFpH are tandem FRET pairsfor the detection of pH changes in the cytosol and nucleus of living cells. GFpH combines GFPuv, which has low pH sensitivity, with pH sensitive EGFP and YFpH combines GFPuv and EYFP [58, 69].Not all tandem GFP biosensors are FRET pairs, however. pHusion is a ratiometric tandem GFP biosensor in whichmRFP (pH insensitive) is tethered to EGFP (pH sensitive) via a linker.pH measurements are determined from the ratio of EGFP to mRFP fluorescence. pHusion biosensor was developed for analysis of intracellular and extracellular pH in developing plants [60].

3.2. In vivoFRET-based biosensors

Genetically-encoded FRET-based biosensors can be applied in a variety of capacities to visualize intracellular spatiotemporal changes in real time. The evolution of these applications has progressed from cell culture systems that transiently express FRET biosensors to transgenic mouse models that express them in a heritable manner [57]. Production of transgenic mice with FRET biosensors arose in an effort to enhance our understanding of the differences that exist between tissue culture and living systems. Transgenic FRET GFP biosensor systems are very efficient and their fluorescence signals are easily distinguished from autofluorescence, which is analyte-independent fluorescence. The sensors themselves can be used to probe a variety of pathways for the activity of signaling enzymes as well as a number of post translational modifications.

3.2.1. Detection of enzyme activity

In transgenic animal models, FRET biosensors can be used to study PKA activation by cAMP, ERK activation by TPA and their association with various physiological changes [57]. PKA and ERK areenzymes that transfer the γ-phosphate of ATP to a number of protein substrates thereby affecting a conformational change. Kinase induced conformational changes are important because they are involved in the control a number of critical cellular processes that include glycogen synthesis, hormonal response, and ion transport [70]. A number of signaling cascades that involve kinases require a means of dynamic control and spatial compartmentalization of the kinase activity; a requirement highlights the need for a mechanism to continuously track kinase activity in different compartments and signaling microdomains *in vivo*.

Traditional methods of assaying kinase activity fail to capture its dynamicity; a void that is filled by genetically encoded FRET-based biosensors. These sensors are constructed so that the substrate protein of the kinase of interest is flanked with a fluorescent protein pair in such a way that the conformational change imparted by phosphorylation translates into a change in the FRET signal (Figure7) [70]. These biosensors can be localized to particular sites of interest with the aid of appropriate targeting signal sequences, allowing the imaging of site-specific kinase activity.G-protein coupled receptors, when used in a biosensor, provide a mechanism for transducingdrug mediated effects on PKA activity into a light signal. Transgenic mice expressing FRET based biosensors provide an ideal system for studying the pharmacodynamics of these drugs.

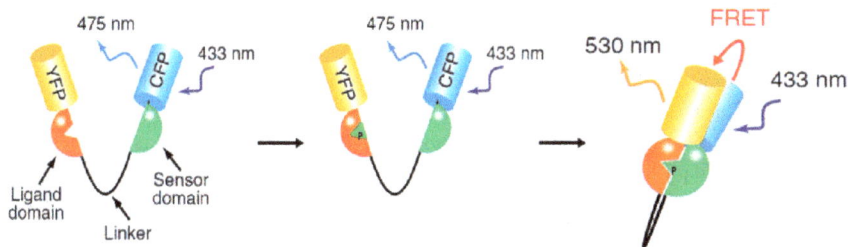

Figure 7. Representation of the mode of action of an intramolecular FRET biosensor containing a molecular switch. The sensor domain and ligand domain of the construct are connected by a flexible linker with CFP and YFP serving as the donor and acceptor for the FRET pair. This switch can perceive various molecular events, such as protein phosphorylation, through binding to the ligand domain. This in turn induces an interaction between the ligand and sensor domains that facilitates a global change in the conformation of the biosensor, which serves to increase the FRET efficiency from the donor to the acceptor (CFP to YFP in this case) [71].

When used to study the signaling events in wound healing, the strength and duration of the fluorescent signals that are generated by these biosensors are dependent on the location within the tissue (tissue depth has a negative impact on the intensity of the fluorescent signal), its vicinity in relation to the site of injury, as well as the contributions made by chemical mediators (drugs) in sustaining kinase activity [57]. These model systems provide a means of visualizing in real-time the agonist/antagonist pharmacodynamics associated with a plethora of signaling molecules that do not necessarily have to be limited to PKA and ERK activity. They also provide a tool for resolving the maze of upstream signaling pathways that contribute to chemotaxis in the animals.

Genetically encodable FRET GFP biosensors have proven to be useful in characterizing the dynamic phosphorylation dependent regulation of small GTPases [70]. Ras GTPases play essential roles in regulating cell growth, cell differentiation, cell migration, and lipid vesicle trafficking. Upon binding GFP, the G-protein Ras recruits the serine/threonine kinase Raf. FRET biosensors for GTPase activity such as Raichu-Ras (Ras and Interacting protein CHimeric Unit for RAS) use this Ras-Raf interaction as the basis for the molecular switch. Raichu-Ras functions by using H-Ras as the sensor domain and the Ras Binding Domain (RBD) of Raf as the ligand domain in constructing a molecular switch that in turn is sandwiched by the FRET pair CFP/YFP (Figure 7). Such a design allows for the monitoring of Ras activation in living cells on the basis of fluctuations in the FRET signals generated.

3.2.2. Detection of antioxidant activity and reactive oxygen species

FRET-based GFP biosensors can also be employed in *in vitro* applications as an alternative tool for high throughput screening assays. These assays are simple, inexpensive, reproducible and highly specific. A good example can be observed in the use of bacterial cell-based assays for screening antioxidant activity of various substances for biological activity [72]. To achieve this objective *E.coli* biosensor strains that carry the plasmid that fuses sodA (manganese superoxide dismutase) and fumC (fumarase C) promoters with GFP genes, called so-dA::gfp and fumC::gfp respectively, were produced and used to evaluate antioxidant activity of a number of phenolic and flavonoid compounds in comparison with two DPPH radical scavenging and SOD activity assays (two more conventional assays). After paraquat treatment of *E. coli* cultures to induce oxidative stress, the putative antioxidant compounds were added and both the GFP fluorescence and cell culture density readings were taken to determine the role played by the respective compounds in reducing the free radical accumulation and intracellular oxidative stress.Genes sodA and fumC are turned on by SoxR and OxyR, respectively, which are the two main regulatory proteins involved in oxidative stress sensing. GFP fluorescence is therefore diminished by successful antioxidants. These constructs are important because they function as alternative screening tools that can be utilized to assess the activity of compounds with therapeutic potential against oxidative stress. Antioxidants have been shown to play a role in disease prevention.

3.2.3. Detection of calcium ions

FRET-based and single domain Ca^{2+} sensors have been constructed using the allosteric effect of calcium binding to receptors calmodulin or troponin [73]. In one construct, the CFP/YFP pairing is separated by a linker containing a calmodulin domain and a calmodulin ligand peptide called M13.When Ca^{2+} is present, it binds to the calmodulin domain, inducing a conformational change and binding of the proximal M13 peptide sequence. The M13 binding results in shortening of the linker, bringing CFP within FRET distance of YFP and changing the emission wavelength from cyan to yellow. The Ca^{2+} binding affinity was found to be highly variable, around 0.3 uM with a Hill coefficient of n=4, depending on conditions. When used *in vivo*, the calmodulin-based biosensors suffered from endogenous interference by host proteins and did not always work [73]. To remedy this, the calmodulin/M13 linker was replaced with troponin C, whose N-to-C distance is shortened by Ca^{2+} binding, resulting in FRET.Using another strategy, calmodulin and M13 peptide sequences were separated by a circularly-permuted EGFP, which was quenched in the absence of Ca^{2+} but recovered fluorescence upon Ca^{2+}-induced binding of the calmodulin to M13. Improved genetically encoded Ca^{2+} indicators have been used *in vivo* to trace action potentials in neurons, with response times in the millisecond range [73, 74], becoming competitive with synthetic indicators and recording electrodes.

4. *In vitro* applications

GFP has great potential to work as an *in vitro* biosensor.Because of its remarkable stability, it can be used and manipulated in multiple ways to impart sensor functionality to the protein.Several approaches are described here, including creating a chimeric protein with antibody fragments, linking fluorescent proteins to quantum dots, manipulating the amino acid sequence to create analyte pores, as well as sequence manipulation that provides increased halide ion and/or pH sensitivity.

4.1. GFP-antibody chimeric proteins

The goal of GFP-antibody chimeric proteins (GFPAbs) is to convert a multi-step experimental process for locating molecules via antibodies and enzyme-linked secondary antibodies, into a one-step process using a GFPAbs.This molecule could then work as a detection reagent in flow cytometry, for intracellular targeting, or fluorescence-based ELISAs [38].However, in order to replace antibodies in these techniques, it is important to achieve the same nanomolar sensitivity that is found in the natural antibodies.To do this, [38] inserted two antigen-binding loops into the GFP structure, counting on cooperativity in binding to enhance affinity.

It became clear that adding loops impinges on the integrity of the native GFP structure.The binding loops must be placed such that their presence in the fluorescent protein does not jeopardize its structural fidelity, or that of the chromophore.There are only a few locations in the molecule that are amenable to such insertions:turn regions β4/β5 (residue 102), β7/β8 (residue 172) and β8/β9 (residue 157).The latter two are too far apart in three-dimensional space to provide for cooperative binding (see Figure 5). The β4/β5 and β8/β9 loop regions are in close proximity, but these do not easily accommodate random loop insertions.

[38] used directed evolution with yeast surface display [75] to find sequences that stabilized the folded conformation in the context of loop insertions.The yeast secretory pathway does not allow unfolded protein to reach the surface of the cell, thus only mutants that yield fully folded GFP were displayed by yeast cells. Directed evolution revealed several mutations that conferred additional stability and increased fluorescence in the context of inserted loops: D19N, F64L, A87T, Y39H, V163A, L221V, and N105T. The F64L mutation has been shown to increase fluorescence of GFP and also to shift the excitation maximum to 488 nm.Y39H and N105T have been shown to improve refolding kinetics and refolding stability, respectively.V163A is linked to improved folding as a result of its increased expression in yeast surface display [38]. These mutations accommodated the insertions of antigen-binding loops from antibodies raised against streptavidin-phycoerythrin, biotin-phycoerythrin, TrkB, or GADPH, all while maintaining 40% of the fluorescence and 60% of the expression of wild type GFP.With dual loop insertion, dissociation constants as low as 3.2 nM have been achieved [38]. The success of this construct means that molecules such as GADPH can be located within cells without having to engineer a second round of antibodies, saving both time and resources.

4.2. A chimeric fluorescent biosensor based on allostery

A general method for developing a biosensor for a specific receptor-ligand interaction has been described [76] in which a receptor protein is inserted into the GFP sequence between strand 8 and strand 9. The insertion puts enough of a strain on GFP that its fluorescence is reduced. Binding of the ligand to the GFP-receptor chimera may then impart enough of a change in its conformation that it causes a change in fluorescence, since the b8/b9 loop is fairly close in space to the chromophore. This change may be found by plate screening for fluorescence. In [76], the receptor Bla1 was cloned into the loop, and random mutations were made to this construct. Mutant constructs that detected the Bla1 ligand BLIP were identified by a visual screen of colonies before and after the induced expression of BLIP. Using this method, a double mutant was found that was shown to detect BLIP *in vitro* with micromolar affinity. In principle, this method could be used to generate a sensor for any ligand that can be expressed in bacteria or added exogenously, as long as a receptor protein exists that can be inserted into the GFP loop.

4.3. FRET-based biosensors using quantum dots

FRET-based in vitro biosensors may be constructed by linking fluorescent proteins to quantum dots (QDs).QDs are inorganic molecular nano-crystals whose absorption and emission spectra are dictated by the size of the QD.For example, a QD may be engineered to absorb ultraviolet light and emit light at 550 nm, which overlaps well with the excitation spectrum of mCherry, a variant of GFP [77], and produces FRET when the two fluorophores are in close proximity.

In order to make the FRET emission analyte-dependent, the QD was linked to the mCherry via an N-terminal linker peptide that contained a protease cleavage site and a 6 histidine tag.The imidazole side chains of the histidines electronically coordinate with the zinc atoms of the CdSe−ZnS core-shell semiconductor of the QD [77]. Multiple mCherry molecules can be coordinated with each QD. Splitting of mCherry from the QD by a protease may be detected by the loss of FRET.By placing the caspase-3 cleavage sequenceinto the linker between GFP and the QD, the FRET complex becomes a biosensor for the presence of caspase-3, glowing red at 610 nm in the absence of the protease, and reverting to the yellow fluorescence of the QD at 550 nm when the protease is present (Figure 8).

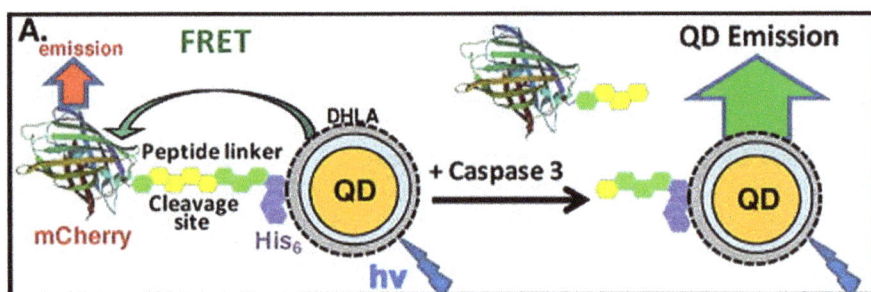

Figure 8. QD-FRET, showing emission of the chromophore only when in close proximity to the QD. When the two are split by caspase activity, FRET is lost. Figure used with permission from [78].

GFP/QD FRET emission may be also be manipulated by pH-induced changes in the spectral overlap, without having to spatially separate the QD from the fluorescent protein.It has been shown that fluorescent proteins such as GFP and mOrange experience a shift in excitation and emission spectra with changes in pH [78].At a slightly acidic pH, there is very little spectral overlap between the QD emission and the mOrange excitation, which means that the QD emission is seen, in this case around 520 nm.However, as the pH increases, the excitation spectrum of mOrange shifts such that there is more overlap with the QD emission, which subsequently causes an increase in FRET.The result is an upward shift in the emission wavelength with increasing pH.It is important to note that since there is a fluctuating hydrogen ion concentration, the histidine-QD coordination complex becomes unstable.In order to remedy this problem, a covalently linked quantum dot must be used.

4.4. Fluorescent proteins as intrinsic ion sensors

Fluorescent proteins, especially E²GFP, have been shown to be sensitive not only to pH changes but also to the concentration of certain ions, particularly chloride ions.E²GFP provides an avenue for single domain ratiometric analysis of pH because it contains two excitation and emission peaks. Only the longer wavelength emission peak is pH dependent [68].Therefore analysis of pH based on the ratio of green fluorescence to cyan.By coupling E²GFP to another fluorescent protein in a fusion construct, it is also possible to measure other intracellular chloride ion concentration.For example, DsRed is neither pH nor chloride ion sensitive, so it can be used to measure chloride ion concentration based on the ratio of its fluorescence to the cyan emission of E²GFP.

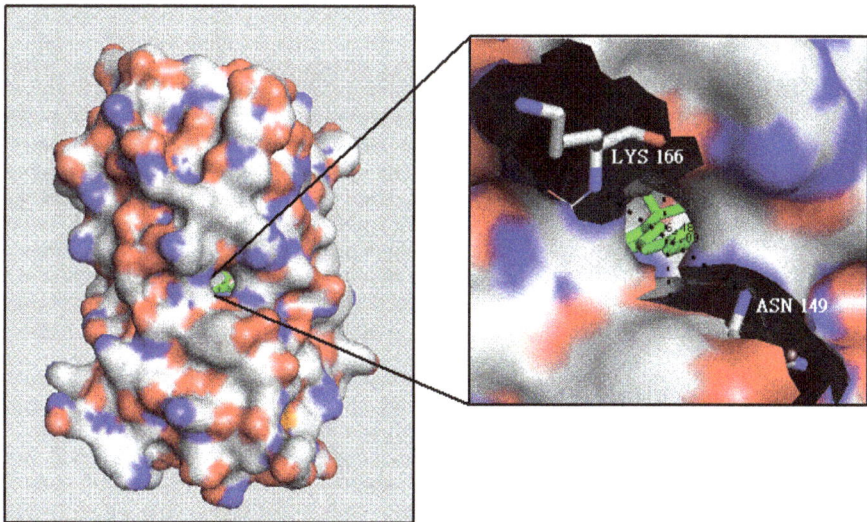

Figure 9. The analyte channel through which copper ions can pass through to the interior of the barrel structure and quench the fluorescence of the chromophore. Used with permission from [80].

Making a few modifications can make GFP sensitive to the concentration of other ions.For example, superfolder GFP can be made sensitive to copper ions by mutating the arginine at position 146 to a histidine, which, as previously mentioned, coordinates well with metal ions [79]. GFP can also become sensitive to ions by creating channels in the structure through

which small molecules can pass through and access the chromophore (Figure 9).By mutating position 165 from a phenylalanine to a glycine, a channel is opened that is about 4 Å
wide.This allows small molecules such as copper ions to enter the hydrophobic core of the
protein and quench fluorescence [80]. GFP, thanks to its stability, has shown a remarkable
ability to be modified, and thus shows great promise in visualizing a large variety of intracellular and extracellular substances.

5. Computationally designed LOO-GFPs

Recent work in the Bystroff lab has focused on programming GFP to accept any desired protein as a binding partner, like an antibody, and to switch on fluorescence only when the targeted protein is bound. The strategy combines Leave-One-Out split reconstitution with
computational design and high throughput screening.

Leave-One-Out (LOO) was described earlier (**"Leave-One-Out" GFP**) as a technique for developing split proteins that spontaneously reconstitute function. Fluorescence is recovered
in LOO-GFP when the left-out piece is encountered in the analyte. A promising application
of LOO-GFP, knowing that it binds to the left-out segment and fluoresces [39, 40], is to engineer novel LOO-GFP molecules that recognize and sense desired peptides derived from other sources such as virus, bacteria and parasites. By modifying the sites of one of the eleven
β-strands to complement shapes of given target peptides, the engineered LOO-GFP molecules will report the presence of specific target proteins, and therefore their host organism,
through simple fluorescence readout (Figure 10).LOO-GFP biosensors can be engineered by
generating mutations that accommodate the shape and charge of a desired target peptide.
The target peptide may be made available for binding by denaturing the target protein.

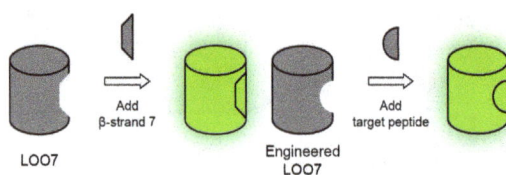

LOO7   Add β-strand 7   Engineered LOO7   Add target peptide

Figure 10. LOO-GFP peptide biosensors. Engineering LOO-GFP molecules to accommodate desired target peptides
create specific sensing tools where fluorescence can be reconstituted upon adding back the left-out peptides and signals the detection.

Theoretically, this goal could be achieved by random mutation followed by high throughput
screening to find mutants that glow in the presence of a peptide. However, random mutation would be extremely inefficient. Computational protein design methods offer a much
better alternative for rationally generating sequence diversity before the labor-intensive experimental screen.

5.1. Computer-aided protein design

Computational protein design predicts protein sequences that fold into predefined protein structures. Proteins are described as a set of atoms with 3D spatial coordinates and physical/chemical properties [81-84]. Instead of mutating residues experimentally, mutations are explored *in silico* and selected using a computed goodness of fit (Figure 11). Mutations predicted to cause collisions between atoms, leave unsatisfied hydrogen bonding partners, cause charge-charge repulsion, or employ rare amino acid side chain conformations are downweighted by assigning them a higher energy value. To facilitate the search for the best mutations, amino acid side chains are discretized into rotational isomers (called rotamers) [85-87]. Protein sequences that preserve the desired functionalities, such as the binding of a ligand, are obtained by searching the space of all side chain rotamers for the minimum free energy. There are few reviews of the methods used [88].

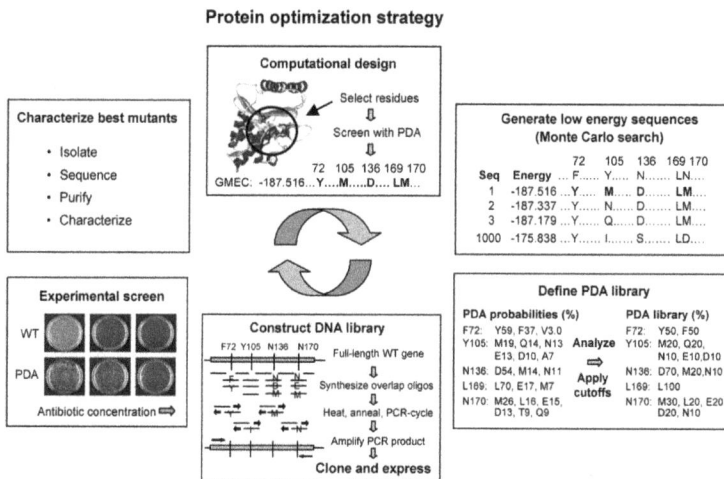

Protein optimization strategy

Figure 11. Computational protein design coupled with design library generation [89]. The entire designed sequence space of selected residues is computationally screened to determine the global minimum energy configuration (GMEC) for the given structure. Starting from the GMEC, sequence space is explored to obtain sub-optimal sequences that are also potentially predicted to be functional. A DNA library is constructed to cover all predicted sequences, and candidates are screened experimentally to select clones with desired functions. Information from analyzing obtained mutants is utilized to validate and improve the computational protein design strategy, and provides a better starting model for iterative optimization.

5.2. Protein biosensors versus other methods for detecting pathogens

Biosensors for specific proteins and pathogens offer potential advantages over the current state of the art, notably speed and simplicity. Laboratory diagnostics of infections commonly includes pathogen isolation using culture, direct antigen detection, or detection of pathogen specific DNA and/or RNA by polymerase chain reaction (PCR). The isolation method requires a culture system to inoculate a specimen, followed by the examination of specific characteristics produced by pathogens, such as the cytopathogenic effect of virus and the distinct metabolism of bacteria. Although culture-based methods have higher detection sensitivity, they generally take three to ten days for diagnosis. Alternatively, immunoassays utilize pathogen specific antibodies and secondary anti-antibodies to detect and report a pathogen. Most of the rapid diagnostic tests only take 15 to 30 min for diagnosis, but raising specific antibodies against pathogens is time-consuming and expensive. Thirdly, molecular diagnosis using PCR takes the advantage of the gene amplification and provides a highly sensitive detection in diagnosis from minute amounts of pathogen genome within a short time. However, the need for real-time PCR and gel electrophoresis apparati and reagents means it will not be possible in all settings, where a simple biosensor test would be possible. PCR assumes that DNA is present, but some pathogens such as anthrax toxin, snake venom and bovine spongiform encephalopathy contain no genetic material. All these point to a need for developing a diagnostic tool for proteins that is fast and easy to use, and suitable for rural, point-of-care facilities in developing nations.

The following describes how the computer-aided design of LOO-GFP was done, and the encouraging but preliminary results. The process has three steps: (1) the selection of a target peptide sequence from the genome of the pathogen, (2) the computational design of the LOO-GFP• target complex, and (3) the experimental screening of a library of potential biosensor sequences.

5.3. Target peptide selection

A target peptide for detection must be unique in order to avoid false positives, and must be conformable to the LOO-GFP binding site, which is the site of one of the eleven β-strands of GFP. From the examination of GFP and homolog fluorescent protein structures and sequences, we defined a set of signature patterns for each β-strand. These patterns define the limits of mutation. For example, no position within a target peptide may be a proline, since it must be hydrogen bonded on both sides to the neighboring β-strands. Cysteines are also disallowed, for experimental reasons.Target peptides are selected by searching the sequences of the target organism for a match to the signature pattern.Other considerations including the location of protease recognition sites, cellular location, and protein expression levels.

In the case study described here, a twelve-residue peptide (SSHEVSLGVSSA) was selected from hemagglutinin (HA) sequence of avian influenza virus H5N1, using the signature pattern of GFP β-strand 7. The target peptide retains the sequence pattern of the wild type β-strand 7, and it can be released by the chymotrypsin digestion of HA protein. A BLAST search of all known protein sequences confirmed that the HA target sequence occurs only in hemagglutinin from influenza virus type A.

5.4. Computational pre-screening of candidate biosensor sequences

To engineer customized LOO-GFP biosensors that sense a given peptide we developed a set of software called DEEdesign. DEEdesign uses a combination of physical properties and statistical knowledge to energetically evaluate the fitness of rotamers in protein structures, along with sampling algorithms to search the space of all possible mutations. The parameters used in the fitness scoring system are trained by a machine learning technique to reproduce the true sidechain conformations in high-resolution crystal structures [90]. Sequence space is searched using one of two methods, either using Monte Carlo [91], with random mutations accepted or rejected based on the calculated energy, or using the dead-end elimination theorem (DEE), which holds that if energies can be decomposed into pairwise terms, then a solution to the problem of finding the lowest energy set of mutations can be found by a process of successive elimination [92].

However, inaccuracies in design due to the imperfect scoring system, the use of discretized side chains, and the lack of precise modeling of backbone flexibility, affect the reliability of the method. Therefore, instead of relying on the accuracy of the single lowest energy protein sequence, DEEdesign provides an ensemble of plausible mutants, all with reasonably low calculated energy scores. These are assembled into a single amino acid profile, from which a library of nucleotide sequences is derived, employing degenerate codons for those positions in the sequence that have more than one possible amino acid.

In our case study, residues 143-154 NSHNVYITADKQ of β-strand 7 were mutated *in silico* to the target peptide sequence SSHEVSLGVSSA from HA. All residues within 7Å of the target were mutated to all amino acids within the constraints of the evolutionary history of GFP, where the latter was derived from a multiple sequence/structure alignment of 34 fluorescent proteins, augmented by additional homologous sequences. If an amino acid was found at a given position in the evolutionary history of GFP, then that amino acid was allowed in the course of the sequence space search, otherwise it was disallowed. DEE and Monte Carlo were used to search this sequence space, identifying an ensemble of low-energy sequences such that the total complexity of the sequence space of the ensemble was only about ten thousand unique sequences, a number that can be efficiently screened on petri plates. The ensemble of sequences was back-translated to DNA and divided into overlapping degenerate-codon oligonucleotides of 60 bases each by the program DNAWorks [93]. The set of mixed oligos was assembled by PCR into a gene library for screening, using the protocols of gene assembly mutagenesis [94].

5.5. Experimental screening and diversity generation by *in vitro* evolution

The computationally generated library for the H5N1 LOO-GFP biosensor had a complexity of around 10000 sequences and was relatively easy to screen in low to medium-throughput manner by looking for colonies that were fluorescent when co-expressed with its target peptide sequence. We fused the target peptide to intein [95] so that it would be cleaved immediately after expression and would exist as a free peptide.

However, potential mutations that are distant from the binding site of the target peptide (i.e. >10Å away from the binding site) may still have indirect effects on the binding of the target, or influence on LOO-GFP folding, are not easily captured in the computational design process because of time and memory limitations. To expand the screening, candidate mutant genes can be subjected to rounds of *in vitro* evolution, such as error-prone PCR [96] and/or DNA shuffling [24].

We demonstrated the first proof-of-concept for designing LOO-GFP biosensors by combining computational protein design and *in vitro* evolution. DEEdesign was used to create a set of degenerate oligonucleotide primers for gene assembly. DNA shuffling was performed directly on this set of genes to further increase the diversity of the constructed library, since gene assembly mutagenesis does not ensure complete representation of all possible anticipated sequences [94]. DNA shuffling also introduces random mutagenesis beyond the predicted mutations on the gene variant.

Potential candidates for LOO-GFP biosensors were plate-screened in *E. coli* that co-expressed the biosensor gene library and the HA peptide fused to a carrier, intein. Expression of both peptide and biosensor library were induced simultaneously, and the intensity of fluorescence was monitored under excitation of 488 nm wavelength after the induction of 24 hours at room temperature. Two potential LOO-GFP biosensors, DS1 and DS2, that produced elevated fluorescence intensity in the presence of the HA peptide were found (Figure 12). There were nine and sixteen mutations found in DS1 and DS2 respectively, and seven of those mutations were from DEEdesign prediction and the remainder were from *in vitro* evolution.

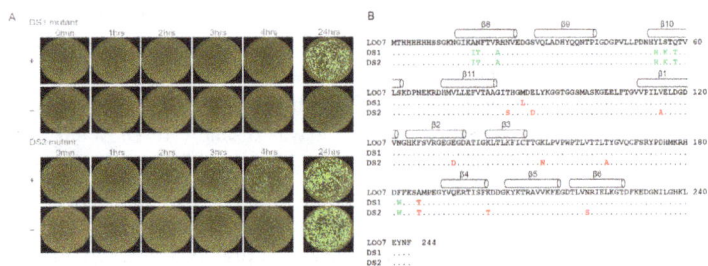

Figure 12. Potential LOO-GFP biosensors against HA target peptides of influenza virus. (A) Time course study of fluorescence recovery upon expression of biosensor variants with [+] and without [-] HA peptides. Protein expression was induced with 0.5mM IPTG and under room temperature. Fluorescence was record every hour for 4 hours and after 24 hours. All pictures were taken with the same setting of digital camera. (B) Multiple sequence alignment of LOO7, DS1 and DS2 mutant. Mutations introduced by computational design (green) and *in vitro* evolution (red) in DS1 and DS2 mutants are shown.

When co-expressed with the HA peptide, the DS1 mutant exhibited target-dependent maturation of chromophore, while in the absense of the peptide it showed barely detectable fluorescence even after 24 hours, indicating a specific interaction between DS1 mutant and the HA peptide. DS2 mutant showed faster recovery of fluorescence within four hours in the presence of the HA peptide; however, a higher degree of nonspecific auto-fluorescence was also observed after 24 hours. The DS1 mutant chromophore formation showed a greater dependency on the left-out peptide (i.e. the HA peptide), implying better folding of designed LOO-GFP molecule, than DS2 mutant *in vivo*, showing DS1 mutant as a better HA-specific LOO-GFP biosensor.

6. Conclusions

The unique physical properties of GFP have made it a gold mine for the development of biosensors and biomarkers. GFP is kinetically super-stable. Its sequence may be readily permuted and mutated. Its engineered variants fluoresce at wavelengths across the visual spectrum, and some pairs of variants can interact via FRET. GFP is quenched by unfolding, by certain ions, and sometimes by light, and variants of GFP are pH sensitive. With many ways of generating a signal, it is no surprise that many types of biosensors have been developed that use GFP and its homolog fluorescent proteins.GFP and its variants can be immobilized and even dried while retaining structure and biosensor function, leading to the promise of future GFP-biosensor microarrays capable of detecting a wide variety of analytes in a single assay. In addition to being broadly useful, such material should be very cheap to produce, and would also be easily stored, used, and read.Arrays of GFP-based biosensors on paper or film may someday become available for household use, so that infections may be rapidly diagnosed without a trip to the hospital, or may become integral parts of devices that continuously monitor the water and air, making the world a healthier and safer place.

Author details

Donna E. Crone, Yao-Ming Huang, Derek J. Pitman, Christian Schenkelberg, Keith Fraser, Stephen Macari and Christopher Bystroff

Department of Biology, Rensselaer Polytechnic Institute, Troy, New York, USA

References

[1] Ormö M, Cubitt AB, Kallio K, Gross LA, Tsien RY, Remington SJ. Crystal Structure of the *Aequorea victoria* Green Fluorescent Protein. Science 1996;273(5280): 1392-5. http://www.sciencemag.org/content/273/5280/1392.long (accessed 15 July 2012)

[2] Osamu Shimomura - Nobel Lecture: Discovery of Green Fluorescent Protein, GFP: Nobelprize.org http://www.nobelprize.org/nobel_prizes/chemistry/laureates/2008/shimomura-lecture.html (accessed 15 July 2012)

[3] Tsien RY. The Green Fluorescent Protein. Annual Reviews of Biochemistry 1998;67: 509-544. http://www.annualreviews.org/doi/pdf/10.1146/annurev.biochem.67.1.509 (accessed 15 July 2012)

[4] Crameri A, Whitehorn EA, Tate E, Stemmer WPC. Improved Green Fluorescent Protein by Molecular Evolution Using DNA Shuffling. Nature Biotechnology 1996;14(3): 315-319. http://www.nature.com/nbt/journal/v14/n3/full/nbt0396-315.html (accessed 15 July 2012)

[5] Enoki S, Saeki K, Maki K, Kuwajima K. Acid Denaturation and Refolding of Green Fluorescent Protein. Biochemistry 2004;43(44): 14238-48. http://pubs.acs.org/doi/abs/ 10.1021/bi048733%2B (accessed 15 July 2012)

[6] Fukuda H, Arai M, Kuwajima K. Folding of Green Fluorescent Protein and the Cycle3 Mutant. Biochemistry 2000;39(39): 12025-32. http://pubs.acs.org/doi/abs/10.1021/ bi0005431 (accessed 15 July 2012)

[7] Heim R, Prasher DC, Tsien RY. Wavelength mutations and posttrasnlational autoxidation of green fluorescent protein. Proceedings of the National Academy of Science of the United States of America 1994;91(26): 12501-4. http://www.pnas.org/content/ 91/26/12501.long (accessed 15 July 2012)

[8] Heim R, Tsien RY. Engineering green fluorescent protein for improved brightness, longer wavelengths and fluorescence resonance energy transfer. Current Biology 1996;6(2): 178-82. http://dx.doi.org/10.1016/ S0960-9822(02)00450-5 (accessed 15 July 2012)

[9] Rosenow MA, Huffman HA, Phail ME, Wachter RM. The Crystal Structure of the Y66L Variant of Green Fluorescent Protein Supports a Cyclization-Oxidation-Dehydration Mechanism for Chromophore Maturation. Biochemistry 2004;43(15): 4464-72. http://pubs.acs.org/doi/abs/10.1021/bi0361315 (accessed 15 July 2012)

[10] Henderson NJ, Ai HW, Campbell RE, Remington SJ. Structural basis for reversible photobleaching of a green fluorescent protein homologue. Proceedings of the National Academy of Sciences of the United States of America 2007;104(16): 6672-7. http:// www.pnas.org/content/104/16/6672.long (accessed 15 July 2012)

[11] Ward WW, Prentice, HJ, Roth AF. Spectral Perturbations of the *Aequorea* Green-Fluorescent Protein. Photochemistry and Photobiology 1982;35(6): 803-808. DOI: 10.1111/j. 1751-1097.1982.tb02651.x

[12] Ward WW, Cody CW, Hart RC, Cormier MJ. Spectrophotometric Identity of the Energy Transfer Chromophores in *Renillia* and *Aequorea* Green-Fluorescent Proteins. Photochemistry and Photobiology 1980;31(6): 611-615. DOI: 10.1111/j. 1751-1097.1980.tb03755.x

[13] Shaner NC, Campbell RE, Steinbach PA, Giepmans BNG, Palmer AE, Tsien RY. Improved monomeric red, orange and yellow fluorescent proteins derived from *Discosoma sp.* red fluorescent protein. *Nature Biotechnology* 2004;22(12): 1567-72. doi: 10.1038/nbt1037 http://www.nature.com/nbt/journal/v22/n12/full/nbt1037.html (accessed 16 July 2012).

[14] Petersen J, Wilmann PG, Beddoe T, Oakley AJ, Devenish RJ, Prescott M, Rossjohn J. The 2.0-Å Crystal Structure of eqFP611, a Far Red Fluorescent Protein from the Sea Anemone *Entacmaea quadricolor*. The Journal of Biological Chemistry2003;278(45): 44626-31. http://www.jbc.org/content/278/45/44626.full (accessed 16 July 2012).

[15] Gurskaya NG, Fradkov AF, Terskikh A, Matz MV, Labas YA, Martynov VI, Yanushevich YG, Lukyanov KA, Lukyanov SA. GFP-like chromoproteins as a source of far-red fluorescent proteins. FEBS Letters2001;507(1): 16-20. http://www.sciencedirect.com/science/article/pii/S0014579301029301 (accessed 16 July 2012).

[16] Shkrob MA, Yanushevich YG, Chudakov DM, Gurskaya NG, Labas YA, Poponov SY, Mudrik NN, Lukyanov S, Lukyanov KA. Far-red fluorescent proteins evolved from a blue chromoprotein from *Actinia equina*. Biochemical Journal 2005;392(Pt 3): 649-54. http://www.biochemj.org/bj/392/0649/bj3920649.htm (accessed 16 July 2012).

[17] Subach OM, Patterson GH, Ting L-M, Wang Y, Condeelis JS, Verkhusha VV. A photoswitchable orange-to-far-red fluorescent protein, PSmOrange. Nature Methods2011;8(9): 771-7. http://www.nature.com/nmeth/journal/v8/n9/full/nmeth.1664.html (accessed 16 July 2012).

[18] Chica RA, Moore MM, Allen BD, Mayo SL. Generation of longer emission wavelength red fluorescent proteins using computationally designed libraries. Proceedings of the National Academy of Sciences of the United States of America 2010;107(47): 20257-62. http://www.pnas.org/content/107/47/20257 (accessed 16 July 2012).

[19] Pollok BA, Heim R. Using GFP in FRET-based applications. Trends in Cell Biology1999;9(2): 57-60. http://www.sciencedirect.com/science/article/pii/S0962892498014342 (accessed 16 July 2012).

[20] Markwardt ML, Kremers G-J, Kraft CA, Ray K, Cranfill PJ, Wilson KA, Day RN, Wachter RM, Davidson MW, Rizzo MA. An Improved Cerulean Fluorescent Protein with Enhanced Brightness and Reduced Reversible Photoswitching. PLoS One2011;6(3): e17896. http://www.plosone.org/article/info%3Adoi%2F10.1371%2Fjournal.pone.0017896 (accessed 16 July 2012).

[21] Albertazzi L, Arosio D, Marchetti L, Ricci F, Beltram F. Quantitative FRET Analysis With the EGFP-mCherry Fluorescent Protein Pair. Photochemistry and Photobiology 2009;85(1): 287-97. http://onlinelibrary.wiley.com/doi/10.1111/j.1751-1097.2008.00435.x/full (accessed 16 July 2012).

[22] Day RN, Davidson MW. Fluorescent proteins for FRET microscopy: monitoring protein interactions in living cells. Bioessays2012;34(5): 341-50. http://onlinelibrary.wiley.com/doi/10.1002/bies.201100098/full (accessed 16 July 2012).

[23] Pritchard L, Corne D, Kell D, Rowland J, Winson M. A general model of error-prone PCR. Journal of Theoretical Biology 2005;234(4): 497-509. http://www.sciencedirect.com/science/article/pii/S0022519304006071 (accessed 15 July 2012)

[24] Stemmer WP. DNA shuffling by random fragmentation and reassembly: in vitro recombination for molecular evolution. Proceedings of the National Academy of Sciences of the United States of America 1994;91(22): 10747-51 http://www.pnas.org/content/91/22/10747.long (accessed 15 July 2012).

[25] Cormack BP, Valdivia RH, Falkow S. FACS-optimized mutants of the green fluorescent protein. Gene1996;173: 33-8 http://www.sciencedirect.com/science/article/pii/0378111995006850 (accessed 16 July 2012).

[26] Waldo GS, Standish BM, Berendzen J, Terwilliger TC. Rapid protein-folding assay using green fluorescent protein. Nature Biotechnology 1999;17(7): 691-5. http://www.nature.com/nbt/journal/v17/n7/full/nbt0799_691.html (accessed 15 July 2012)

[27] Pédelacq JD, Cabantous S, Tran T, Terwilliger TC, Waldo GS. Engineering and characterization of a superfolder green fluorescent protein. Nature Biotechnology 2006;24(1): 79-88. http://www.nature.com/nbt/journal/v24/n1/abs/nbt1172.html (accessed 16 July 2012).

[28] Lawrence MS, Phillips KJ, Liu DR. Supercharging Proteins Can Impart Unusual Resilience. *Journal of the American Chemical Society* 2007;129(33): 10110-2. http://pubs.acs.org/doi/full/10.1021/ja071641y (accessed 16 July 2012).

[29] Melnik BS, Povarnitsyna TV, Glukhov AS, Melnik TN, Uversky VN. SS-Stabilizing Proteins Rationally: Intrinsic Disorder-Based Design of Stabilizing Disulphide Bridges in GFP. Journal of biomolecular structure and dynamics2012;29(4): 815-24. DOI:10.1080/07391102.2012.10507414

[30] Rosenman D, Huang YM, Xia K, Colon W, Van Roey P, Bystroff C. Green-lighting GFP folding by loop remodeling to eliminate a cis-trans peptide isomerization. (in preparation), 2012.

[31] Evdokimov AG, Pokross ME, Egorov NS, Zaraisky AG, Yampolsky IV, Merzlyak EM, Shkoporov AN, Sander I, Lukyanov KA, Chudakov DM. Structural basis for the fast maturation of Arthropoda green fluorescent protein. *EMBO Reports* 2006;7(10): 1006-12. http://www.nature.com/embor/journal/v7/n10/full/7400787.html (accessed 16 July 2012).

[32] Dai M, Fisher HE, Temirov J, Kiss C, Phipps ME, Pavlik P, Werner JH, Bradbury AR. The creation of a novel fluorescent protein by guided consensus engineering. *Protein Engineering Design & Selection* 2007;20(2): 69-79. http://peds.oxfordjournals.org/content/20/2/69.full (accessed 16 July 2012).

[33] Kiss C, Temirov J, Chasteen L, Waldo GS, Bradbury AR. Directed evolution of an extremely stable fluorescent protein. *Protein Engineering Design & Selection* 2009;22(5): 313-23. http://peds.oxfordjournals.org/content/22/5/313.full (accessed 16 July 2012).

[34] Topell S, Hennecke J, Glockshuber R. Circularly permuted variants of the green fluorescent protein. *FEBS Letters* 1999;457(2): 283-9. http://www.sciencedirect.com/science/article/pii/S0014579399010443 (accessed 16 July 2012).

[35] Baird GS, Zacharias DA, Tsien RY. Circular permutation and receptor insertion within green fluorescent proteins. Proceedings of the National Academy of Sciences of the United States of America 1999;96(20): 11241-6. http://www.pnas.org/content/96/20/11241.long (accessed 16 July 2012).

[36] Reeder PJ, Huang Y-M, Dordick JS, Bystroff C. A Rewired Green Fluorescent Protein: Folding and Function in a Nonsequential, Noncircular GFP Permutant. *Biochemistry* 2010;49(51): 10773-9. http://pubs.acs.org/doi/full/10.1021/bi100975z (accessed 16 July 2012).

[37] Gilbert D, Westhead D,Nagano N, Thornton J. Motif-based searching in TOPS protein topology databases. Bioinformatics 1999;15(4): 317-326 http://bioinformatics.oxfordjournals.org/content/15/4/317.abstract%C2%B5%EF%A3%B9%CE%B0z (accessed 16 July 2012)

[38] Pavoor TV, Cho YK, Shusta EV. Development of GFP-based biosensors possessing the binding properties of antibodies. *Proceedings of the National Academy of Sciences of the United States of America* 2009;106(29): 11895-900. http://www.pnas.org/content/106/29/11895.long (accessed 15 July 2012)

[39] Huang YM, Bystroff C. Complementation and Reconstitution of Fluorescence from Circularly Permuted and Truncated Green Fluorescent Protein. *Biochemistry* 2009;48(5): 929-40. http://pubs.acs.org/doi/full/10.1021/bi802027g (accessed 16 July 2012).

[40] Huang YM, Nayak S, Bystroff C. Quantitative *in vivo* solubility and reconstitution of truncated circular permutants of green fluorescent protein. *Protein Science* 2011;20(11): 1775-80. http://onlinelibrary.wiley.com/doi/10.1002/pro.735/full (accessed 16 July 2012).

[41] Kent KP, Oltrogge LM, Boxer SG. Synthetic Control of Green Fluorescent Protein. *Journal of the American Chemical Society* 2009;131(44): 15988-9. http://pubs.acs.org/doi/full/10.1021/ja906303f (accessed 16 July 2012).

[42] Nemethy G, Scheraga HA. A Possible Folding Pathway of Bovine Pancreatic RNase. *Proceedings of the National Academy of Sciences* 1979;76(12): 6050-6054. http://www.ncbi.nlm.nih.gov/pmc/articles/PMC411798/?tool=pubmed (accessed 16 July 2012)

[43] Kent KP, Boxer SG. Light-Activated Reassembly of Split Green Fluorescent Protein. *Journal of the American Chemical Society* 2011;133(11): 4046-52. http://pubs.acs.org/doi/full/10.1021/ja110256c (accessed 12 June 2012).

[44] Do K, Boxer SG. Thermodynamics, Kinetics, and Photochemistry of β-Strand Association and Dissociation in a Split-GFP System. *Journal of the American Chemical Society* 2011;133(45): 18078-81. http://pubs.acs.org/doi/full/10.1021/ja207985w (accessed 4 June 2012).

[45] Chalfie M, Tu Y, Euskirchen G, Ward WW, Prasher DC. Green fluorescent protein as a marker for geneexpression. Science. 1994;263(5148): 802-5. http://www.sciencemag.org/content/263/5148/802.long (accessed 15 July 2012)

[46] Prasher DC, Eckenrode VK, Ward WW, Prendergast FG, Cormier MJ. Primary structure of the Aequorea victoria green-fluorescent protein. Gene. 1992;111(2): 229-33.

[47] Chudakov DM, Matz MV, Lukyanov S, Lukyanov KA. Fluorescent proteins and their applications inimaging living cells and tissues. Physiological Reviews 2010;90(3): 1103-63. http://physrev.physiology.org/content/90/3/1103.long (accessed 15 July 2012)

[48] Kremers GJ, Gilbert SG, Cranfill PJ, Davidson MW, Piston DW. Fluorescent proteins at a glance. Journal of Cell Science 2011;124(Pt 2): 157-60. doi: 10.1242/jcs.072744 http://jcs.biologists.org/content/124/2/157.long (accessed 15 July 2012)

[49] Miyawaki A. Proteins on the move: insights gained from fluorescent protein technologies. Nature revews. Molecular cell biology 2011;12(10): 656-68. doi: 10.1038/ nrm3199. http://www.nature.com/nrm/journal/v12/n10/full/nrm3199.html (accessed 15 July 2012)

[50] Cabantous S, Terwilliger TC, Waldo GS. Protein tagging and detection with engineered self-assemblingfragments of green fluorescent protein. Nature Biotechnology. 2005;23(1): 102-7. http://www.nature.com/nbt/journal/v23/n1/full/nbt1044.html (accessed 15 July 2012)

[51] Goedert M. Tau protein and the neurofibrillary pathology of Alzheimer's disease. Annals of the New York Academy of Sciences 1996;777: 121-31. DOI:10.1111/j. 1749-6632.1996.tb34410.x (accessed 15 July 2012).

[52] Chun W, Waldo GS, Johnson GV. Split GFP complementation assay: a novel approach to quantitatively measure aggregation of tau in situ: effects of GSK3beta activation and caspase 3 cleavage. Journal of neurochemistry 2007;103(6): 2529-39. DOI: 10.1111/j.1471-4159.2007.04941.x   http://onlinelibrary.wiley.com/doi/10.1111/j. 1471-4159.2007.04941.x/abstract (accessed 15 July 2012).

[53] Chun W, Waldo GS, Johnson GV. Split GFP complementation assay for quantitative measurement of tau aggregation in situ. In: Roberson, ED. (ed.) Alzheimer's Disease and Frontotemporal Dementia Methods and Protocols. CliftonNJ: Springer. 2011. p109-23. Available from http://www.springerlink.com/content/l8444n2g5116g5t7/? MUD=MP (accessed 15 July 2012)

[54] Rothbauer U, Zolghadr K,Tillib S, Nowak D, Schermelleh L, Gahl A, Backmann N, Conrath K, Muyldermans S, Cardoso MC, Leonhardt H. Targeting and tracing antigens in live cells with fluorescent nanobodies. *Nature methods 2006;*3(11): 887-889. http://www.nature.com/nmeth/journal/v3/n11/full/nmeth953.html (accessed 15 July 2012)

[55] Betzig E; Patterson GH, Sougrat R, Lindwasser OW, Olenych S, Bonifacino JS, Davidson MW, Lippincott-Schwartz J, Hess HF. Imaging intracellular fluorescent proteins at nanomolor resolution. *Science 2006;*313(5793): 1642-1645. http://www.sciencemag.org/content/313/5793/1642.long (accessed 16 July 2012)

[56] Ries J, Kaplan C, Platonova E, Eghlidi H, Ewers H. A simple, versatile method for GFP-based super-resolution microscopy via nanobodies. *Nature methods 2012;* 9(6): 582-584. doi: 10.1038/nmeth.1991. http://www.nature.com/nmeth/journal/v9/n6/full/ nmeth.1991.html (Accessed 15 July 2012).

[57] Kamioka Y, Sumiyama K, Mizuno R, Sakai Y, Hirata E, Kiyokawa E, Matsuda M. Live imaging of protein kinase activities in transgenic mice expressing FRET biosensors. *Cell structure and function* 2012;37(1): 65-73. http://dx.doi.org/10.1247/csf.11045 (accessed 15 July 2012).

[58] Bizzarri R, Serresi M, Luin S, Beltram F. Green fluorescent protein based pH indicators for in vivo use: a review. Analytical and bioanalytical chemistry 2009;393(4): 1107-22. DOI: 10.1007/s00216-008-2515-9 http://www.springerlink.com/content/y73v37381q82v3h6/ (accessed 15 July 2012).

[59] Choi WG, Swanson SJ, Gilroy S. High-resolution imaging of Ca2+ , redox status, ROS and pH using GFP biosensors. The Plant journal 2012;70(1): 118-28. doi: 10.1111/j.1365-313X.2012.04917.x. http://onlinelibrary.wiley.com/doi/10.1111/j.1365-313X.2012.04917.x/abstract (accessed 15 July 2012).

[60] Gjetting KS, Ytting CK, Schulz A, Fuglsang AT. Live imaging of intra- and extracellular pH in plants using pHusion, a novel genetically encoded biosensor. Journal of experimental botany. 2012;63(8): 3207-18. http://jxb.oxfordjournals.org/content/63/8/3207.long (accessed 15 July 2012).

[61] Patterson GH, Lippincott-Schwartz J. A photoactivatable GFP for selective photolabeling of proteins and cells. Science 2002;297(5588): 1873-7. http://www.sciencemag.org/content/297/5588/1873.long (accessed 15 July 2012).

[62] Brakemann T, Stiel AC, Weber G, Andresen M, Testa I, Grotjohann T, Leutenegger M, Plessmann U, Urlaub H, Eggeling C, Wahl MC, Hell SW, Jakobs S. A reversibly photoswitchable GFP-like protein with fluorescence excitation decoupled from switching. Nature Biotechnology 2011;29(10): 942-7. doi: 10.1038/nbt.1952. http://www.nature.com/nbt/journal/v29/n10/full/nbt.1952.html (accessed 15 July 2012)

[63] Kneen M, Farinas J, Li Y, Verkman AS. Green fluorescent protein as a noninvasive intracellular pH indicator. Biophysical journal 1998;74(3): 1591-9. http://www.sciencedirect.com/science/article/pii/S0006349598778701 (accessed 7 July 2012)

[64] Miesenböck G, De Angelis DA, Rothman JE. Visualizing secretion and synaptic transmission with pH-sensitive green fluorescent proteins. Nature 1998;394(6689): 192-5. http://www.nature.com/nature/journal/v394/n6689/full/394192a0.html (accessed 15 July 2012)

[65] Hanson GT, McAnaney TB, Park ES, Rendell ME, Yarbrough DK, Chu S, Xi L, Boxer SG, Montrose MH, Remington SJ. Green fluorescent protein variants as ratiometric dual emission pH sensors. 1. Structural characterization and preliminary application. Biochemistry 2002;41(52): 15477-88. http://pubs.acs.org/doi/abs/10.1021/bi026609p (accessed 15 July 2012)

[66] Elsliger MA, Wachter RM, Hanson GT, Kallio K, Remington SJ. Structural and spectral response of green fluorescent protein variants to changes in pH. Biochemistry 1999;27;38(17): 5296-301. http://pubs.acs.org/doi/abs/10.1021/bi9902182 (accessed 15 July 2012)

[67] Bizzarri R, Arcangeli C, Arosio D, Ricci F, Faraci P, Cardarelli F, Beltram F. Development of a novel GFPbased ratiometric excitation and emission pH indicator for intracellular studies. Biophysical journal 2006;90(9): 3300-14. http:// www.sciencedirect.com/science/article/pii/S0006349506725127 (accessed 15 July 2012)

[68] Arosio D, Ricci F, Marchetti L, Gualdani R, Albertazzi L, Beltram F. Simultaneous intracellular chloride and pH measurements using a GFP-based sensor. Nature methods 2010;7(7): 516-8. http://www.nature.com/nmeth/journal/v7/n7/full/nmeth. 1471.html (accessed 15 July 2012)

[69] Awaji T, Hirasawa A, Shirakawa H, Tsujimoto G, Miyazaki S. Novel green fluorescent protein-based ratiometric indicators for monitoring pH in defined intracellular microdomains. Biochemical and biophysical research communications 2001;289(2): 457-62. http://www.sciencedirect.com/science/article/pii/S0006291X01960048 (accessed 15 July 2012)

[70] Zhou X, Herbst-Robinson K, Zhang J. Visualizing dynamic activities of signaling enzymes using genetically encodable FRET-based biosensors. From design to applications. In: Conn PM (ed.) Methods in enzymology:Imaging and Spectroscopic Analysis of Living CellsOptical and Spectroscopic Techniques. San Diego, Calif. : Elsevier Inc. *2012.* 504: 317-340. Available from http://www.sciencedirect.com/ science/article/pii/B9780123918574000161 (accessed 16 July 2012)

[71] Komatsu N, Aoki K, Yamada M, Yukinaga H, Fujita Y, Kamioka Y, Matsuda M. Development of an optimized backbone of FRET biosensors for kinases and GTPases. Molecular biology of the cell 2011;22(23): 4647-56. http://www.molbiolcell.org/ content/22/23/4647.long (accessed 16 July 2012)

[72] Eiamphungporn W, Prachayasittikul S, Isarankura-Na-Ayudha C, Prachayasittikul V. Development of bacterial cell-based system for intracellular antioxidant activity screening assay using green fluorescence protein (GFP) reporter. *African Journal of Biotechnology 2012;*11(27): 6934-45. DOI: 10.5897/AJB11.3790 http://www.academicjournals.org/ajb/PDF/pdf2012/3Apr/Eiamphungporn%20et%20al.pdf (accessed 15 July 2012)

[73] Hires, S.A., L. Tian, and L.L. Looger, Reporting neural activity with genetically encoded calcium indicators. Brain cell biology 2008;36(1-4): 69-86. http://www.springerlink.com/content/a145q526472454q4/ (accessed 15 July 2012).

[74] Tian L,Hires SA,Mao T,Huber D,Chiappe ME,Chalasani SH,Petreanu L,Akerboom J,McKinney SA, Schreiter ER. Imaging neural activity in worms, flies and mice with improved GCaMPcalcium indicators. Nature methods 2009;6(12): 875-81. http:// www.nature.com/nmeth/journal/v6/n12/full/nmeth.1398.html (accessed 15 July 2012)

[75] Boder ET, Wittrup KD. Yeast surface display for screening combinatorial polypeptide libraries. Nature Biotechnology 1997;15(6): 553-7. http://www.nature.com/nbt/ journal/v15/n6/full/nbt0697-553.html (accessed 15 July 2012)

[76] Doi N,Yanagawa H. Design of generic biosensors based on green fluorescent proteins with allosteric sites by directed evolution. *FEBS letters 1999;453*(3): 305-7. http://dx.doi.org/10.1016/S0014-5793(99)00732-2 (accessed 15 July 2012)

[77] Boeneman K, Mei BC, Dennis AM, Bao G, Deschamps JR, Mattoussi H, Medintz IL. Sensing caspase 3 activity with quantum dot-fluorescent protein assemblies. *Journal of the American Chemical Society 2009;131*(11); 3828-9. doi:10.1021/ja809721j http://pubs.acs.org/doi/abs/10.1021/ja809721j (accessed 15 July 2012)

[78] Dennis, A. M., Rhee, W. J., Sotto, D., Dublin, S. N., & Bao, G. (2012). Quantum dot-fluorescent protein FRET probes for sensing intracellular pH. *ACS nano 2012;6*(4): 2917-24. doi:10.1021/nn2038077 http://pubs.acs.org/doi/abs/10.1021/nn2038077 (accessed 15 July 2012)

[79] Tansila N, Becker K, Na-ayudhya CI, Prachayasittikul V, Bu L. Metal ion accessibility of histidine-modified superfolder green fluorescent protein expressed in Escherichia coli. *Biotechnology Letters 2008;*30(8): 1391-1396. doi:10.1007/s10529-008-9692-7 http://www.springerlink.com/content/3813241647413002/ (accessed 15July 2012)

[80] Tansila N, Tantimongcolwat T, Isarankura-Na-Ayudhya C, Nantasenamat C, Prachayasittikul V. Rational design of analyte channels of the green fluorescent protein for biosensor applications. *International journal of biological sciences 2007;*3(7): 463-70. http://www.biolsci.org/v03p0463.htm (accessed 15 July 2012)

[81] Schueler-Furman O, Wang C, Bradley P, Misura K, Baker D. Progress in Modeling of Protein Structures and Interactions. Science 2005;310(5748): 638-42. http://www.sciencemag.org/content/310/5748/638.long (accessed 15 July 2012)

[82] Street AG, Mayo SL. Computational Protein Design. Structure 1999;7(5): R105-R109.

[83] Park S, Yang X, Saven JG. Advances in Computational Protein Design. Current Opinion in Structural Biology 2004;14(4): 487-94. http://www.sciencedirect.com/science/article/pii/S0959440X04001009 (accessed 15 July 2012)

[84] Lippow SM, Tidor B. Progress in Computational Protein Design. Current Opinion in Biotechnology 2007;18(4): 305-11. http://www.sciencedirect.com/science/article/pii/S0958166907000778 (accessed 15 July 2012)

[85] Dunbrack RL. Rotamer Libraries in the 21st Century. Current Opinion in Structural Biology 2002;12(4): 431-40. http://www.sciencedirect.com/science/article/pii/S0959440X02003445 (accessed 16 July 2012)

[86] Lovell SC, Word JM, Richardson JS, Richardson DC. The Penultimate Rotamer Library. Proteins 2000;40(3): 389-408. doi: 10.1002/1097-0134(20000815)40:3<389::AID-PROT50>3.0.CO;2-2 (accessed 16 July 2012)

[87] Lovell SC, Davis IW, Arendall WB 3rd, de Bakker PI, Word JM, Prisant MG, Richardson JS, Richardson DC. Structure Validation by Cα Geometry: φ,ψ and Cβ Deviation. Proteins 2003;50(3) 437-450. doi: 10.1002/prot.10286 http://onlinelibrary.wiley.com/doi/10.1002/prot.10286/abstract (accessed 16 July 2012)

[88] Pokala N, Handel TM, Review: protein design--where we were, where we are, where we're going. Journal of structural biology 2001;134(2-3): 269-81. http://www.science-direct.com/science/article/pii/S1047847701943497 (accessed 16 July 2012)

[89] Hayes RJ, Bentzien J, Ary ML, Hwang MY, Jacinto JM, Vielmetter J, Kundu A, Dahiyat BI. Combining Computational and Experimental Screening for Rapid Optimization of Protein Properties. Proceedings of the National Academy of Sciences of the United States of America 2002;99(25): 15926-31 http://www.pnas.org/content/99/25/15926.long (accessed 15 July 2012)

[90] Huang YM, Bystroff C. Exploring Objective Functions and Cross-terms in the Optimization of an Energy Function for Protein Design. (submitted) ACMBCB2012, 7-10 October 2012, Orlando, Florida, USA 2012

[91] Shakhnovich EI, Gutin AM, A new approach to the design of stable proteins. Protein Engineering 1993;6(8): 793-800.

[92] Desmet J,De Maeyer M, Hazes B, Lasters I. The dead-end elimination theorem and its use in protein side-chain positioning. Nature 1992;356(6369): 539-42. doi: 10.1038/356539a0   http://www.nature.com/nature/journal/v356/n6369/pdf/356539a0.pdf (accessed 16 July 2012)

[93] Hoover D. M. and Lubkowski J., DNAWorks: an automated method for designing oligonucleotides for PCR-based gene synthesis. Nucleic Acids Research 2002;30(10): e43. http://nar.oxfordjournals.org/content/30/10/e43.long (accessed 16 July 2012)

[94] Bessette PH, Mena MA, Nguyen AW, Daugherty PS. Construction of Designed Protein Libraries Using Gene Assembly Mutagenesis. Methods in Molecular Biology 2003;231(1) 29-37. DOI: 10.1385/1-59259-395-X:29

[95] Intein – Wikipedia: http://en.wikipedia.org/wiki/Intein. (20 July2012)

[96] Cirino P. C., Mayer K. M., and Umeno D., Generating mutant libraries using error-prone PCR. In: Arnold FH, Georgiou G. (eds.) Methods In Molecular Biology: Directed Evolution Library Creation: Methods and Protocols. TotowaNJ:Humana Press; 2003. 231: p. 3-10. Available from http://www.springerlink.com/content/978-1-58829-285-8/ (accessed 16 July 2012)

Chapter 2

Polymers for Biosensors Construction

Xiuyun Wang and Shunichi Uchiyama

Additional information is available at the end of the chapter

1. Introduction

The unprecedented interest in the development and exploitation of analytical devices for detection, quantification and monitoring of specific chemical species has led to the emergence of biosensors. Electrochemical biosensors have gained ever-increasing acceptance in the field of medical diagnostics, health care, environmental monitoring, and food safety due to high sensitivity, specificity, and ability for real-time analysis coupled with speed and low cost and polymers are promising candidates that can facilitate a new generation of biosensors [1-7]. A biosensor is a device having a biological sensing element either intimately connected to or integrated within a transducer. The aim is to produce a digital electronic signal, which is proportional to the concentration of a specific chemical or set of chemicals (Fig. 1). A definition of biosensor is proposed by IUPAC as: "a self-contained integrated device, which is capable of providing specific quantitative or semi-quantitative analytical information using a biological recognition element (biochemical receptor), which is retained in direct spatial contact with a transducer element." The life time of the enzyme electrode, the rate of electron transfer be-tween the enzymatic redox reaction and electrode, and the miniaturization of enzyme elec-trode are some of the critical points appeared as central to this interdisciplinary research. Biosensor has been pursued extensively in a wide range for their unparalleled selectivity and mild reaction conditions. As a coin has two sides, enzymes which are the key biological recognition element are usually costly and easy to inactivate in their free forms. The immobilization of enzymes is the main approach to optimizing the in-service performance of an enzyme, particularly in the field of non-aqueous phase catalysis. However, the immobilization process for enzymes will inevitably result in some loss of activity, improving the activity retention of the immobilized enzyme is critical. To some extent, the performance of an immobilized enzyme is mainly governed by the supports used for immobilization, thus it is important to fully understand the properties of supporting materials and immobilization processes [8-10]. The properties of immobilized enzymes are governed by the properties of both the enzyme and the support material. The interaction between the two lends an immobilized enzyme specific physico-chemical and kinetic properties that may be decisive for its practical application, and thus, a support judiciously chosen can significantly enhance the operational performance of the immobilized system. It is widely acknowledged that analytical sensing at electrodes modified with polymeric materials results in low detection limits, high sensitivities, lower applied potential, good stability, efficient electron transfer and easier immobilization of enzymes on electrodes. In recent years, there has been growing concern in using polymeric materials as supports for their good mechanical and easily adjustable properties [11]. Of the many carriers that have been considered and studied for immobilizing enzymes, conducting polymer (CP), redox polymer (RP), sol-gel and hydrogel materials, chitin and chitosan are of interest in that they offer most of the above characteristics [12, 13].

The proper use of different compositions of binder and immobilization matrix, electron transport mediators, biomaterials and biocatalysts, and solid supports as electron collectors in the construction of enzyme electrode is critical to generate optimum current from the enzymatic redox reactions. In essence, design and fabrication of advanced materials coupled with good understanding of their behaviors when incorporated as interfacial or transducer elements would be of paramount importance. The recent advancement of polymer materials is greatly influencing the redox reactions and electron transport kinetics of the enzyme electrodes. To achieve high specificity, high sensitivity, rapid response and flexibility of use, it is clear that the research continues to focus on new assembly strategies. Polymers are becoming inseparable from biomolecule immobilization strategies and biosensor platforms. Their original role as electrical insulators has been progressively substituted by their electrical conductive abilities, which opens a new and broad scope of applications. This chapter highlights recent contributions in the incorporation of promising polymeric materials within biosensors, special emphasis was placed on different classes of polymeric materials such as nanomaterials, sol-gel and hydrogel materials, conducting polymers, functional polymers and biomaterials that have been used in the design of sensors and biosensors. We want to remind our readers that this chapter is not intended to provide comprehensive coverage of electrochemical biosensor development but rather to provide a glimpse of the incorporation of polymers within biosensors. These materials have attracted much attention to their potentials for interesting applications, broad applicability as well as tunable properties according to applications needs. In addition, the critical issues related to the fabrication of enzyme electrodes and their application for biosensor applications are also highlighted in this article. Effort has been made to cover the recent literature on the advancement of polymers to develop enzyme electrodes and their potential applications for the construction of biosensors [14-15].

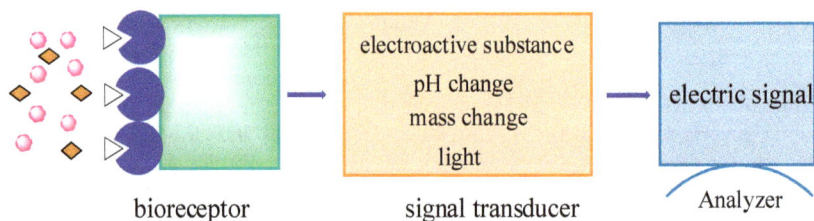

Figure 1. Principle of biosensor

2. Construction of enzyme-based biosensors

2.1. Immobilization methods of enzyme for biosensors

Enzyme immobilization is one of the most important subjects for any enzyme-based biosensor research. Considerable efforts have been invested in this topic for a number of years [16-18]. The biosensing process of the immobilized enzyme is regarded as a heterogeneous phase reaction; thus, the main consideration for enzyme immobilization is to achieve stable and high enzymatic activity with low mass-transfer resistance. Enzymes electrodes have the longest tradition in the field of biosensors. They are one of the most intensely investigated biosensors due to highly selective and fast response. One of the key factors in developing a reliable biosensor is the immobilization of enzymes at transducer surfaces. The method of enzyme immobilization plays an important role on the performance of an enzyme electrode, such as lifetime, linear range, sensitivity, selectivity, response time, stability and anti-interferent.

Enzymes may be immobilized by a variety of methods (Fig. 2), which may be broadly classified as physical approaches and chemical approaches. To the physical methods belong: (i) physical adsorption on a water-insoluble matrix based on hydrophobic, electrostatic and van der Waals attractive forces; (ii) entrapment enzyme in sol-gel, or hydrogel, or a paste, confined by semi-permeable membranes; (iii) microencapsulation with a solid membrane; (iv) encapsulation, containment of an enzyme within a membrane reactor; (v) formation of enzymatic Langmuir-Blodgett films or self-assemble monolayer which are the spontaneous and uninstructed structural reorganizations that form from a disordered system.

The chemical immobilization methods include: (i) covalently binding enzyme to support materials immobilizing enzyme into a membrane matrix or directly onto the surface of the transducer; (ii) crosslinking enzyme employing a multifunctional, low molecular weight reagent based on the formation of strong covalent binding between the transducer and the biological material using a bifunctional agent and (iii) electrochemical polymerization based on electrochemical oxidation of a given monomer from a solution containing the enzyme obtaining a conducting or non-conducting polymer layer and (iv) Micelle: The molecule must have a strongly polar/ hydrophilic "head" and a non-polar/hydrophobic "tail". When this type of molecule is added to water, the hydrophilic head of the molecule presents itself for inter-action with the water molecules on the outside of the micelle, and the hydrophobic tails of the molecules clump into the center of a ball like structure, called a micelle. Enzyme micelle membrane presented here is an innovative way and will be a well-developed biosensor technology to provide rapid and reliable measurements of food, water pollution and clinical analysis.

2.2 Recent development of biosensor researches

Electrochemical biosensors are the oldest and most widely available group in the solid-state chemical sensor field. Electrochemical sensors provide a crucial analytical tool as demand for

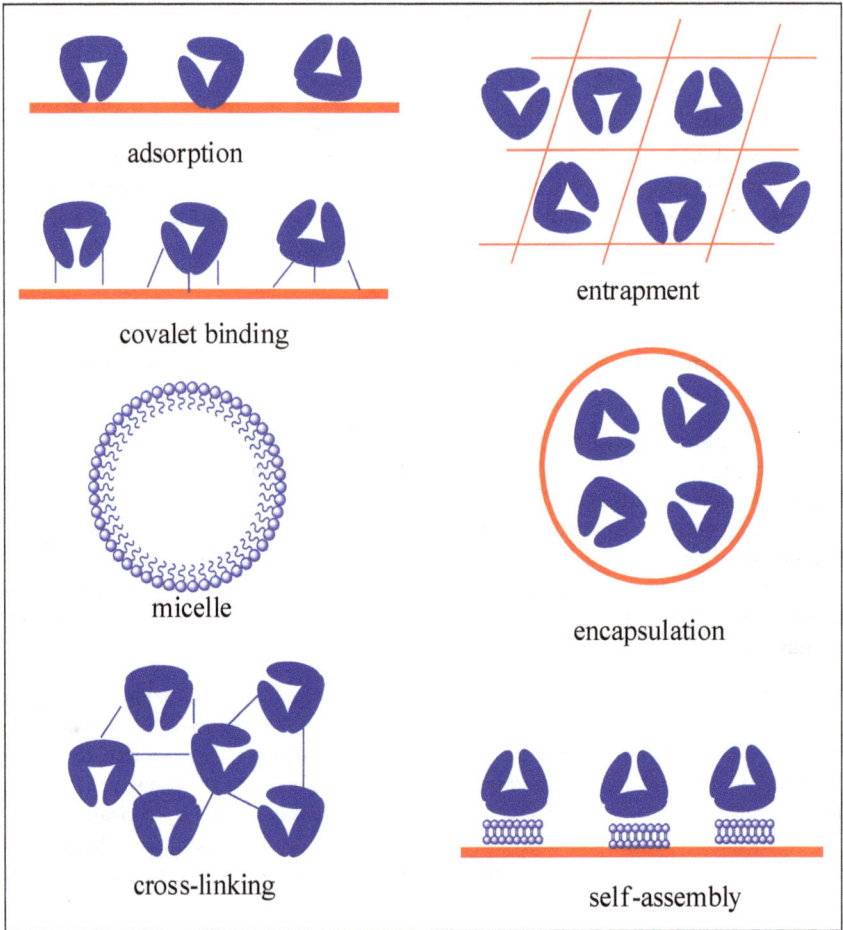

Figure 2. Illustration of enzyme immobilization methods

sensitive, rapid, and selective determination of analytes increases. Over the past two decades we have witnessed a tremendous amount of activity in the area of biosensors. Enzyme based electrodes require the immobilization of an enzyme onto an electrode surface for the quantification of an analyte and hold a leading position among biosensor systems presently available. Biosensors have found promising applications in various fields such as biotechnology, food and agriculture product processing, health care, medicine and pollution monitoring. Recent development has focused on improving the immobilization and stability of the enzymes. There have been a number of recent review articles that have focused on the development of various materials, techniques, and applications of biosensors [19-24]. Since there have been a wealth of biosensor developments in the past years, new approaches and materials for enzyme based sensors have been primarily focused. Strategies for incorporating materials to enhance speed, sensitivity, and stability of these sensors has been of particular interest. Major advancements in biosensors revolve around immobilization and interface capabilities of the biological material with the electrode. The use of polymer and nanomaterials has provided a means for increasing the signal response from these types of sensors. Moreover, the combination of various nanomaterials into composites in order to explore their synergistic effects has become an interesting area of research. The ability to incorporate biomaterials with the potential for direct electron transfer is another growing research area in this field [25-31]. In general, we believe that the field of electrochemical sensors will focus on the incorporation and interaction of unique materials, both nano and biological, in the coming years.

3. Polymers coating in biosensors

Enzyme immobilization using supports have been of great interest for many researches. Various supporting films on electrode surface have been developed to immobilize proteins or enzymes and many polymeric materials were used for enzyme immobilizations. Polystyrene (PS) membrane is a very promising support for the immobilization of enzymes due to its excellent biocompatibility, no toxicity, high affinity, strong adsorption ability, low molecules permeability, physical rigidity and the chemical inertness in biological processes. Its molecular structure is shown in Fig. 3. It is popular for the immobilization in enzyme-linked immuno-sorbent assays (ELIA) by adsorption, however, and few methods for immobilizing proteins on the PS surface by covalent bonding have been proposed, because the complicated multi-step methods must be employed for introduction of functional groups that react with proteins and their procedures are tedious and time consuming. In order to solve the difficulty of introduction of functional groups, we adopted polymaleimidostyrene (PMS) to introduce maleimide group in the bulk of PS, and the coating of enzyme containing PS membrane on the electrode surface under mild conditions opens up enormous possibilities for the immobilization of biomolecules [32-40].

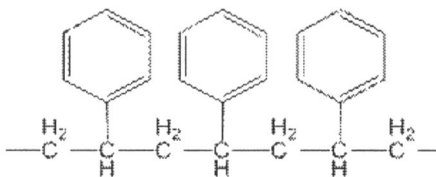

Figure 3. Molecular structure of polystyrene

The interferents and biofouling are two major problems which can affect the performance of a biosensor. Interference from electroactive substances is especially problematic when electrochemical measurements are being made in vivo. Biocompatible membranes are preferable both as a selective barrier as well as for enhancing biocompatibility within electro-chemical biosensors [41]. The cellulose acetate layer permits only small molecules, such as hydrogen peroxide to reach the electrode, eliminating many electrochemically-active com-pounds that could interfere with the measurement. Fig. 4 depicts the molecular structure of cellulose.

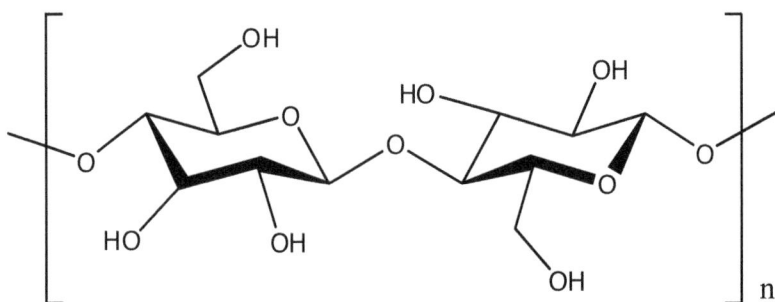

Figure 4. Molecular structure of cellulose

Nafion, a perfluorinated sulfonated cation exchanger (see Fig. 5), has been widely used as an electrode modifier due to the chemical inertness, thermal stability, mechanical strength, and antifouling properties. Nafion coated electrodes have been applied in the analysis of phenol, for the determination of parathion [42, 43]. Nafion has also been widely utilised as a coating material. The polymer displays the advantages of being chemically inert and easily cast from solution. The polymer is anionic and upon casting forms a structure with hydrophilic channels contained within a hydrophobic matrix. Films formed from this material are reasonably robust, show strong exclusion of anionic interferents and display enhanced biocompatibility.

As functional materials, chitin and chitosan offer a unique set of characteristics: biocompati-bility, biodegradability to harmless products, nontoxicity, physiological inertness, antibacte-rial properties, heavy metal ions chelation, high affinity to proteins, gel forming properties and hydrophilicity, remarkable affinity to proteins, availability of reactive functional groups for direct reactions with enzymes and for chemical modifications, mechanical stability and rigidity, and ease of preparation in different geometrical configurations that provide the system with permeability and surface area suitable for a chosen biotransformation [44]. Owing to these characteristics, chitin and chitosan offer a unique set of these characteristics and are predicted to be widely exploited in the near future especially in enzyme immobilization supports. The most distinguishing chitosan properties are its biodegradability and biocom-patibility, which makes it a green polymer. The increasing importance of materials from renewable sources has put chitosans in the spotlight, especially due to their biological

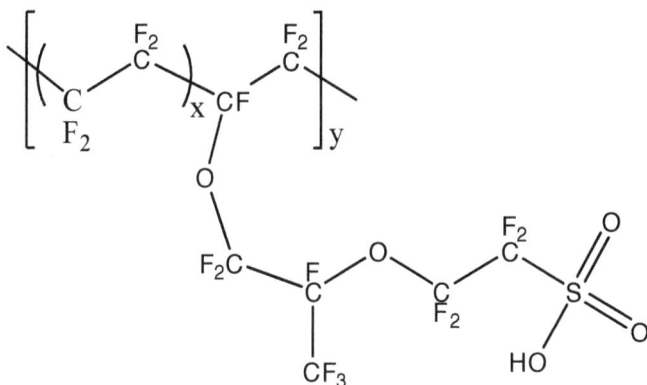

Figure 5. Molecular structure of Nafion

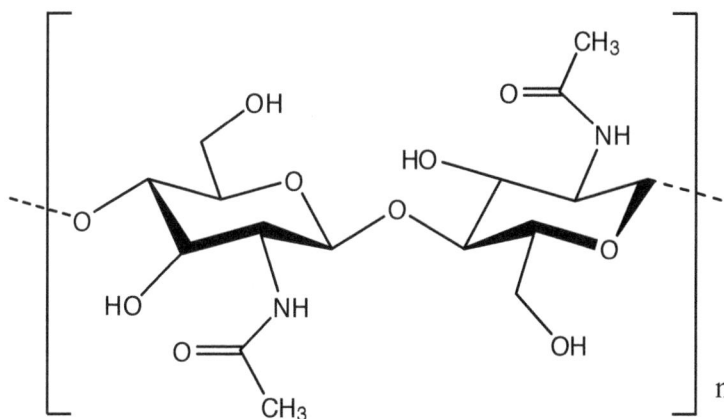

Figure 6. Molecular structure of chitin

properties, which have been exploited in many applications [45, 46]. The molecular structure of chitin and chitosan are shown in Fig. 6 and Fig. 7, respectively.

4. Polymaleimidostyrene in biosensors

Polymers are becoming inseparable from biomolecule immobilization strategies and biosensor platforms. Their original role as electrical insulators has been progressively substituted by their electrical conductiveabilities, which opens a new and broad scope of applications in both the physical adsorption and chemical coupling methods, protein molecules are immobilized on the

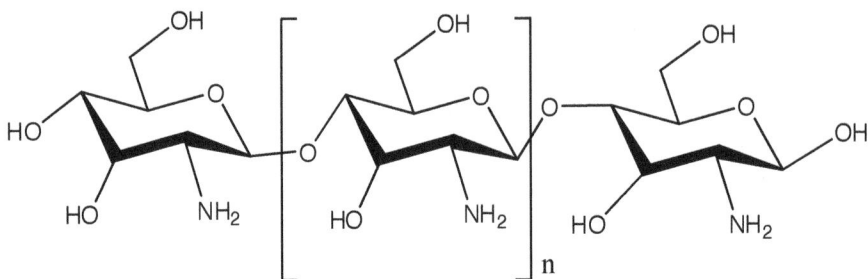

Figure 7. Molecular structure of chitosan

surface with random orientations and are likely to be denatured. In order to increase the lifetime stability of enzyme electrode, it is necessary that there should be a strong and an efficient bonding between the enzymes and immobilizing material. Hence, covalent binding of enzyme on a stabilizer or on the transducer is an efficient method of immobilization. Recently, we have developed an advanced design and preparation of enzyme-based amperometric biosensors using enzyme reverse micelle membrane, as well as the functional structure and principle. Particular emphasis is directed to the discussion and exploration of electrochemical biosensors based on novel functional polymer, polymaleimidostyrene (PMS), as effective immobilization stabilizer as support. It can be expected to be a common method for the immobilization of enzymes to fabricate various bioelectrochemical sensors [32-40].

4.1. Synthesis, structure and properties

PMS which is a polymerization of N-4-vinylphenylmaleimide (N-VPMI) is a new type of polymer mixible with polystyrene (PS). It was synthesized by Prof. Hagiwara's group in 1991 [34]. The compound was prepared by a modified method for synthesis of N-phenylmaleimide (N-PMI). N-VPMI purified by recrystallization was used as a monomer after thoroughly dried below 20 °C under reduced pressure. Polymerization was started by the addition of a tetrahydrofuran (THF) solution of initiator to the monomer solution at a determined temperature with stirring. The synthesis scheme and the molecular structure of PMS are depicted in Scheme. 1.

PMS possesses two polymerizable carbon-carbon double bonds with different reactivities, one of which is the vinylene group of the maleimide moiety and the other the vinyl group of the styrene moiety. The vinylene groups of a maleimide moiety react easily with sulphydryl or amino groups of enzyme by covalent bonds which prevent the unfolding of enzyme. PMS is a very effective, important and useful reagent to immobilize enzyme strongly via covalent bond, because high density of maleimide groups of PMS can catch not only exposed SH groups but also buried SH groups forming enzyme micelles [35–36]. To model the enzyme micelle structure, an illustration is displayed in Fig. 8. The hydrophobic PMS groups are outside the structure shielding the hydrophilic enzyme inside the interior. Therefore, the structure is named as reverse micelle. The reverse micelle structure tends to evolve to a lower-energy configuration under equilibrious conditions.

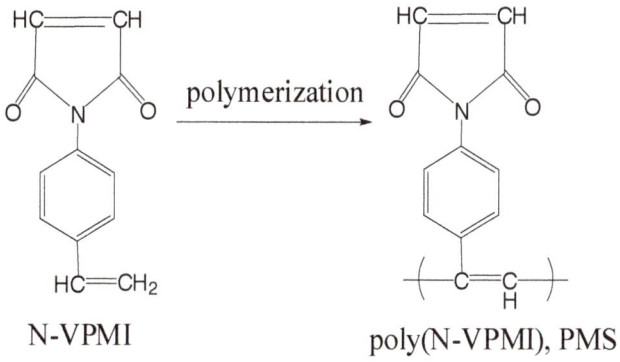

Scheme 1. Synthesis of polymaleimidostyrene

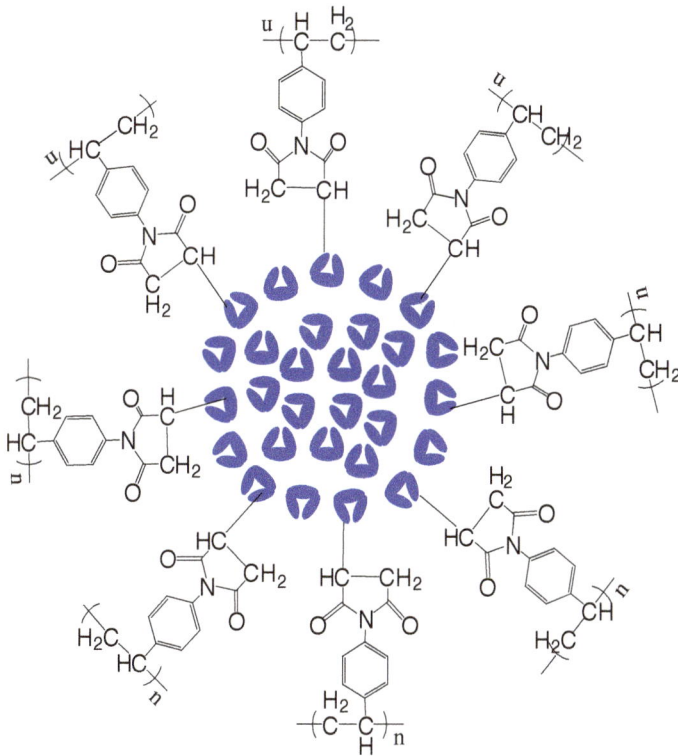

Figure 8. Illustration of enzyme reserve micelle in polystyrene film

Although, it should be mentioned that the free enzyme exists in the outer surface of micelle may lose its activity due to the hydrophobility of chloroform reagent, which was used to dissolve PS and PMS, thus there is partly loss of enzyme activity during the immobilization, the strong adsorption ability of PS membrane makes the enzyme micelle based biosensors particularly attractive for on-line analytical systems which need high stability and good endurability due to the long-time and continuous determinations in a flow system during experimentation.

Furthermore, PMS exhibits a strong affinity specific to a variety of hydrophobic materials such as polystyrene, polyethylene and polyetheretherketone due to the styrene moiety. Then PMS is considered to be an ideal stabilizer for both covalent bonding enzyme and hydrophobic affinity to PS film [30–32]. The application of PMS as a convenient immobilization reagent is summarized in table 1. PMS appears to be effective and promising for the maintenance of biological activity as well as long time stability.

4.2. Application of PMS to biosensors

It has been found that the urease reverse micelle membrane exhibits good sensitivity after stored in a phosphate buffer solution (0.1 M, PH 5.5) for one month compared with its initial sensitivity. The ideal immobilization process should be easy, quick, and enzyme friendly, result in high loading surface. In our previous study, enzyme micelle membrane, which is one-step immobilization, has been proved to be an excellent immobilization method to preserve enzyme conformation resulting in good activity and stability. To form enzyme micelle, hydrophilic PMS covalently bonded-enzyme (via both exposed and buried sulfhydryl groups and amino groups of enzyme and maleimide groups of PMS) was dispersed in hydrophobic PS solution. PMS-bonded enzyme aggregated to form micelle with enzyme inside the structure and PMS outside. The enzyme micelles were immobilized on the surface of glassy carbon electrode (GCE) utilizing PS, which has admirable properties of biocompatibility and strong adsorption ability. PMS also possess good biocompatibility, the PMS-bonded enzyme exhibits good activity due to PMS is a significantly excellent stabilizer for enzyme. On the other hand, it is amazing that the free enzyme exists in the inner part of micelle enhanced the stability of enzyme. The applications of PMS as a function polymer have been published [34-40] and were summarized in table 1.

5. Intrinsically conducting polymers in biosensors

5.1. Types, structures and features

It is generally recognized that the modern study of electric conduction in conjugated polymers began in 1977 with the publication describing the doping of polyacetylene (PA). Conductive polymers or, more precisely, intrinsically conducting polymers (CPs) having unique conjugated π-electron backbone system, are organic conjugated polymers that conduct electricity and are one of the more promising biocompatible materials [48]. Professor Alan Heeger along with Prof. Alan G. MacDiarmid and Prof. Hideki Shirakawa shared the 2000 Nobel Prize in

Electrode configuration	Enzyme	PH	Potential (V)	Linear range	Response time (min.)	Stability (days)	Ref.
OE/PCS	GOD	5.50		0.1– 3.0 mM	12	< 5	36
	ASOD	7.00	−0.7	0 – 1.5 mM	4	150	35
AGCE/PCS	Urease	7.00	1.2	0.5 – 21 mM	< 10	3	37
Gold/ERMM	Uricase	8.5	−0.7	0.005 – 0.105 mM	< 1	7	39
AGCE/ERMM	ASOD	5.50	−0.30	0.01 –0.6mM	< 1	15	38
	GOD	6.50	−0.45	0.2 – 2.6 mM	< 2	"/> 60	40
	Urease	7.00	1.2	0.5 – 16 mM	< 10	"/> 60	34

OE: oxygen electrode; PCS: porous carbon sheet; AGCE: aminated glassy carbon electrode; ERMM: enzyme reverse micelle membrane; GOD: glucose oxidase; ASOD: ascorbate oxidase.

Table 1. Amperometric sensors using PMS as a stabilizer

Chemistry for their seminal contribution to the discovery and development of conductive polymers. A substantial account of the electronic properties was studied. From the point of versatility of synthesis techniques, properties, and broadness of the scope of application, CPs have raised a great deal of scientific and technological interest and have led the research in materials science in a new direction. The last comprehensive reviews devoted to CP were published and are excellent summary of earlier work [49-52].

Since the beginning of conductive polymer research, it has witnessed the emergence of CPs as an intriguing class of organic macromolecules that offer high electrical conductivity and optical properties of metals and semiconductors and, in addition, have the processability advantages and mechanical properties of polymers, in particular, are especially amenable to be further exploited to develop a new form of electrochemical biosensor either as sensitive components or as a matrix for providing biomolecule immobilization, signal amplification and for rapid electron transfer for the fabrication of efficient biosensor devices. A variety of monomers can be electropolymerised on an electrode surface and under correct conditions form stable conductive films. Structures of some CPs commonly used in biosensors are described in Fig. 9.

CPs like polypyrrole (PPy), polyaniline (PANI), polythiophene (PT) can be obtained by electrochemical polymerization either potentiostatically, galvanostatically or by means of multi-sweep experiments. The thickness of the polymer films can be defined by measuring the charge transferred during the electrochemical polymerization process and by controlling parameters like temperature, monomer concentration, polymerization potential or current, as well as the concentration and nature of the supporting electrolyte. Moreover, CPs films exhibt interesting properties concerning the decrease of the influence of interfering compounds due to their size-exclusion and ion-exchange characteristics.

The π-electron backbone which is an extended conjugated system having single and double bonds alternating along the polymer chain is responsible for their unusual electronic properties such as electrical conductivity, low energy optical transitions, low ionization potential and high electron affinity [53]. Scientists from many disciplines are now combining expertise to study organic solids that exhibit remarkable conducting properties. A key requirement for a polymer to become intrinsically electrically conducting is that there should be an overlap of molecular orbitals to allow the formation of delocalized molecular wave function. Besides this, molecular orbitals must be partially filled so that there is a free movement of electrons throughout the lattice [54]. The electronic conductivity of conducting polymers changes over several orders of magnitude in response to changes in pH and redox potential of their environment. The electrical properties can be fine-tuned using the methods of organic synthesis and by advanced dispersion techniques. Polymeric material containing interesting electrical properties is a step forward for research in materials. CPs has the ability to efficiently transfer electric charge produced by the biochemical reaction to electronic circuit. Moreover CPs can be deposited over defined areas of electrodes. The study of unique property of CPs has resulted in fundamental insights into the understanding of the chemistry and physics of this novel class of materials, and it has been exploited for the fabrication of amperometric biosensors.

5.2. Application of CPs in biosensors

Polymers are being discarded for their traditional roles as electric insulators to literally take charge as conductors with a range of novel applications. The electronic CPs has an organised molecular structure on metal substrates, which serves as proper and functional immobilizing platforms for biomolecules. These matrices provide a suitable environment for the immobilization and preserve the activity for long duration. This property of the conducting polymer together with its functionality as a membrane has provided opportunities to investigate the development of biosensors. Application of organic CPs in biosensors has recently aroused much interest as potential candidates to enhance speed, sensitivity and versatility for electrochemical biosensors due to their easy preparation methods along with attractive unique properties such as high stability at room temperature, good conductivity output and facile polymerization and being compatible with biological molecules in a neutral aqueous solution. Moreover, the CP film provides a suitable environment for the immobilization of biomolecules. Thus, CPs have been studied extensively for the development of biosensors. The electrochemically prepared conducting polymers used for the biomolecule immobilization are polyacetylene (PA), polypyrrole (PPy), polythiophene (PT), polyaniline (PANI) etc because of their good electrical properties, environmental stability. Many applications of conducting polymers including analytical chemistry and biosensing devices have been reviewed by various researchers [55, 56].

Enzyme immobilization onto the electrode surface is a crucial step in assembling amperometric biosensors. The CPs have attracted much interest as suitable matrices for biomolecules due to that the extended conjugation along the polymer backbone provides unusual electrochemical properties such as low energy optical transitions, high electrical conductivity, low ionization potential, high electronic affinities. Polymer matrices can be used either in the sensing mechanism or in the immobilization of the bioelement responsible for sensing the analyte. The

poly(acetylene)   poly(pyrrole)   polt(thiophene)
       PA   PPy   PT

poly(para-phenylene)   poly(3,4-ethlenedioxythiophene)
        PPP   PEDOT

poly(phenylene sulfide)   poly(para-phenylene vinylene)
        PPS   PPV

polyaniline
PANI

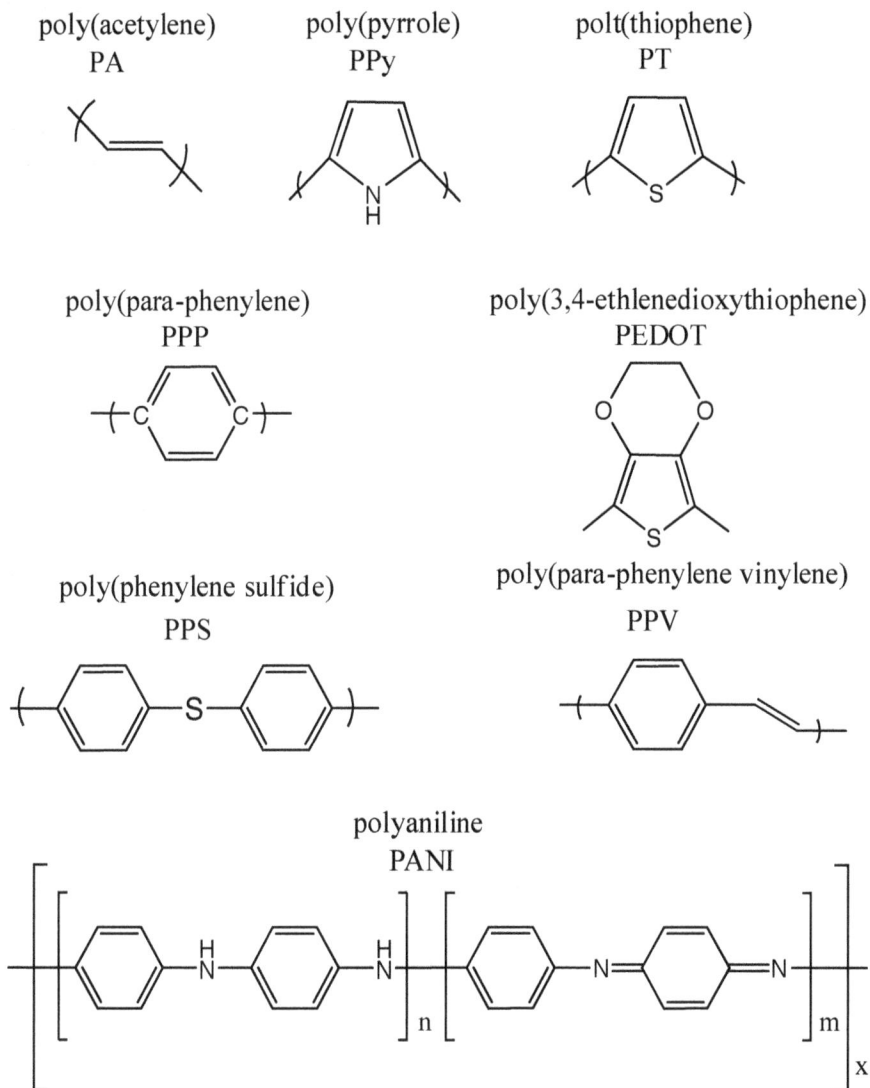

Figure 9. Structures of some conducting polymers commonly used in biosensors.

empolyment of promising CPs in electron transfer as an appropriate surface for enzyme immobilization provides rapid response encourages the coexistences of biomolecules and raises the stability of the biosensors. Numerous papers have been published indicating organic

CPs as a convenient component, forming an appropriate environment for the immobilization of enzyme at the electrode surface. Stable immobilization of macromolecular biomolecules on conducting microsurfaces with complete retention of their biological recognition properties is a crucial problem for the commercial development of miniaturized biosensor. Most of the conventional procedure for biomolecule immobilization such as cross-linking, covalent binding and entrapment in gels or membrane suffer from a low reproducibility and a poor spatially controlled deposition.

Due to that CPs have considerable flexibility in the available chemical structure, which can be modified as required, CPs have attracted much interest to serve as good matrices for the immobilization of enzymes. The techniques of incorporating enzymes into electro-depositable conducting polymeric films permit the localization of biologically active molecules on electrodes of any size or geometry and are particularly appropriate for the elaboration of multi-analyte micro-amperometric biosensors. Another advantage offered by CPs is that the electrochemical synthesis allows the direct deposition of the polymer on the electrode surface, while simultaneously trapping the protein molecules. In addition, the electrochemically prepared CPs can be grown with controlled thickness using lower potential and they also provide an excellent enzyme-entrapping property. The polymerization through electrochemical oxidation provides greater control over the process and enables control over the thickness of the polymer layer and even small electrode substrate to be coated. This technique offers a suitable way to make a homogeneous film that adheres strongly with the electrode surface.

Another important advantage of using CPs is that the biomolecules can be immobilized onto the nanowire structure in a single step rather than the multiple steps that are required when other non-polymeric materials are used. Nanostructured conjugated polymers and their nanocomposites represent new advanced materials that are key issues for the development of new devices and structures offering the association of the various properties required in advanced applications. As conducting polymer nanomaterials are light weight, have large surface area, adjustable transport properties, chemical specificities, low cost, easy processing and scalable productions, they are used for applications in nanoelectric devices, chemical and biological sensors [57]. There are extensive studies in the literature concerning the synthesis, characterization, and application of these CPs.

Among the CPs, PPy is one of the most extensively used conducting polymers in design of bioanalytical sensors. PPy and its derivatives play a leading role due to several interesting properties such as electroactivity, ionic exchange properties, supercapacitors for energy storage, secondary batteries, and elastic textile composites of high electrical conductivity, as well as good stability. PPy have the most versatile applicability for the construction of different types of bioanalytical sensors. The background presented illustrates that PPy is a very attractive, versatile material, suitable for preparation of various catalytic and affinity sensors and biosensors. The immobilization of biologically active molecules into PPy can be obtained during electrochemical deposition during which either some undesirable electrochemical interactions can be prevented or the electron transfer from some redox enzymes can be facilitated. The developments in nano-structured conducting polymers and polymer nano-composites have large impact on biomedical research. Significant advances in the fabrications of nanobiosensors/sensors using nano-structured CPs have been reviewed [58]. Recent advances in application of PPy in immunosensors and DNA sensors and recent progress and problems in development of molecularly imprinted PPy have been presented. The use of PPy in conjunction with bioaffinity reagents has provident to be a powerful route that has expanded the range of applications of electrochemical detection and its future development is expected to continue [59].

Among various CPs, PANI which can be directly and easily deposited on the sensor electrode and has controlled, high surface area, chemical specificities, long term environmental stability and tuneable properties has attracted much attention to be a suitable candidate to be used in various applications in biosensors due to its unique and controllable chemical and electrical properties, its environmental, thermal and electrochemical stability, and its interesting electrochemical, electronic, optical and electro-optical properties. PANI has gained much popularity in biosensor applications, partially due to its favourable storage stability, simple synthetic procedures with good processibility, rapid electron transfer and direct communication to produce a range of analytical signals and new analytical applications. Efforts have been made to discuss and explore various characteristics of PANI responsible for direct electron transfer leading towards fabrication of biosensor interfaces and can also be used as a suitable matrix for immobilization of biomolecules [60, 61]. Moreover, PANI exhibits two redox couples in right potential range to facilitate an enzyme–polymer charge transfer and thereby acts as self-contained electron transfer mediator. In particular, PANI's transport properties, electrical conductivity or rate of energy migration, provide enhanced sensitivity. In addition, Nano-structures of PANI can offer the possibility of enhanced performance and also helps to overcome the processibility issues associated with PANI. In a conclusion, the various remark-able characteristics of PANI matrix make it a novel platform for fabrication of variety of biosensors interface.

6. Sol-gel and hydrogel material in biosensors

The terminology sol-gel is used to describe a broad class of processes in which a solid phase is formed through gelation of a colloidal suspension. A typical hydrogel network involve poly (vinyl alcohol) (PVA) or poly(acrylic acid) (PAA). Sol-gel is gradually attracting the attention of the electrochemical community as a versatile way for the preparation of modified electrodes and solid electrolytes. Sol-gel electrochemistry is rather young, and often researchers are still excited by the mere feasibility of realizing an application by sol-gel technologies due to its better processibility, improved diffusion rate, large ion-exchange capacity, fast proliferation of organic-inorganic hybrids, and other application specific chemical properties. The types of sol-gel materials that are useful for electrochemistry and the recent advances in the various fields of sol-gel electrochemistry have been reviewed by a few research groups [62, 63]. The ease of preparation and the wide ranging flexibility of incorporating desired functionalities by careful selection and design can be readily applied to different types of electrode materials without being restricted by electrode shapes and designs, which are suited for the immobilization of biomolecules. However, sol–gel matrices, despite being also biocompatible, are traditional fragile nature, similarly to biological membranes, and suffer from low sensitivity and reproducibility which hampered their application in biosensor. The organic–inorganic material prepared by sol–gel method can yield a highly sensitive, robust and stable biosensor.

Hydrogels have also been extensively investigated as coatings support for immobilization of enzymes. Enzymes can often denature and lose their efficiency; however this effect can be mitigated by encapsulating it inside a hydrogel, because hydrogel is a type of waterswollen and cross-linked polymer formed by the gelling process and features a highly hydrophilic structure of three dimensional networks. And the consequent swelling of the polymer matrix provides a biocompatible microenviroment for the enzyme to maintain its natural configuration, thus is an ideal matrix for the massive entrapment of cell and enzyme. Moreover, the biosensors based on the hydrogel have high sensitivities. Although it exhibits a high affinity for water, it does not dissolve, and provides sufficient permeability for both solvent and substrate molecules so that they are capable of diffusing quickly through the water-swollen polymer which has reasonably high water content. In addition, the external hydrophobic organic solvent is unable to distort the native conformation of the entrapped enzyme in the sol-gel, and the hydrogel is soft and of rubbery consistence which closely resemble living tissues. Consequently, the widely utilised application for hydrogels has been as enzyme stabilising agents [64-68].

7. Conclusion and future perspectives

Proper electrode fabrication using different materials for efficient electron transport has recently aroused much interest as a versatile tool for the constructing biosensors. Biosensors designed employing polymeric materials results in low detection limits, high sensitivities, lower applied potential, reduction of background, efficient electron transfer and easier immobilization of enzymes on electrodes. Application of organic CPs in biosensors has recently aroused much interest as potential candidates to enhance speed, sensitivity and versatility for electrochemical biosensors due to their easy preparation methods along with attractive unique properties such as high stability at room temperature, good conductivity output and facile polymerization and being compatible with biological molecules in a neutral aqueous solution. However, there are verious of challenges to be addressed in order to fulfill the applications of polymers. In addition, long laboratory synthetic pathways and costs are also involved in the production of the functional polymers. The aforementioned disadvantages of the above polymeric materials call for search for low cost biomaterials as alternative for the development of novel electrochemical sensors and biosensors. Those focuses towards designing smart polymers such as nanostructure doped polymers are the most promising for polymers to be further investigated. By combination of the unique properties of nanostructured material and various polymers, it is possible to develop novel enzyme-based bioelectronic devices with particular advantages. In particular, the integration of nanotechnology, with novel polymeric materials should lead to very sensitive and fast assays.

Acknowledgements

I have benefited over the years from many publications with the authors of the papers cited. I thank them for their explanations and inspiration. I apologize to those whose work I have seemed to overlook.

Financial support from the the financial support from the National Natural Science Foundation of China (21205008), the Fundamental Research Funds for the Central Universities (DUT12LK31) and the Beijing National Laboratory for Molecular Sciences (BNLMS) are gratefully acknowledged.

Author details

Xiuyun Wang[1] and Shunichi Uchiyama[2]

*Address all correspondence to: xiuyun@dlut.edu.cn

*Address all correspondence to: uchiyama@sit.ac.jp

1 School of Chemistry, Dalian University of Technology, Dalian, China

2 Saitama Institute of Technology Fukaya, Saitama, Japan

References

[1] Bilitewski U. and Turner A. P. F. Turner, Biosensors for environmental monitoring: 4. water analysis. The Netherlands, Harwood Academic Publishers. 2000, 137 – 213.

[2] Buerk D. G., Biosensors: theory and applications. USA, Technomic Publishing Company, Inc. 1993, 208 – 209.

[3] Sara R., Marco M., Alda M., Damia Barcelo. *Anal. Bioanal. Chem.* 2004, 378, 588 – 598.

[4] Wang J., *Biosens. Bioelectron.*, 2006, 21, 1887 – 1892.

[5] Wilson G. S. and Hu Y., *Chem. Rev.* 2000, 100, 2693-2704.

[6] Cai Q., Zeng K., Ruan C., Desai T. A., and Grimes C., *Anal. Chem.*, 2004, 76 (14), 4038–4043.

[7] Sara R. and Alba M. J L., *Anal. Bioanal. Chem.*, 2006, 386, 1025 – 1041.

[8] Barton C. S., Gallaway J. and Atanassov P., *Chem. Rev.*, 2004, 104, 3239 – 3265.

[9] Nakamura H. and Karube I., *Anal. Bioanal. Chem.*, 2003, 377, 446 – 468.

[10] Bakker E., *Anal. Chem.*, 2004, 76, 3285 – 3298.

[11] Murphy L., *Curr. Opin. Chem. Biol.*, 2006, 10, 177 – 184.

[12] Higgins M. J., Molino P. J., Yue Z., and Wallace G. G., *Chem. Mater.* 2012, 24, 828–839.

[13] Vidal J., Esperanza G., and Castillo J., *Microchim. Acta* 2003, 143, 93–111.

[14] Vankelecom IFG, *Chem. Rev.* 2002, 102, 3779 –3810.

[15] Anish K. M., Soyoun J. and Taeksoo J., *Sensors* 2011, 11, 5087–5111.

[16] Guilbault G. G., Analytical uses of immobilized enzymes: 3. Principles of immobilized enzymes. New York, Marcel Dekker, Inc., 1984, 78 – 93.

[17] Moyo M., Okonkwo J. O. and Agyei N. M., *Sensors* 2012, 12, 923-953.

[18] Prodromidis M. I. and Karayannis M. I., *Electroanalysis* 2002, 14, No. 4.

[19] Kimmel D., LeBlanc G., Meschievitz M., Cliffel D., *Anal. Chem.* 2012, 84, 685–707.

[20] Heller A. and Feldman, B., *Chem. Rev.* 2008, 108, 2482–2505.

[21] Joo S and Brown RB, *Chem. Rev.* 2008, 108, 638–651.

[22] Wang J., *Chem. Rev.* 2008, 108, 814-825.

[23] Dongen S., Hoog H., Peters R., Nallani M., Nolte R. and Hest J., *Chem. Rev.* 2009, 109, 6212–6274.

[24] Kobayashi S. and Makino A., *Chem. Rev.* 2009, 109, 5288–5353.

[25] Shao Y.; Wang J., Wu H., Liu J., Aksay I. A., Lin Y., *Electroanalysis* 2010, 22, 1027–1036.

[26] Zhao Z.; Lei W., Zhang X., Wang B., Jiang H., *Sensors* 2010, 10, 1216–1231.

[27] Siqueira J. R., Caseli L., Crespilho F. N., Zucolotto V., Oliveira O. N., *Biosens. Bioelectron.* 2010, 25, 1254–1263.

[28] Harper A. and Anderson M. R. *Sensors* 2010, 10, 8248–8274.

[29] Singh R. P., Oh B. K., Choi J. W., *Bioelectrochem.*, 2010, 79, 153–161.

[30] Park B. W., Yoon, D. Y., Kim D. S., *Biosens. Bioelectron.* 2010, 26, 1–10.

[31] Su L., Jia W., Hou C., Lei Y., *Biosens. Bioelectron.* 2011, 26, 1788–1799.

[32] Uchiyama S., Watanabe H., Yamazaki H., Kanazawa A., Hamana H., and Okabe Y., *J. Electrochem. Soc.*, 2007, 154 (2), F31 – F35.

[33] Hagiwara T., Suzuk, I., Takeuchi K., Hamana H. and Narita T., *Macromolecules.* 1991, 24, 6856 – 6858.

[34] Wang X., Uchiyama S., *Anal. Lett.*, 41(7), 1173-1183 (2008).

[35] Tomita R., Kokubun K., Hagiwara T. and Uchiyama S., *Anal. Lett.*, 2007, 40, 449 – 458.

[36] Uchiyama S., Tomita R., Sekioka N., Imaizumi E., Hamana H. and Hagiwara T., *Bioelectrochemistry*. 2006, 68, 119 – 125.

[37] Wang X., Watanabe H., Sekioka N., Hamana H. and Uchiyama S., *Electroanalysis*, 2007, 12, 1300 – 1306.

[38] Wang X., Watanabe H., and Uchiyama S., *Talanta*, 2008, 74 (5), 1681–1685.

[39] Wang X., Hagiwara T., and Uchiyama S., *Anal. Chim. Acta*, 2007, 587, 41 – 46.

[40] Wang X. and Uchiyama S., *ITE Lett.*, 2007, 8 (3).

[41] Davis F. and Higson S., Biomedical Polymers, Woodhead Publishing, 2007.

[42] Tsai Y., Li S., and Chen J., *Langmuir* 2005, 21, 3653-3658.

[43] Norouzi P., Faridbod F., Larijani B., Mohammad Reza Ganjali1, *Int. J. Electrochem. Sci.*, 5 (2010) 1213 – 1224.

[44] Krajewska B., *Separation and Purification Technology* 2005, 41 (3), 305–312.

[45] Krajewska B., *Enzyme and Microbial Technology* 2004, 35, 126–139.

[46] Cosnier S., *Biosen. Bioelectron.*, 1999, 14, 443–456.

[47] Scott C., Nanostructured Conductive Polymers, John Wiley & Sons, Ltd., 2010.

[48] Liu J., Lam J. W. Y., Tang B. Z., *Chem. Rev.* 2009, 109, 5799–5867.

[49] Teles F., Fonseca L., *Mater. Sci. Engineer. C* 2008, 28, 1530–1543.

[50] Peter S. Heeger†‡ and Alan J. Heeger, *PNAS*, 1999, 96 (22), 12219–12221.

[51] McQuade D. T., Pullen A. E., Swager T. M., *Chem. Rev.* 2000, 100, 2537–2574.

[52] Lu J. and Toy P., *Chem. Rev.*, 2009, 109, 815–838.

[53] Adam K. Wanekaya, Lei Y., Bekyarova E., Chen W., Haddon R., Mulchandani A., Nosang V. Myung, *Electroanalysis*, 2006, 18 (11),1047 – 1054.

[54] Bakhsgum A., Kaur A. and Arora V., *Indian J. Chem.* 2012, 5, 57-68.

[55] Higgins M., Molino P., Yue Z., and Wallace G., *Chem. Mater.*, 2012, 24 (5), 828–839.

[56] Leon A.P., and Wallace G., *Chem. Rev.* 2010, 39, 2545–2.

[57] Li X., Huang M., Duan W., Yang Y., *Chem Rev.* 2002, 102(9), 2925–3030.

[58] Dhand C., Das M., Datta M., Malhotra B.D., *Biosen. Bioelectron.*, 2011, 26, 2811–2821.

[59] Odaci D., Kayahan S. K., Timur S., Toppare L., *Electrochim. Acta*, 2008, 53, 4104–4108.

[60] Hatchett D, Josowicz M., *Chem. Rev.* 2008, 108, 746–769.

[61] Murat A. A., and Saracb S., *Progress in Organic Coatings* 2009, 66, 337–358.

[62] Blackman C S., Parkin I P, *Chem. Mater.*, 1997, 9, 2354-2375.

[63] Vlierberghe S., Dubruel P., and Schacht E., *Biomacromolecules*, 2011, 12, 1387–1408.

[64] Wang B., Li B., Deng Q., Dong S., *Anal. Chem.*, 1998, 70, 3170–3174.

[65] Wang C., Yu B., Knudsen B., Harmon J., Moussy F., Moussy Y., *Biomacromolecules* 2008, 9, 561–567.

[66] Andersson O., Larsson A., Ekblad T. and Liedberg B., *Biomacromolecules* 2009, 10, 142–148.

[67] Amanda K. Andriola Silva, Cyrille Richard, Michel Bessodes, Daniel Scherman and Otto-Wilhelm Merten, *Biomacromolecules* 2009, 10 (1), 9–18.

[68] Gray K. M., Liba B. D, Wang Y., Cheng Y., Rubloff G., Bentley W., Montembault A., Royaud I., David L., and Payne G., *Biomacromolecules* 2012, 13, 1181–1189.

Chapter 3

Layered Biosensor Construction

Joanna Cabaj and Jadwiga Sołoducho

Additional information is available at the end of the chapter

1. Introduction

Biosensors for last two decades make ideal sensing systems to monitor the effects of pollu‐ tion on the environment, in the food or textile industry as well as medical diagnostic due to their biological base, ability to operate in complex matrices, short response time and small size.

In enzymatic devices, efforts have been concentrated on the control over enzyme activity, which is highly dependent on the interface between the electrode and the enzyme. Such con‐ trol has led to immobilization techniques suitable for anchoring the enzyme close to elec‐ trode with preservation of its biological activity. In these type of devices, where retaining of the enzyme activity at the electrode/enzyme interface is the key to design efficient electrode, charge transfer between enzyme and electrode should be fast and reversible. Moreover, the charge transfer may also be optimized with some mediating particles (i.e. conducting units) being used in conjunction with the biological molecules at the electrode surface. To the use of conducting polymers for the fabrication of various biosensors have been dedicated exten‐ sively study due to their redox, optical, mechanical and electrical properties, as well as to their unique capability to act both, as transducers, and an immobilization matrices for en‐ zyme retention [1].

It is essential for the sensitivity of the system that the recognition units have optimized sur‐ face density, good accessibility, long-term stability and minimized non-specific interactions with compounds other than the analyte. Such model molecular assemblies can be prepared by Langmuir-Blodgett (LB) and Langmuir-Schaefer (LS) techniques [2,3], layer-by-layer (LbL) or by employing self-assembly monolayers (SAMs) or electrolytic deposition (Fig. 1) [4,5]. The main advantage of using thin films to build a biosensor is the possibility to de‐ crease dramatically the response time of the device. Langmuir-Blodgett type technology al‐ lows building up i.e. lamellar lipid stacking at an air/water interface, which can be easy transported onto a solid support. When all parameters are optimized, this technique corresponds to one of the most promising for preparing thin films of amphiphilic molecules. Based on the self-assembled properties of amphiphilic biomolecules at the air/water interface, LB technology offers the possibility to prepare biomimetic layers suitable for adsorptive immobilization of bioactive molecules [6].

Proteins are more challenging to prepare for the different microarray than i.e. DNA, and protein functionality is often dependent on the state of proteins. Since enzymes often significantly reduce their activity during immobilization, the optimized adsorption methods seem to be optimal for the retaining of conformational states of proteins on solid surfaces. Among the various immobilization techniques available, adsorption may have a higher commercial potential than other methods because the adsorption process is simpler, less expensive, retains a high catalytic activity, and most importantly, the support could be repeatedly reused after inactivation of the immobilized enzyme.

Enzyme-based biosensors play an important role in various industries, such as food, manufacturing, clinic, and environment. Recently, mediators have been employed in enzyme-based biosensors in order to shuttle electrons between the redox enzyme and the electrode surface. Solution-phase mediators may cause electrode contamination and operation inconvenience. In order to overcome the above-mentioned drawbacks and improve the performances of the biosensors, the immobilization of the mediator with protein on a solid support provides a new way to construct reagentless biosensors [7].

However, there is many techniques of biocatalysts immobilization and much research is dedicated to fabricate the biosensing elements, the construction of novel type of biosensor is challenge for new technologies and the key problem is modification of electrode by enzyme using thin film preparation methods.

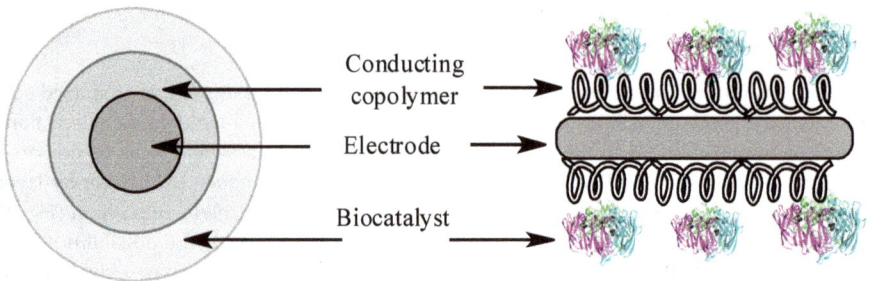

Figure 1. Layered biosensor system

2. Langmuir-Blodgett, Langmuir-Schaefer, layer-by-layer assembly multilayers of proteins

The concept of using biomolecules as an elementary structure to develop self-assembled structures of defined geometry has thus received considerable attention. In this way, the self-assembly ability of amphiphilic biomolecules such as lipids, to spontaneously organize into nanostructures mimicking the living cell membranes, appears as a suitable concept for the development of biomimetic membrane models. The potential of two-dimensional molecular self-assemblies is clearly illustrated by Langmuir monolayers of lipid-like molecules, which have been extensively used as models to understand the role and the organization of biological membranes and to acquire knowledge about the molecular recognition process [6,8]. Langmuir-Blodgett technology allows building up lamellar lipid stacking by transferring a monomolecular film formed at an air/water interface – named Langmuir monolayer or Langmuir film– onto a solid support (Figure 2).

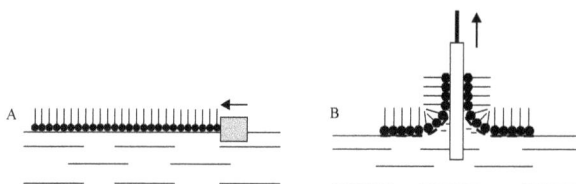

Figure 2. Langmuir – Blodgett deposition

Lipid-based phases are particularly attractive because they can be nanostructurally customized, for instance, to closely resemble cellular components, or to formulate delivery vehicles for biomolecules and drugs. In the presence of water or aqueous buffers, lipid molecules can self-assemble into a wide range of nanostructures [9]. The intrinsically low degree of nonspecific adsorptivity of supported membranes makes them interesting as an interface between the nonbiological materials on the surface of a sensor or implant and biologically active fluids [10,11]. Potential applications include the acceleration and improvement of medical implant acceptance, programmed drug delivery, production of catalytic interfaces, as a platform to study transmembrane proteins and membrane-active peptides, and as biosensors [12-14].

Although the LB method does not solve all problems associated with engineering the structure of condensed phases, it does provide a level of control over the orientation and placement of molecules in monolayer and multilayer assemblies that is not otherwise available. When all parameters are optimized, this technique corresponds to one of the most promising for preparing thin films of amphiphilic molecules as it enables an accurate control of the thickness, an homogeneous deposition of the monolayer over large areas compared to the dimension of the molecules, as well as the possibility to transfer monolayers on almost any kind of solid substrate. Based on the self-assembled properties of amphiphilic biomolecules at the air/water interface, LB technology offers the possibility to prepare biomimetic layers suitable for immobilization of bio-active molecules [8].

Systems mimicking natural membranes appear promising in the field of bioelectronic devices and represent useful models in basic research on membrane behavior in life science. For such purposes, the interest of LB films is now largely recognized, and several enzyme sensors based on the LB technology have been reported (Table 1).

Protein	Immobilization method	Thin ordered film	Stability	References
Glucose oxidase	Lipid - coating	LB – two layers film of lipids	3 month	[15]
Catalase	Adsorption	LB – one layer film of phospholipids	>3 months	[16]
Laccase	Adsorption	LB – five layers film of benzothiadiazole copolymers in mixture with linoleic acid	>3 months	[2]
Invertase	Adsorption	LS – one layer film of phospholipids and octadecylamine	>4 months	[17]
Cellulase	Adsorption	LS film of cellulose	No data	[18]
Glucose oxidase	Adsorption	LbL films of alternate layers of poly(allylamine) hydrochloride and glucose oxidase	20 days	[19]
Urease	Adsorption and covalent grafting	LbL multilayer films of alternate charged polysaccharides, chitosan and polyaniline	3 weeks	[20]
β-Galactosidase, glucose oxidase, peroxidase	Adsorption	self-assembled monolayer of a 3-mercaptopropionic acid	4 weeks	[21]
Cholesterol oxidase	Adsorption	LB layers of octadecyltrimethylammonium and nano-sized Prussian blue clusters	No data	[22]

Table 1. Selected biosensor based on thin ordered films

In particular LB films are offering a possibility of obtaining extended two–dimensional π-electron systems [23]. Although, majority of conventional conducting polymers are not soluble in common solvents, making any LB deposition impossible, one can increase their solubility by attachment of side groups (usually n–alkyl ones) to the main chains. According to this Langmuir-Blodgett, horizontal lifting or other self–assembled method is employed for obtaining molecular films of conducting structures. This type of material is popular in designing of sensor devices. Product in any solid-state sensor, analyte molecules have to diffuse into and react with the acting sensing component and any product of the reaction must diffuse out. It therefore follows that the thinner the sensing layer is, the less time this will take and thereby speed and reversibility being improved.

Device preparation requires use of facilitative methodologies for the organization of biological components in a particular configuration. Self-assembled monolayer and LB methods have been used for organization of functional elements in two-dimensional or layered structures, respectively. These methods offer opportunities for immobilization of functional components into well-organized structures. As a convenient methodology alternate layer-by-layer adsorption has been paid much attention as an emerging methodology.

Recent research has proved the great applicability of the LbL technique not only for preparation of bio-related devices but also for producing various device structures, including sensors [24], photovoltaic devices [25], electrochromic devices [26], fuel cells [27].

Materials that can be used in LbL methods cover a wide range including conventional polyelectrolytes, conductive polymers, dendrimers, proteins, nucleic acids, saccharides, virus particles, inorganic colloidal particles, quantum dots, clay plates, nanosheets, nanorods, nanowires, nanotubes, dye aggregates, micelles, vesicles, LB film, and lipid membranes [28]. In most cases, the LbL assembly is carried out based on electrostatic interactions. As illustrated in Figure 3, adsorption of counterionic species at relatively high concentrations leads to excess adsorption of the substances, as a result of charge neutralization and resaturation, finally resulting in charge reversal.

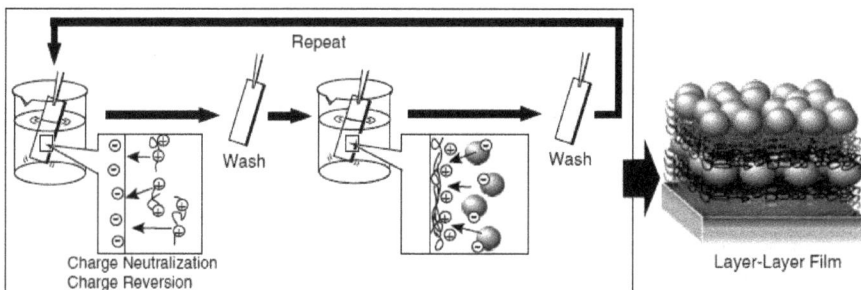

Figure 3. Layer-by-Layer assembly [28]

The forces for LbL assembly are not limited to electrostatic interactions alone. Various interactions including metal-ligand interaction, hydrogen-bonding, charge transfer, supramolecular inclusion, bio-specific recognition, and stereo-complex formation can be used for LbL assembly [28]. Biocompatibility is the most prominent advantage of the LbL assembly because this procedure requires mild conditions for film construction. Most proteins, especially those soluble in water, have charged sites on their surface, and so the electrostatic LbL adsorption is useful for the construction of various protein organizations.

In order the aim to develop models mimicking biomembranes usable for applications in the biosensor field, studies of biological activities of membraneous proteins after incorporation in a phospholipidic bilayer were widely investigated [29,30]. Well known is the sensing system built by the incorporation of glutamate dehydrogenase or choline oxidase into fatty acid LB films through an adsorption or an inclusion process which consists of sandwiching the protein molecules between two LB layers [31]. The molecules of glutamate dehydrogenase adsorbed on the surface of behenic acid LB films work as a protective screen against the rearrangement of the multilayers induced by diffusion of the alkaline buffer inside the structure; on the contrary, the choline oxidase molecules operate as an accelerating factor of the structural lipidic reorganization induced in the same conditions.

The physiological activity of transmembrane proteins, however may depend on the physico-chemical properties of neighboring phospholipids. Such dependence has been demonstrated in the case of, among others, hydroxybutyrate dehydrogenase, Ca^{2+}-ATPase and melibiose-permease. Moreover, all integral membrane proteins are surrounded by a layer of phospholipids, the annular region, which provides the adequate lateral pressure and fluidity to seal the membrane during the changes in the protein during transport events [32].

2.1. Main membrane lipids

All the amphiphilic molecules are potentially surface active agents and substantially monolayer-forming materials. It is possible to find a discussion on the range of a large variety of amphiphile compounds able to form insoluble monomolecular films [33]. Due to the synthesis of biomimetic membranes, the most important types of amphiphilic molecules are fatty acids, phospholipids and glycolipids. Cholesterol as a type of steroid extremely abundant in the cell membrane, can also form insoluble monolayers but it is generally more studied mixed with other phospholipids [34-36] in order to its implication in the formation of lipid microdomains. Figure 4 presents the examples of the principal structures of these different types of lipids. The amphiphilic nature of biological surfactants is responsible for their aggregation at the air/water interface. Their affinity for the air/water interface is determined by the physico-chemical properties of the hydrophilic and hydrophobic parts. The monolayer forming abilities of the amphiphiles is dependent on the balance between these two opposite forces, which are determined by the size of the hydrophobic tail group (i.e. the alkyl chain) and the strength of the hydrophilic head group (i.e. size, polarity, charge, hydration capacity) [6]. If the equilibrium between hydrophilic and hydrophobic part of molecule is disturbed, the material dissolve in the subphase and is not able to form a stable monolayer.

Most of the lipidic cell membrane components are composed of a zwitterionic head group at pH 7.0 (phospholipids) or contain a highly hydrophilic polar group (glycolipids), and a hydrophobic part which is constituted by two hydrocarbon chains per molecule and drastically reduces the water solubility of the complete lipidic membrane molecule. Consequently, many components of cell membranes form insoluble monolayers at the air/water interface since the lipid concentration in the aqueous subphase is negligible, and some of them may be built up into multilayer films by Langmuir-Blodgett deposition.

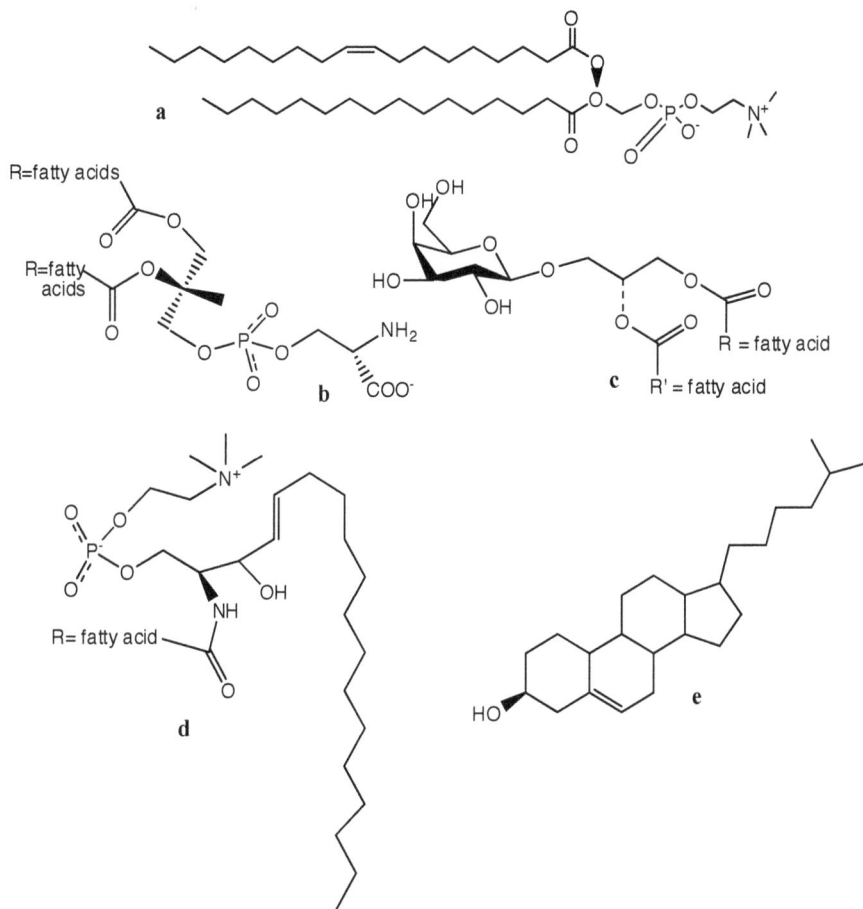

Figure 4. Examples of the main membrane lipids; 1) phospholipids: a, phosphatidylcholine; b, phophatidylserine; 2) glycolipids: c, monogalactosyldiglyceride; 3) sphingolipids: d, sphingomyelin; e, cholesterol

3. Protein adsorption at the solid/liquid interface

Protein adsorption at solid/solution interface has been a research focus for more than three decades due to its importance in the development of biocompatible materials, various bio-technological processes, food and pharmaceutical industries, and promising new areas such as biosensor, gene microarray, biochip, biofuel cell and so on. In order to control and manip-ulate protein adsorption, the mechanisms which govern the adsorption process need to be well understood.

The effect of variables like pH, temperature, the ionic strength, the properties of the protein and the surface, the nature of the solvent and other components on protein adsorption have been studied. Protein adsorption is a very complex process (Figure 5), which is driven by different protein-surface forces, including van der Waals, hydrophobic and electrostatic forces. Attention is also paid to the structural rearrangements in the protein, dehydration of the protein and parts of the surfaces, redistribution of charged groups in the interfacial layer and the role of small ions in the overall adsorption process. Protein adsorption also depends on the chemical and physical characteristic of the surface. Conformational changes in the protein can greatly contribute to the driving force for adsorption. Proteins are highly ordered structures (i.e., states of low conformational entropy). Partial or complete unfolding of the protein on the sorbent surface leads to an increase in conformational entropy, which can be the driving force for protein adsorption. To assess the tendency of proteins to unfold on surfaces, it is important to have a clear picture of protein stability.

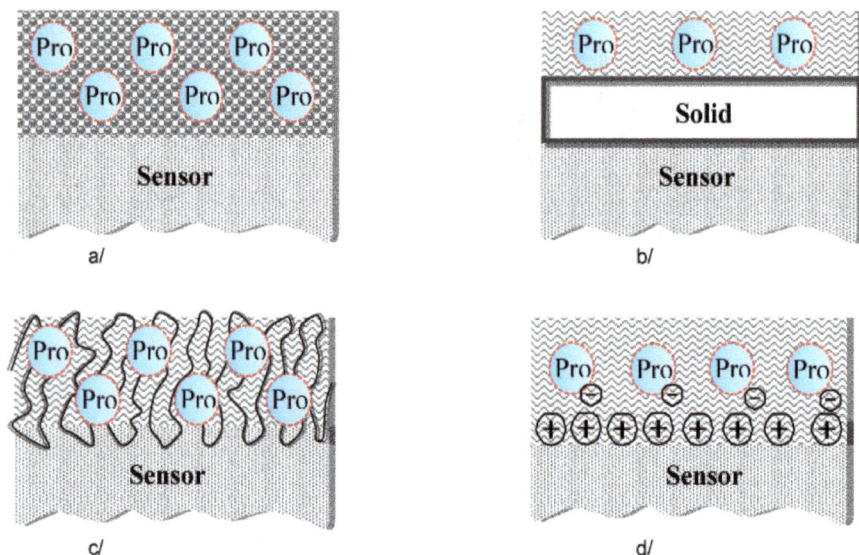

Figure 5. Protein adsorption; a/ adsorption in membrane, b/ classic adsorption on solid, c/ encapsulation in porous material, d/ protein with LbL assembly

Now is generally accepted that the adsorption behavior of proteins at relatively high concentrations often does not follow the true equilibrium isotherm because the slow relaxation of nonequilibrium structure leads to multilayer build-up [37, 38]. Such behavior can be monitored by atomic force microscopy (AFM) measurements, neutron reflection, dual polarization interferometry, circular dichroism, and Fourier transform infrared attenuated total reflectance (FTIR/ATR) [39] as well as other techniques.

The process of macromolecular multilayer adsorption, is still too complicated to be effectively modeled by kinetic models. At low concentrations, an interfacial cavity kinetic model has been used to characterize monolayer or submonolayer protein adsorption with surface-induced structural transitions [39].

Many studied projects have focused on the effect of various modifications in the adsorption systems, including surface modification [40], protein modification [41], the use of saccharides [42] and surfactants [43], and adjustment of solvent conditions such as ionic strength and pH [38, 44], for the purpose of either reducing or promoting protein adsorption.

Adsorption capacity of cytochrome c on chelated Cu^{2+} bead was demonstrated to be dependent on the buffer type with the observed adsorption in the order phosphate >N-(2-hydroxyethyl)-piperazine-N'-2-ethanesulfonic acid >morpholinopropane sulfonic acid >morpholinoethane sulfonic acid >tris(hydroxymethyl)-aminomethane hydrochloride (Tris-HCl) [45]. Vasina and Dejardin reported that the adsorption of α-chymotrypsin on muscovite mica was depressed by increasing the concentration of Tris-HCl buffer at pH 8.6, close to α-chymotrypsin's isoelectronic point [46]. Phosphate buffered saline (PBS) is the most commonly used buffer at the pH range close to 7, since it is reported to be able to stabilize protein structure in bulk solution environment in most cases [47]. The behavior of PBS buffer is particularly complex in adsorption studies due to the various types of phosphate ions present and the tendency of these ions to adsorb competitively and/or to form complexes either with the proteins or with the surfaces.

It is also well known that protein size and net charge have significant effects on adsorption. Changes in protein secondary structure are frequently monitored as indications of denaturing. Denaturing upon adsorption is important in many applications such as implants, biofueling. Quantification of secondary structure is sensitive to the peak assignments.

Adsorption of a protein to a surface may induce conformational changes in the protein. The degree of conformational changes is determined by a combination of the native stability of a protein, the hydrophobicity and the charges of the protein and the sorbent surface. Protein adsorption can be driven by a conformational entropy gain especially if adsorption is endothermic. This entropy gain can arise from the release of the solvent molecules from hydrophobic patches on the protein surface. Loses of translation entropy of the protein may play a minor rule [48].

Norde et al. [49] study the thermodynamics for adsorption of human serum albumin (soft) and ribonuclease (hard) on polystyrene surface (hydrophobic). That was reported that a net increase in entropy on a like-charged polystyrene surface drives the adsorption process for both proteins. The entropy increases because hydrophobic parts of the polystyrene surface gets dehydrated and structural changes in the proteins allowed the molecule feel free. Proteins with low Gibbs energy of denaturation (i.e., a protein with low native-state stability is called soft protein) are driven by entropy gains associated with the breaking down of secondary and tertiary protein structure.

4. Immobilization of protein monolayers on planar solid supports

The concept of using self-assembled biomolecules as an elementary units to develop super-structures of defined geometry has thus received considerable attention. In this contents, the self-assembly ability of amphiphilic biomolecules such as lipids, to spontaneously organize appears as a suitable concept for the development of membrane models. The concept is clearly illustrated i.e. by Langmuir monolayers, which have been extensively used as models to understand the role and the organization of biological membranes [50] and to acquire knowledge about the molecular recognition process [51,52]. Langmuir-Blodgett technology allows to build lamellar lipid stacking by transferring a monomolecular film formed at an air/water interface onto a solid support. When all parameters are optimized, this technique corresponds to one of the most promising for preparing thin films of amphiphilic molecules [6]. The sensitive element produced by LB technology has higher sensitivity and faster response time, can work in room temperature.

The optimal value of the surface pressure to produce the best results depends on the nature of the monolayer and is often established empirically [53]. However, the LB/LS deposition is traditionally carried out in the condensed phase since it is generally believed that the transfer efficiency increases when the monolayer is in a close-packed state. In that condition the surface pressure is sufficiently high to ensure a strong lateral cohesion in the monolayer, so that the monolayer does not fall apart during the transfer process. Although the optimal surface pressure depends on the nature of the material constituting the film, biological amphiphiles can seldom be successfully transferred at surface pressures lower than 10 mN/m and at surface pressures above 40 mN/m, where collapse and film rigidity often pose problems [6].

Moreover, the main advantage of the adsorption of the enzyme onto pre-formed LB films lies in the possible interaction of the enzyme with a hydrophobic or hydrophilic surface depending on the number of the deposited layers, thus allowing the control of the enzyme environment. Likewise, this approach allows the control of the thickness and the homogeneity of the LB films harboring the enzymes. Nevertheless, the release of protein molecules due to the weakness of their association with the surface is often the main reason which explains the poor reproducibility of responses of LB membrane-based sensors. Due to avoid desorption, some authors have proposed to covalently immobilize the enzyme on LB film surfaces by the use of cross-linking agents [54].

Electrostatic layer-by-layer assembly was first proposed by Decher in 1990s and proved to be possible to build-up ordered multilayer structures by consecutive adsorption of polyanions and polycations [55]. This film assembly approach has great advantages because of the simplicity preparation of ultrathin films with defined composition and uniform thickness in nanoscale in which synergy between distinct materials may be achieved in a straightforward, low-cost manner. With the LbL technique a wide diversity of materials may be employed, and film fabrication is performed under mild conditions, which is particularly important for preserving activity of biomolecules. The fundamental concepts and mechanisms involved in the LbL technique have been detailed in a series of papers [56]. In most cases adsorption in LbL films is governed by electrostatic interactions between species bearing opposite charges, but secondary interactions have also been shown to be important. The LbL technique is versatile with regard to the substrates that may be used, which include hydrophilic and hydrophobic glass, mica, silicon, metals, quartz, and polymers [57]. In addition, LbL films may be deposited directly onto colloidal suspensions [58].

Several attempts have been made to fabricate hybrid enzyme electrodes with the method [59]. In 1995 this new method was applied to immobilize negatively charged glucose oxidase (GOx) in a polyethyleneimine based multilayer structure [60] and proved to be one of the most perspective methods for preparing amperometric enzyme biosensors. One year later, an oxygen mediated glucose biosensor based on GOx and poly(L-lysine) co-adsorbed onto a negatively charged monolayer of mercaptopropionic acid, deposited on an Au electrode was described [60]. Hodak *et al*. introduced LbL assembly technique to construct reagentless biosensor with glucose oxidase and ferrocene modified with poly(allylamine). Sun *et al*. [61] and Chen *et al*.[62] fabricated peroxidase and glucose oxidase biosensors with Os-based redox polymer and enzymes. Also known isreagentless biosensor built of organic dye methylene blue with peroxidase [63]. Vossmeyer and co-workers investigated the optical and electrical properties of layer-by-layer self-assembly of gold nanoparticle/alkanedithiol films [64]. Though gold nanoparticles or enzymes have been widely used to form multilayer films by layer-by-layer technology.

5. Biorecognition elements

Bioreceptors or biorecognition elements are the key to specification of biosensor technologies. They are responsible for binding the analyte of interest to the sensor for the measurement. These bioreceptors can take many forms and the different bioreceptors that have been used are as numerous as the different analytes that have been monitored using biosensors. However, bioreceptors can generally be classified into a few different major categories (Figure 6). These categories include: antibody/antigen, enzymes, nucleic acids/DNA, cellular structures/cells (i.e. microorganisms), and biomimetics. The specificity of molecular recognition makes these molecules very attractive as tools for therapeutic diagnostic and other analytical applications.

5.1. Enzymes

Enzymes are the most commonly used bio-receptors in bioassays. The analyte can be the enzyme, whose enzymatic activity is determined, or the substrate or the enzyme cofactors. Enzymatic assays are mainly based on either inhibition of the enzyme activity or catalysis. According to the fact, variety of enzymes such as organophosphorous hydrolase, alkaline phosphatase, ascorbate oxidase, tyrosinase and acid phosphatase have been employed in design of pesticide bioassays and biosensors [65].

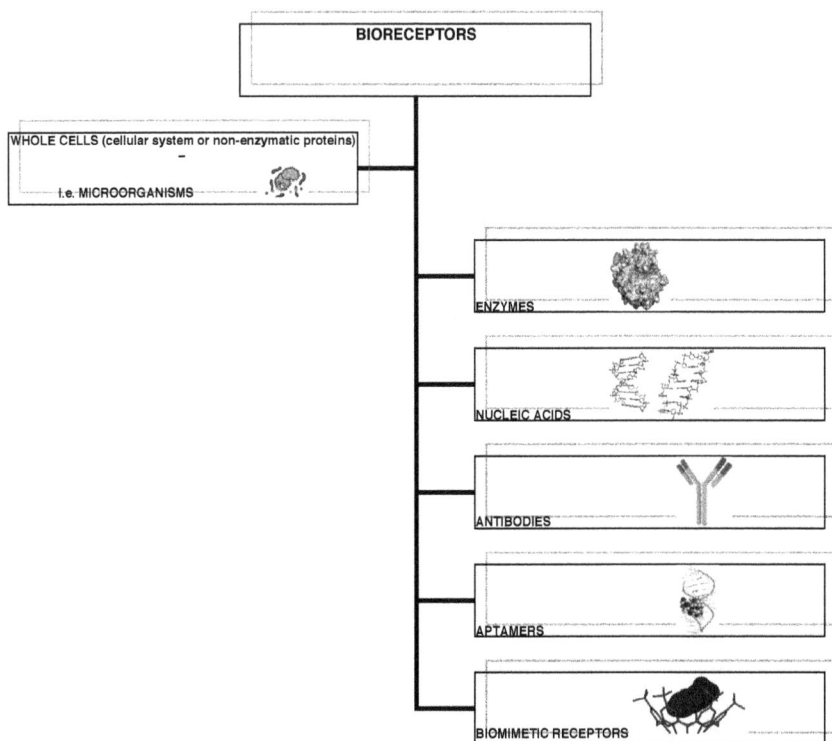

Figure 6. Classification of bioreceptors

Enzymes are often chosen as bioreceptors based on their specific binding capabilities as well as their catalytic activity. In biocatalytic recognition mechanisms, the detection is amplified by a reaction catalyzed by macromolecules - biocatalysts. With the exception of a small group of catalytic ribonucleic acid molecules, all enzymes are proteins. Some enzymes require no chemical groups other than their amino acid residues for activity. Others require an additional cofactor, which may be either one or more inorganic ions (Fe^{2+}, Mg^{2+}, Mn^{2+}, or Zn^{2+}), or a more complex coenzyme. The catalytic activity provided by enzymes allows for much lower limits of detection than would be obtained with common binding techniques. The catalytic activity of enzymes depends upon the integrity of their native protein conformation. Enzyme-coupled receptors can also be used to modify the recognition mechanisms. In example, the activity of an enzyme can be modulated when a ligand binds at the receptor. This enzymatic activity is often greatly enhanced by an enzyme cascade, which leads to complex reactions in the cell [66].

The use of enzymes as the recognition element was very popular in the first generation of biosensor due to their availability. Among various oxidoreductases, glucose oxidase, horse-radish peroxidase and alkaline phosphatase have been employed in most biosensor studies [67-69]. Some amperometric based methods use duel enzyme systems such as acetylcholine esterase and choline oxidase. In example, organophosphorous hydrolase catalyzes the hydrolysis of a wide range of organophosphate pesticides, and as a result of its versatility, this enzyme has been incorporated into a number of assays and sensors for the detection of this type of compounds. Additional enzymes can be used to detect other environmental and food contaminants such as nitrate, nitrite, sulfate, phosphate, heavy metals and phenols. Tyrosinase is frequently used to determine phenols, chlorophenols, cyanide, carbamates and atrazine.

Enzymes offer many advantages connected with high sensitivity, possibility of direct visualization and stability, but there are still some problems, which include multiple assay steps and the possibility of the interference from endogenous enzymes. Many enzyme detection procedures are visual eliminating the need for expensive equipment, but the enzyme stability is still problematic and the ability to maintain enzyme activity for a long time [70].

5.2. Nucleic acids

Recently, advances in nucleic acid recognition have enhanced the using of DNA biosensors and biochips [71]. In the case of nucleic acid bioreceptors for pathogen detection, the identification of analyte's nucleic acid is achieved by matching the complementary base pairs. Since each organism has unique DNA sequences, any self-replicating microorganism can be easily detect [70].

Grabley and coworkers have reported on the use of DNA biosensors for the monitoring of DNA-ligand interactions [72]. Surface plasmon resonance was used to monitor real-time binding of low molecular weight ligands to DNA fragments that were irreversibly bound to the sensor surface *via* coulombic interactions. The detection of specific DNA sequences has been employed for detecting microbial and viral pathogens [73] and food pathogen like *E. coli* [74], *Salmonella sp.*

Recent advances in nucleic acid recognition, like the introduction of Peptide Nucleic Acid (PNA) has opened new opportunities for DNA biosensors. PNA is a synthesized DNA in which the sugar-phosphate backbone is replaced with a pseudopeptide. PNA as a probe molecule has several advantages, i.e. superior hybridization characteristics, detection of single-based mismatches, better stability compared to enzymes [70].

5.3. Antibodies

Antibodies are common bioreceptors used in biosensor technologies. Antibodies are biological molecules that exhibit very specific binding capabilities for specific structures. This is very important due to the complex nature of most biological systems. An antibody is a complex biomolecule, made up of hundreds of individual amino acids arranged in a highly ordered sequence. For an immune response to be produced against a particular molecule, a certain molecular size and complexity are necessary: proteins with molecular weights greater than 5000 Da are generally immunogenic.

The way in which an antigen and its antigen-specific antibody interact is analogues of a lock and key fit [66], by which specific geometrical configurations of a unique key enables it to open a lock.

An antigen-specific antibody fits its special antigen in a highly specific manner, according to that the three-dimensional structures of antigen and antibody molecules are matching [70]. Antibody biosensors are interested wide and interesting group of sensing devices, which includes i.e. Surface plasmon resonance [74], fiber-optic biosensor [75], magnetoelastic resonance sensor [76] and immunosensor [77].

5.4. Aptamers

Aptamers are folded single stranded DNA or RNA oligonucleotide sequences with the capacity to recognize various target molecules. They are generated in the systematic evolution of ligands by exponential enrichment process which was first time reported by Ellington [78] and Tuerk [79]. In this way suitable binding sequences are first isolated from large oligonucleotide libraries and subsequently amplified. The main application for aptamers is in biosensors. While antibodies are used in ELISA, the similar process for aptamers is called ELONA (enzyme linked oligonucleotide assay). They have many advantages over antibodies such as easier deposition on sensing surfaces, higher reproducibility, longer shelf life, easier regeneration and a higher resistance to denaturation. As antibodies, they are characterized by both, their high affinity and specificity to their targets [80].

5.5. Biomimetic receptors

A receptor which is designed and fabricated and to mimic a bioreceptor is often defined as biomimetic receptor. According to the phenomena several different techniques have been developed over the years for the construction of biomimetic receptors [81,82]. These procedures include: genetically engineered molecules, artificial membrane fabrication and molecular imprinting method. The molecular imprinting method has existed as an attractive and accepted tool in developing an artificial recognition agents.

Artificial membrane fabrication for bioreception has been performed for many different applications. Stevens *et al.* has developed an artificial membrane by incorporating gangliosides into a matrix of diacetylenic lipids [83]. The lipids were allowed to self-assemble into Langmuir-Blodgett layers and were then photopolymerized *via* ultraviolet irradiation into polydiacetylene membranes. However, molecular imprinting has been used for the construction of a biosensor based on electrochemical detection of morphine [84].

5.6. Cellular bioreceptors

Cellular structures and cells have been used in the development of biosensors and biochips. These bioreceptors are either based on biorecognition by an entire cell/microorganism or a specific cellular component that is capable of specific binding to certain species. There are presently three major subclasses of this category: a) cellular systems, b) enzymes and c) non-enzymatic proteins. Due to the importance and large number of biosensors based on enzymes, these have been given their own classification [85]. Microorganisms offer a form of bioreceptor that often allows a whole class of compounds to be monitored. Generally these microorganism biosensors rely on the uptake of certain chemicals into the microorganism for digestion. Often, a class of chemicals is ingested by a microorganism, therefore allowing a class-specific biosensor to be created. Microorganisms such as bacteria and fungi have been used as indicators of toxicity or for the measurement of specific substances. For example, cell metabolism (e.g., growth inhibition, cell viability, substrate uptake), cell respiration or bacterial bioluminescence have been used to evaluate the effects of toxic heavy metals [85].

A microbial biosensor has been developed for the monitoring of short-chain fatty acids in milk [86]. *Arthrobacternicotianae* microorganisms were immobilized in a calciumalginate gel on an electrode surface. By monitoring the oxygen consumption of the *Anthrobacter* electrochemically, its respiratory activity could be monitored, thereby providing an indirect means of monitoring fatty acid consumption.

6. Quantitative detection of protein binding to the solid surface

Proteins adsorb in differing quantities, densities, conformations, and orientations, depending on the chemical and physical characteristics of the surface [87]. Protein adsorption is a complex process involving van der Waals, hydrophobic and electrostatic interactions, and hydrogen bonding. Although surface-protein interactions are not well understood, surface chemistry has been shown to play a fundamental role in protein adsorption. Moreover, the properties of protein over-layers can be altered by the underlying chemistry, which directly impinges on control of conformation and/or orientation [88].

During the past decade substantial progress has been made in understanding the mechanism of protein adsorption. Authors have reported a numerous of techniques, e.g. QCM [89, 90], surface plasmon resonance (SPR) [91,92], ellipsometry [91], FTIR [93], atomic force microscopy (AFM) [94].

6.1. Quartz crystal microbalance measurements

QCM (Quartz Crystal Microbalance, Fig. 7) technology enables studies of molecular interactions by measuring the weight of the molecules, much like a very sensitive scale or balance. When molecules are added to or removed from the sensor surface, it is detected as a change in the oscillation frequency of the sensor crystal; the change in resonance frequency is correlated to the change in mass on the surface. QCM technology does not have the same limitations with regard to surface proximity as other biosensor technologies, making it possible for the instrument to measure binding to large structures such as cells.

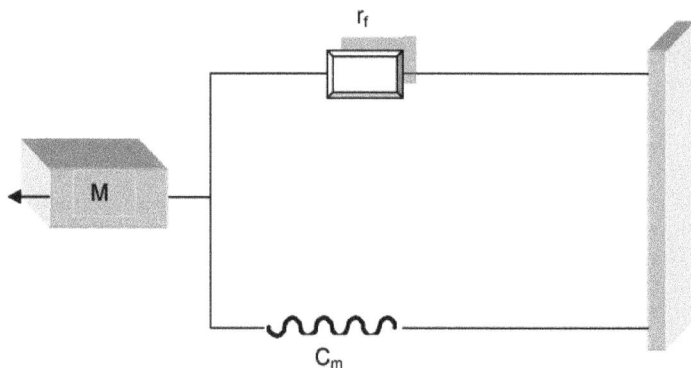

Figure 7. Quartz Crystal Microbalance - equivalent mechanical model; mass (M),a compliance (Cm), and a resistance. (rf). The compliance represents energy stored during oscillation and the resistance represents energy dissipation during oscillation

The method is very useful for monitoring the rate of deposition in thin film deposition systems under vacuum. In liquid, it is highly effective at determining the affinity of molecules (i.e. proteins) to surfaces functionalized with recognition sites. Larger entities such as viruses or polymers are investigated, as well. QCM has also been used to investigate interactions between biomolecules.

Upon protein adsorption to the crystal surface, the oscillatory motion of the crystal was dampened, causing a decrease in the resonant frequency. The frequency shift of the QCM is due to a change in total coupled mass, including water interaction within the protein layer. The *Sauerbrey* equation relates the measured frequency shift (Δf) and the adsorbed mass (m) [95].

$$\Delta f = -\frac{2\Delta m f_0^2}{A\sqrt{\rho_q \mu_q}} = -\frac{2 f_0^2}{A\sqrt{\rho_q \mu_q}} \Delta m \tag{1}$$

where

- f_0 – Resonant frequency (Hz)

- Δf – Frequency change (Hz)

- Δm – Mass change (g)

- A – Area between electrodes (cm^2)

- ρ_q – Density of quartz (ϱ_q = 2.648 g/cm^3)

- μ_q – Shear modulus of quartz (μ_q = 2.947 x 10^{11} g/cm.S^2)

The sensor can be used for the direct, marker-free measurement of specific interactions between immobilized molecules and analytes in solution. Binding of a soluble analyte to the immobilized ligand causes a shift in the resonance frequency, and this signal can be recorded using a frequency counter with high resolution. This method, despite its existence for four decades, has only recently been developed for immunological measurements in a flow through system [96].

In contrast to the optical techniques, which are not sensitive to water associated with adsorbed proteins, the f-shift of the QCM is due to the change in total coupled mass, including hydrodynamically coupled water, water associated with the hydration layer of e.g. proteins and/or water trapped in cavities in the film [89].

A recent extension of the technique, called QCM-D, to simultaneously measure changes in the frequency, Δf, and in the energy dissipation, ΔD, of the QCM provides new insight into e.g. protein adsorption processes [89] as well as other surface-related processes.

6.2. Surface plasmon resonance

Surface plasmon resonance (Figure 8) can be applied as a convenient, sensitive and label-free technique to study various surface phenomena. SPR is a surface sensitive, spectroscopic method which measures change in the thickness or refractive index of biomaterials at the interface between metal surfaces, usually a thin gold film (50–100 nm) coated on a glass slide, and an ambient medium. In SPR the test proteins are immobilized on a gold-surface, unlabelled query protein is added, and change in angle of reflection of light caused by binding of the probe to the immobilized protein is measured to characterize biomolecular interactions in the real-time [97].

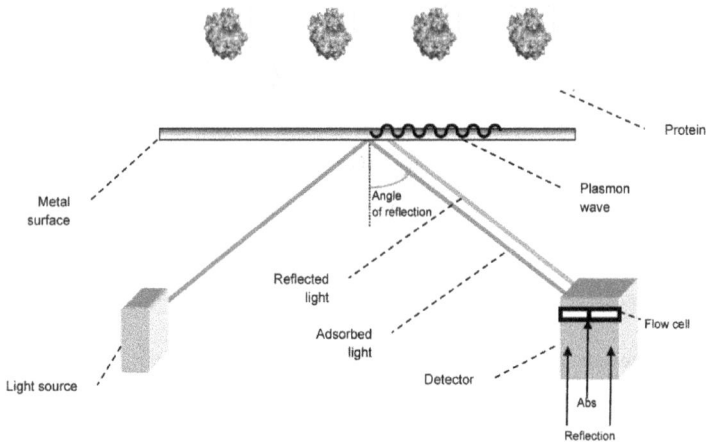

Figure 8. Surface Plasmon Resonance

SPR has been used for many biomedical, food and environmental applications [98]. In example, Hiep *et al.* [99] developed a localized SPR immunosensor for detection of casein allergen in raw milk. There was also generated a unique SPR-based microarray using natural glycans for rapid screening of serum antibody profiles [100]. SPR microarrays was utilized in combination with HT antibody purification technologies for rapid and proper affinity ranking of antibodies [101]. SPR-based biosensors are in great demand as they provide label-free, real-time detection of the biomolecular interactions.

6.3. Ellipsometry

Ellipsometry (ELM, Fig. 9) is an optical method that has been used extensively for protein adsorption studies. The method is based on the change upon protein adsorption of the state of polarization of elliptically polarized light reflected at a planar surface. From the changes in the ellipsometric angles (Δ, ψ), the refractive index and the thickness, morphology or roughness of the surface of layers can be deduced and used to determine e.g. the amount of adsorbed protein on a surface. Since the refractive index of adsorbed protein films is always close to $n=1.5$, which the film thickness can be calculated with quite good accuracy [102]. Moreover, the clear advantage of this technique is that the proteins under investigation require no chemical treatments with markers etc. before use. Also, the measurement procedure is quite fast (on the order of a few seconds).

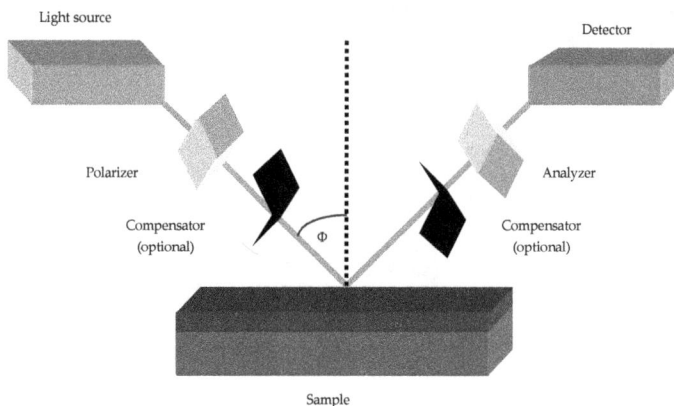

Figure 9. Ellipsometry setup

Since ellipsometry can be performed on most reflective substrates it can be easily conducted on an electrode surface and combined with electrochemistry. The ellipsometry and electrochemical methods have been used to study protein adsorption on metal surfaces and specifically human serum albumin on gold surfaces [103]. Chronoamperometry and ellipsometry were combined for the study of immunosensor interfaces based on methods of Immunoglobulin G adsorption onto mixed self-assembled monolayers [104]. The combined imaging ellipsometry with electrochemistry have been employed to investigate the influence of electrostatic interaction on fibrinogen adsorption on gold surfaces [105].

6.4. Atomic force microscopy

Since its invention atomic force microscopy (AFM) has been a useful tool for imaging a wide class of biological specimens such as nucleic acids, proteins, and cells at nanometer resolution in their native environment [106]. AFM can also be applied to measure intermolecular force based on the deflection signal of the AFM probe (cantilever) caused by the force between the cantilever modified with a molecule of interest and a complementary molecule immobilized on a substrate.

Atomic force microscopy (AFM) is a very high-resolution type of scanning probe microscopy, with demonstrated resolution on the order of fractions of a nanometer, more than 1000 times better than the optical diffraction limit. Because the atomic force microscope relies on the forces between the tip and sample, knowing these forces is important for proper imaging. The force is not measured directly, but calculated by measuring the deflection of the lever, and knowing the stiffness of the cantilever. Hook's law gives:

$$F = -kz \tag{2}$$

where F is the force, k is the stiffness of the lever, and z is the distance the lever is bent.

The major advantage of AFM is to carry out measurements on non-conducting substrates, and to determine particle height as well, with atomic level precision and flexibility [107]. AFM is a useful device not only to trace topography of biological samples with molecular resolution under physiological conditions but also to study the interaction force between bio-molecular pairs and the mechanical properties of proteins at the single molecular level [108]. Moreover, the AFM is used as a manipulator to obtain DNA from chromosomes [109], or mRNA from local regions of living cells [110].

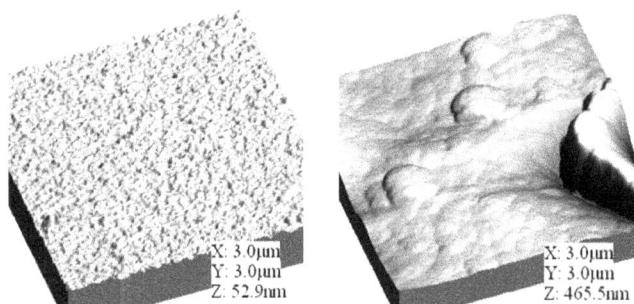

Figure 10. AFM image of immobilized invertase in LS lipid-like film; left – pure lipid film, right – hybrid protein film. All images are 3 μm x 3 μm [17]

Several measurements of the intermolecular force produced by biomolecular interaction were reported, i.e. the AFM surface topography of phospholipids LB films have shown a smooth surface. In presence of protein in LB film well-defined structures were observed, characterized by domains, globules, grains with different diameters. The images indicates that both enzyme molecules are not only properly entrapped in the composite membrane but also well exposed at the surface, which can be clearly seen in Figure 10. The recorded images show a relatively high homogeneity of the topography, especially in case of lipid film [17].

7. Prospects and future trends

The advances observed in the areas of biochemistry, chemistry, electronics and bioelectronics will markedly influence future of biosensor production. Progresses in biosensors technology focus on two main aspects: transducer technology development and sensing element development [111]. New improved detection systems developed under the areas of microelectronics or even nanoelectronics can be used in biosensors. However, since biosensor sensitivity and selectivity depend basically on the properties of the biorecognition elements, a crucial aspect in future biosensors is the development of improved molecular recognition elements. In this respect, biotechnology and genetic engineering offer the possibility of tailor binding molecules with predefined properties

According to fact, that miniaturization of devices as well as multi-sensor arrays are expected to have a marked impact in biosensors technology, the use of thin film methods for preparation of the recognition layers provides a simple procedure for the functionalization of electrode surfaces using nanogram amounts of material. Different techniques can give either highly ordered or amorphous film, ensured a high level of control of the environment and often resemble the environment found inside biomembranes, thereby guaranteed the stabilization of biomolecules. Biosensors produced using these layer technologies can display high sensitivities, be easily interrogated using electronic, optical or mass-sensitive techniques, can often be regenerated and display good stability.

Another current trend is the combination of physics and biology in the creation of new nanostructures. Nanotechnology comprises a group of emerging techniques from physics, chemistry, biology, engineering and microelectronics that are capable of manipulating matter at nanoscale. This novel technology bridges materials science, and biochemistry/chemistry, where individual molecules are of major interest [112]. Inspired by nature, molecular self-assembly has been proposed for the synthesis of nanostructures capable to perform unique functions. According to that, novel tools that combine different sensing methods can provide also the necessary complementary information that is needed to understand the limitations and to optimize the performance of the new techniques. Therefore, introducing existing methods (e.g., SPR, QCM, ellipsometry) allow parallel complementary investigations of the biochemical processes that take place at the interface between the devices and the biological sample.

At present, biosensor research is not only driving the ever-accelerating race to construct smaller, faster, cheaper and more efficient devices, but may also ultimately result in the successful integration of electronic and biological systems. Thus, the future development of highly sensitive, highly specific, multi-analysis, nanoscale biosensors and bioelectronics will require the combination of much interdisciplinary knowledge from areas such as: quantum, solid-state and surface physics, biology and bioengineering, surface biochemistry, medicine and electrical engineering. Any advancement in this field will have an effect on the future of diagnostics and health care.

Acknowledgments

The authors are gratefully acknowledged to Wroclaw University of Technology and Polish National Centre of Progress of Explorations for the financial support (Grant no. NR05-0017-10/2010).

Author details

Joanna Cabaj and Jadwiga Sołoducho

Faculty of Chemistry, Wrocław University of Technology, Wrocław, Poland

References

[1] Kuwahara T., Oshima K., Shimomura M., Miyauchi S. Glucose sensing with glucose oxidase immobilized covalently on the films of thiophene copolymers. Synthetic Metals 2005, 152, 29-32.

[2] Cabaj J., Sołoducho J., Nowakowska-Oleksy A., Langmuir-Blodgett film based biosensor for estimation of phenol derivatives. Sensors and Actuators B 2010, 143, 508-515.

[3] Cabaj J., Sołoducho J., Świst A., Active Langmuir–Schaefer films of tyrosinase—Characteristic. Sesnors and Actuators B 2010, 150, 505-512.

[4] Cabaj J., Sołoducho J., Chyla A., Jędrychowska A. Hybrid phenol biosensor based on modified phenoloxidase electrode Sensors and Actuators B 2011, 157, 225-231.

[5] Cabaj J., Chyla A., Jędrychowska A., Olech K., Sołoducho J., Detecting platform for phenolic compounds – characteristic of enzymatic electrode. Optical Materials 2012, 34, 1677-1681.

[6] Girard-Egrot AP., Blum LJ., Langmuir-Blodgett Technique for Synthesis of Biomimetic Lipid Membranes in Nanobiotechnology of Biomimetic Membranes, Ed. D. Martin; Springer; 2007.

[7] Zhou X., Xi F., Zhang Y., Lin X. Reagentless biosensor based on layer-by-layer assembly of functional multiwall carbon nanotubes and enzyme-mediator biocomposite. Journal of Biomedicine and Biotechnology 2011, 12(6) 468-476.

[8] El Kirat K., Besson F., Prigent A.F., Chauvet JP., Roux B. Role of calcium and membrane organization on Phospholipase D localization and activity. Competition between a soluble and an insoluble substrate. Journal of Biological Chemistry 2002, 277, 21231–21236.

[9] Kulkarni CV., Tomseice M., and Glatter O. Immobilization of Nanostructured Lipid Particles in Polysaccharide Films. Langmuir 2011, 27 (15) 9541–9550.

[10] Heyse S.; Vogel H., Sanger M., Sigrist H. Covalent attachment of functionalized lipid bilayers to planar waveguides for measuring protein binding to biomimetic membranes. Protein Science 1995, 4 (12) 2532–2544.

[11] Stelzle M., Weissmuller G., Sackmann E. On the application of supported bilayers as receptive layers for biosensors with electrical detection. Journal of Physical Chemistry, 1993, 97(12) 2974–2981.

[12] Castellana ET., Cremer PS. Solid supported lipid bilayers: From biophysical studies to sensor design. Surface Science Reports 2006, 61(10) 429–444.

[13] Cooper MA. Signal transduction profiling using label-free biosensors. Journal of Receptors Signal Transductors, 2009, 29 (3-4) 224–233.

[14] Reimhult E., Kumar K. Membrane biosensor platforms using nano- and microporous supports. Trends in Biotechnology, 2008, 26(2) 82–89.

[15] Okahata Y., Tsuruta T., Ijiro K., Ariga K. Langmuir-Blodgett films of an enzyme-lipid complex for sensor membranes. Langmuir, 1988, 4, 1373-1375.

[16] Goto T., Lopez RF., Oliveira ON., Caseli L. Enzyme activity of catalase immobilized in Langmuir-Blodgett films of phospholipids. Langmuir, 2010, 26, 11135-11139.

[17] Cabaj J., Sołoducho J., Jędrychowska A., Zając D., Biosensinginvertase-based Langmuir–Schaefer films: Preparation and characteristic. Sensors and Actuators B, 2012, 66–67, 75–82.

[18] Ahola S., Turon X., Osterberg M., Laine J., and Rojas OJ. Enzymatic hydrolysis of native cellulose nanofibrils and other cellulose model films: Effect of surface structure. Langmuir 2008, 24, 11592-11599.

[19] Ferreira M., Fiorito PA., Oliveira ON. Jr., Córdoba de Torresi S. Enzyme-mediated amperometric biosensors prepared with the Layer-by-Layer (LbL) adsorption technique. Biosensors and Bioelectronics, 2004, 19, 1611-1615.

[20] Lakarda B., Magninb D., Deschaumeb O., Vanlanckerb G., Glinelb K., Demoustier-Champagneb S., Nystenb B., Jonasb AM., Bertrandb P., Yunus S. Urea potentiometric enzymatic biosensor based on charged biopolymers and electrodeposited polyaniline. Biosensors and Bioelectronics, 2011, 26(10) 4139-4145.

[21] Conzuelo F., Gamella M., Campuzano S., Ruiz MA., Reviejo A J., and Pingarron JM. An integrated amperometric biosensor for the determination of lactose in milk and dairy products. Journal of Agriculture Food Chemistry, 2010, 58, 7141–7148.

[22] Ohnuki H., Honjo R., Endo H., Imakubo T., Izum M. Amperometric cholesterol biosensors based on hybrid organic–inorganic Langmuir–Blodgett films. Thin Solid Films 2009, 518, 596–599.

[23] Cabaj J., Sołoducho J., Nowakowska A., Chyla A. Synthesis, design and electrochemical properties of Langmuir-Blodgett (LB) films built of bis(pyrrolyl)fluorene. Electroanalysis, 2006, 18(8) 801-806.

[24] Ji Q., Yoon SB., Hill JP., Vinu A., Yu J-S., Ariga K. Layer-by-layer films of dual-pore carbon capsules with designable selectivity of gas adsorption. Journal of American Chemical Society, 2009, 131, 4220-4221.

[25] Kniprath R., McLeskey JT. Jr, Rabe JP., Kirstein S. Nanostructured solid-state hybrid photovoltaic cells fabricated by electrostatic layer-by-layer deposition. Journal of Applied Physics 2009, 105,124313-124320.

[26] Jain V., Khiterer M., Montazami R., Yochum HM., Shea KJ., Heflin JR. High-contrast solid-state electrochromic devices of viologen-bridged polysilsesquioxane nanoparticles fabricated by layer-by-layer assembly. ACS Applied Materials and Interfaces 2009, 1, 83-89.

[27] Jiang SP., Liu Z., Tian ZQ. Layer-by-layer self-assembly of composite polyelectrolyte/Nafion membrane for direct methanol fuel cells. Advanced Materials 2006,18,1068-1072.

[28] Ariga K., Ji Q., and Hill JP. Enzyme-encapsulated layer-by-layer assemblies: Current status and challenges toward ultimate nanodevices. Advanced Polymer Science 2010, 229, 51–87.

[29] Puu G., Gustafson E., Artursson E., Ohlsson P-Aj. Retained activities of some membrane proteins in stable lipid bilayers on a solid support. Biosensors and Bioelectronics 1995, 10, 463-476.

[30] Ohlsson P-Aj., Tjarnhage T., Herbai E., Lofas S., Puu G. Liposome and proteoliposome fusion onto solid substrates, studied using atomic force microscopy, quartz crystal microbalance and surface Plasmon resonance. Biological activities of incorporated components. Bioelectrochemistry and Bioenergetics 1995, 38, 137-148.

[31] Girard-Egrot AP., Morelis RM., and Coulet PR. Influence of lipidic matrix and structural lipidic reorganization on choline oxidase activity retained in LB films, Langmuir 1998, 14, 476-482.

[32] Picas L., Suarez-Germa C., Montero MT., and Hernandez-Borrell J. Force spectroscopy study of Langmuir-Blodgett asymmetric bilayers of phosphatidylethanolamine and phosphatidylglycerol. Journal of Physical Chemistry B 2010, 114, 3543–3549.

[33] Robert GG. Langmuir-Blodgett films. New York: Plenum Press; 1990.

[34] Yuan C., Johnston LJ. Phase evolution in cholesterol/DPPC monolayers: atomic force microscopy and near field scanning optical microscopy studies, Journal of Microscopy 2002, 205, 136–146.

[35] McConnell HM., Radhakrishnan A. Condensed complexes of cholesterol and phospholipids. Biochimica et BiophysicaActa 2003,1610, 159–173.

[36] Bonn M., Roke S., Berg O., Juurlink LBF., Stamouli A., Muller M. A molecular view of cholesterol-induced condensation in a lipid monolayer. Journal of Physical Chemistry B 2004,108,19083–19085.

[37] Heinrich L., Mann EK., Voegel JC., Koper GJM., Schaaf, P. Scanning angle reflectometry study of the structure of antigen-antibody layers adsorbed on silica surfaces. Langmuir 1996, 12, 4857-4865.

[38] Wei T., Kaewtathip S., and Shing K. Buffer effect on protein adsorption at liquid/solid interface. Journal of Physical Chemistry C 2009, 113, 2053–2062.

[39] Tie Y., Ngankam AP., Van Tassel PR. Probing macromolecular adsorbed layer structure via the interfacial cavity function. Langmuir 2004, 20, 10599-10603.

[40] Zhang Z., Chen S., Jiang S. Dual-functional biomimetic materials: nonfouling poly(carboxybetaine) with active functional groups for protein immobilization. Biomacromolecules 2006, 7, 3311-3315.

[41] Daly SM., Przybycien TM., Tilton RD. Adsorption of polyethylene glycol modified lysozyme to silica. Langmuir 2005, 21, 1328-1337.

[42] Wendorf JR., Radke CJ., Blanch HW. Reduced protein adsorption at solid interfaces by sugar excipients Biotechnology and Bioenineering 2004, 87, 565-573

[43] Wahlgren M., Arnebrant T. Protein and a mutant with lower thermal stability. Langmuir 1997, 13, 8-13.

[44] Buijs J., Norde W., Lichtenbelt JWTh. Changes in the secondary structure of adsorbed IgG and F(ab')₂ studied by FTIR spectroscopy. Langmuir 1996, 12, 1605-1613.

[45] Emir S., Say R., Yavuz H., Denizli A. A new metal chelate affinity adsorbent for cytochrome C. Biotechnology Progress 2004, 20, 223-228.

[46] Vasina EN., Dejardin P. Adsorption of alpha-chymotrypsin onto mica in laminar flow conditions. Adsorption kinetic constant as a function of tris buffer concentration at pH 8.6. Langmuir 2004, 20, 8699-8706.

[47] Wang W. Instability, stabilization, and formulation of liquid protein pharmaceuticals. International Journal of Pharmaceutics 1999, 185, 129-188.

[48] Nakanishi K., Sakiyama T., and Imaura K. On the adsorption of proteins on solid surfaces, a common but very complicated phenomenon. Journal of Biosience and Bioengineering, 2001.

[49] Anzai JI., Furuya K., Chen CW., Osa T., Matsuo T. Enzyme sensors based on ion-sensitive field effect transistor. Use of Langmuir-Blodgett membrane as a support for immobilizing penicillinase. Analytical Sciences, 1987, 3, 271–272.

[50] Vollhardt D. Supramolecular organization in monolayers at the air/water interface. Materials Science Engineering C, 2002, 22, 121–127.

[51] Siegel S., Kindermann M., Regenbrecht M., Vollhardt D., von Kiedrowski G. Molecular recognition of a dissolved carboxylate by amidium monolayers at the air-water interface. Progress in Colloid and Polymer Science 2000, 115, 233–237.

[52] Vollhardt D., Fainerman VB. Penetration of dissolved amphiphiles into two dimensional aggregating lipid monolayers. Advance in Colloid and Interface Sciences 2000, 86, 103–151.

[53] Hann RA. Molecular structure and monolayer properties. In Robert G.G., ed. Langmuir-Blodgett films. New York: Plenum Press; 1990.

[54] Ganguly P., Paranjape DV., Sastry M., Chaudhari SK., Patil KR. Deposition of yttrium ions in Langmuir-Blodgett films using arachidic acid. Langmuir 1993, 9, 487–490.

[55] Lvov Y., Essler F., Decher G. Combination of polyanion/polycation self-assembly and Langmuir–Blodgett transfer for the construction of superlattice films. Journal of Physical Chemistry, 1993, 97, 13773-13777.

[56] Crespilho FN., Zucolotto V., Oliveira ON. Jr. and Nart FC. Electrochemistry of layer-by-layer films: A review. International Journal of Electrochemical Society 2006, 1, 194-214.

[57] Anzai J., Kobayashi Y., Nakamura N., Nishimura M. and Hoshi T. Layer-by-layer construction of multilayer thin films composed of avidin and biotin-labeled poly(amine)s. Langmuir 1999, 15, 221-226.

[58] Ichinose I., Tagawa H., Mizuki S., Lvov Y. and Kunitake T. Formation process of ultrathin multilayer films of molybdenum oxide by alternate adsorption of octamolybdate and linear polycations. Langmuir 1998, 14, 187-192.

[59] Hodak J., Etchenique R., Calvo EJ., Singhal K., Bartlett PN. Layer-by-layer self-assembly of glucose oxidase with a poly(allylamine)ferrocene redox mediator. Langmuir, 1997, 13, 2708-2716.

[60] Zhao W., Xu J-J., and Chen H-Y. Extended-range glucose biosensor via layer–by-layer assembly incorporating gold nanoparticles. Frontiers in Bioscience 2005, 10, 1060-1069.

[61] Li WJ., Wang ZC., Sun CQ., Xian M., Zhao MY. Fabrication of multilayer films containing horseradish peroxidase and polycation-bearing Os complex by means of elec-

trostatic layer-by-layer adsorption and its application as a hydrogen peroxide sensor. Analytica Chimica Acta 2000, 418, 225-232.

[62] Hou SF., Yang KS., Fang HQ., Chen HY., Amperometric glucose enzyme electrode by immobilizing glucose oxidase in multilayers on self-assembled monolayers surface. Talanta, 1998, 47, 561-567.

[63] Yang SM., Li YM., Jiang XM., Lin XF. A novel reagentless biosensor constructed by layer-by-layer assembly of HRP and Nile Blue premixed with polyanion. Chinese Chemical Letters 2005, 16(7) 983-986.

[64] Joseph Y., Besnard I., Rosenberger M., Guse B., Nothofer HG., Wessels JH., Wild U., Knop-Gericke A., Su D., Schlogel R., Yasuda A. and Vossmeyer T. Self-assembled gold nanoparticle/alkanedithiol films: preparation, electron microscopy, XPS-analysis, charge transport, and vapor-sensing properties. Journal of Physical Chemistry B 2003, 107, 7406-7413.

[65] Kandimalla VB. and Ju HX. New horizons with a multi dimensional tool for applications in analytical chemistry – Aptamer. Analytical Letters 2004, 37, 2215–2233.

[66] Vo-Dinh T., Cullum B., Biosensors and biochips: advances in biological and medical diagnostics, Fresenius Journal of Analytical Chemistry 2000, 366, 540–551.

[67] Wang J. Analytical Electrochemistry, 2nd Edition, New York: Wiley-VCH; 2000.

[68] Laschi S., Franek M., and Mascini M. Screen-printed electrochemical immunosensors for PCB detection, Electroanalysis 2000, 12, 1293-1298.

[69] Luong JHT., Male KB., Glennon JD. Biosensor technology: Technology push versus market pull. Biotechnology Advances 2008, 26, 492-500.

[70] Rahaie M. and Kazemi SS. Lectin-based biosensor: As powerful tools in bioanalytical applications. Biotechnology, 2010, 9, 428-443.

[71] Velusamy V., Arhak K., Korostynska O., Oliwa K. and Adley C. An overview of foodborne pathogen detection: In the perspective of biosensors. Biotechnology Advances 2010, 28, 232-254.

[72] Piehler J., Brecht A., Gauglitz G., Zerlin M., Maul C., Thiericke R., Grabley S. Label-free monitoring of DNA ligand interactions. Analytical Biochemistry 1997, 249, 94–102.

[73] Yang M., McGovern ME., and Thompson M. Genosensor technology and the detection of interfacial nucleic acid chemistry, Analytica Chimica Acta, 1997, 346, 259-375.

[74] Waswa J., Irudayaraj J. and DebRoy C. Direct detection of E. coli O157:H7 in selected food systems by surface plasmon resonance biosensor. LWT-Food Science and Technology 2007, 40, 187-192.

[75] Lim DV., Detection of microorganisms and toxins with evanescent wave fiber-optic biosensors. Proceedings of the IEEE, 2003, 91, 902-907.

[76] Guntupalli R., Hu J., Lakshmanan RS., Huang TS., Barbaree JM. and Chin BA., A magnetoelastic resonance biosensor immobilized with polyclonal antibody for the detection of Salmonella typhimurium. Biosensors and Bioelectronics 2007, 22, 1474-1479.

[77] Tokarskyy O. and Marshall DL. Immunosensors for rapid detection of Escherichia coli O157:H7-perspectives for use in the meat processing industry. Food Microbiology 2008, 25, 1-12.

[78] Ellington AD., Szostak JW. Invitro selection of RNA molecules that bind specific ligands. Nature 1990, 346(6287) 818–822.

[79] Tuerk C., Gold L. Systematic evolution of ligands by exponential enrichment- RNA ligands to bacteriophage-t4 DNA-polymerase. Science 1990, 249(4968) 505–510.

[80] Grieshaber D., MacKenzie R., Voros J. and Reimhult E. Electrochemical biosensors - sensor principles and architectures, Sensors 2008, 8, 1400-1458.

[81] Costello RF., Peterson IR., Heptinstall J., Walton DJ. Improved gel-protected bilayers. Biosensors and Bioelectronics 1999, 14, 265–271.

[82] Cornell BA., Braach-Maksvytis VLB., King LG., Osman PDJ., Raguse B., Wieczorek L., Pace RJ. A biosensor that uses ion-channel switches. Nature 1997, 387, 580 583.

[83] Charych D., Cheng Q., Reichert A., Kuziemko G., Stroh M., Nagy JO., Spevak W., Stevens RC. A 'litmus test' for molecular recognition using artificial membranes. Chemistry and Biology 1996, 3,113–120.

[84] Kriz D., Mosbach K. Competitive amperometric morphine sensor-based on an agarose immobilized molecularly imprinted polymer. Analytica Chimica Acta 1995, 300, 71–75.

[85] Vo-Dinh T., Cullum B. Biosensors and biochips: advances in biological and medical diagnostics. Fresenius Journal of Analytical Chemistry 2000, 366, 540–551.

[86] Schmidt A., Standfuss-Gabisch C., and Bilitewski U. Microbila biosensor for free fatty acids using an oxygen electrode based on thick film technology, Biosensors and Bioelectronics 1996, 11, 1139- 1145.

[87] Ta TC., McDermott MT. Mapping Interfacial Chemistry Induced Variations in Protein Adsorption with Scanning Force Microscopy. Analytical Chemistry 2000, 72, 2627-2634.

[88] Lenk TJ., Horbett TA., Ratner BD., Chittur KK. Infrared spectroscopic studies of time-dependent changes in fibrinogen adsorbed to polyurethanes. Langmuir 1991, 7(8) 1755-1764.

[89] Hook F., Rodahl M., Brzezinski P., Kasemo B. Energy dissipation kinetics for protein and antibody-antigen adsorption under shear oscillation on a quartz crystal microbalance. Langmuir 1998, 14, 729-734.

[90] Caruso F., Furlong DN., Kingshott P. Characterization of ferritin adsorption onto gold. Journal of Colloid and Interface Science 1997, 186, 129-140.

[91] Hook F., Kasemo B., Nylander T., Fant C., Sott K., and Elwing H. Variations in coupled water, viscoelastic properties, and film thickness of a Mefp-1 protein film during adsorption and cross-linking: a quartz crystal microbalance with dissipation monitoring, ellipsometry, and surface plasmon resonance study. Analytical Chemistry 2001, 73, 5796-5804.

[92] Sigal GB., Mrksich M., Whitesides GM. Effect of surface wettability on the adsorption of the proteins and detergents. Journal of American Chemical Society 1998, 120(14) 3464-3473.

[93] Chittur KK. FTIR/ATR for protein adsorption to biomaterial surfaces. Biomaterials 1998, 19(4-5) 357-369.

[94] Kidoaki S., Matsuda T. Adhesion forces of the blood plasma proteins on self-assembled monolayer surfaces of alkanethiolates with different functional groups measured by an atomic force microscope. Langmuir 1999, 15, 7639-7646.

[95] Sauerbrey G. Use of quartz crystal vibrator for weighting thin films on a microbalance. Phys. 1959, 155, 206-222.

[96] Kösslinger C., Drost S., Aberl F., Wolf H., Koch S., and Woias P. A quartz crystal biosensor for measurement in liquids. Biosensors and Bioloectronics 1992, 7, 397-404.

[97] Ray S., Mehta G. and Srivastava S. Label-free detection techniques for protein microarrays: Prospects, merits and challenges. Proteomics 2010, 10, 731–748.

[98] Shankaran DR., Gobi KV., Miura N. Recent advancements in surface plasmon resonance immunosensors for detection of small molecules of biomedical, food and environmental interest. Sensors and Actuators B 2007, 121, 158–177.

[99] Hiep HM., Endo T., Kerman K., Chikae M., Saito M., Tamiya E. A localized surface plasmon resonance based immunosensor for the detection of casein in milk. Science and Technology of Advanced Materials 2007, 8, 331–338.

[100] Wassaf D., Kuang G., Kopacz K., Wu QL. Nguyen Q., Toews M., Cosic J., Jacques J., Wiltshire S., Lambert J., Pazmany CC., Hogan S., Ladner RC., Nixon AE., Sexton DJ. Highthroughput affinity ranking of antibodies using surface plasmon resonance microarrays. Analytical Biochemistry 2006, 351, 241–253.

[101] de Boer RA., Hokke CH., Deelder AM., Wuhrer M. Serum antibody screening by surface plasmon resonance using a natural glycan microarray. Glycoconjugate Journal 2008, 25, 75–84.

[102] Hook F., Voros J., Rodahl M., Kurrat R., Boni P., Ramsden JJ., Textor M., Spencer ND., Tengvall P., Gold J., Kasemo B., A comparative study of protein adsorption on titanium oxide surfaces using in situ ellipsometry, optical waveguide lightmode

spectroscopy, and quartz crystal microbalance/dissipation, Colloids and Surfaces B: Biointerfaces 2002, 24, 155–170.

[103] Ying PQ., Viana AS., Abrantes LM., Jin G. Adsorption of human serum albumin onto gold: a combined electrochemical and ellipsometric study. Journal of Colloid and Interface Science 2004, 279(1), 95–99.

[104] Wang ZH.; Viana AS., Jin G., Abrantes LM. Immunosensor interface based on physical and chemical immunogylobulin G adsorption onto mixed self-assembled monolayers. Bioelectrochemistry 2006, 69(2), 180–186.

[105] Yu Y., Jin G. Influence of electrostatic interaction on fibrinogen adsorption on gold studied by imaging ellipsometry combined with electrochemical methods. Journal of Colloid and Interface Science 2005, 283(2), 477–481.

[106] Wakayama J., Sekiguchi H., Akanuma S., Ohtani T., Sugiyama S., Methods for reducing nonspecific interaction in antibody–antigen assay via atomic force microscopy. Analytical Biochemistry 2008, 380, 51–58.

[107] Wang HT., Kang BS., Chancellor TF. Jr., Lele TP., Tseng Y., Ren F., Pearton SJ., Dabiran A., Osinsky A. and Chow PP. Selective detection of Hg(II) ions from Cu(II) and Pb(II) using AlGaN/GaN high electron mobility transistors. Electrochem. ECS Solid-State Letters 2007, 10, J150–J153.

[108] Sekiguchi H., Hidaka A., Shiga Y., Ikai A., Osada T. High-sensitivity detection of proteins using gel electrophoresis and atomic force microscopy, Ultramicroscopy 2009, 109, 916–922.

[109] Tsukamoto K., Kuwazaki S., Yamamoto K., Ohtani T., Sugiyama S. Dissection and high-yield recovery of nanometer-scale chromosome fragments using an atomic-force microscope. Nanotechnology 2006, 17, 1391-1396.

[110] Uehara H., Osada T., Ikai A., Quantitative measurement of mRNA at different loci within an individual living cell. Ultramicroscopy 2004, 100, 197-201.

[111] Rodriguez-Mozaz S., Lopez de Alda MJ., Marco M-P., Barcelo D., Biosensors for environmental monitoring, a global perspective. Talanta 2005, 65, 291–297.

[112] Kossek S., Padeste C., Tiefenauer LX., Siegenthaler H. Localization of individual biomolecules on sensor surfaces. Biosensors and Bioelectronics 1998,13, 31-43.

Signal Transduction Methods

Chapter 4

Porous Silicon Biosensors

M. B. de la Mora, M. Ocampo, R. Doti, J. E. Lugo and
J. Faubert

Additional information is available at the end of the chapter

1. Introduction

There are a vast number of applications for biosensors ranging from medical monitoring and control, to release of drugs [1], and biosecurity [2]. The goblal market for biosensors in 2012 is estimated to reach 8.5 billion USD and projected to reach 16.8 billion by 2018 [3]. Porous silicon (p-Si) offers several advantages for its use as a biosensor such as a large specific surface area (of the order of 500 m ^{2} cm^{-3}) [4], visible luminescence at room temperature [5] and biocompatibility [6]. The p-Si was accidentally discovered when, in 1956 at the U.S. Bell Laboratories, Arthur Uhlir Jr. and Ingeborg Uhlir observed a red-green film formed on the wafer surface while trying a new technique for polishing silicon (Si) crystalline wafers. At the time however, it was not considered an interesting material. But when Leigh Canham in 1990 [5] discovered its visible luminescence properties, researchers started studying its non-linear optical, electric and mechanical properties. These academic and technological efforts have permitted the fabrication of uniform porous layers with diameters as small as one nanometer, permitting an enormous inner surface density, which is useful for biosensing applications. Several techniques exist to form this structure from a pure silicon crystalline wafer. The most popular is the electrochemical etching of crystalline silicon wafers (c-Si) [5]. Anodization begins when a constant current is applied between the c-Si wafer and the electrolyte by means of an electronic circuit controlling the anodization process [6].

Generally, p-Si is fabricated as shown in figure 1. We have a c-Si wafer (single crystalline) with the top face in contact with a hydrofluoric acid solution (HF) and where an immersed platinum electrode is placed at certain distance and parallel to the wafer. In the bottom face of the wafer we find a flat metallic electrode that is in close electric contact. Between the two electrodes there is a controlled voltage supply with its negative pole connected to the platinum immersed electrode. A current is established from the anodic electrode (back of the wafer) and the catodic electrode (platinum immersed). Modulating four variables: the intensity and interval of application of this current, the HF solution concentration, and the concentration and type of dopant previously applied to the c-Si wafer (type-n, type-p, or highly doped: type-p^+ and type n^+) it is then possible to control the porous size and p-Si layer geometrical parameters, as well as the number of layers. Dopant refers to a different element atom that replaces a percentage of the Si atom inside the wafer and that is crystallographically compatible with it, but that presents an electron in excess (type n) or an electron lack (type p). This introduces a number of properties that modify the material behavior when an electric field is applied, mainly the resistivity, that will influence the etching process performance.

The electric current oxidizes the surface silicon atoms permitting a fluoride ion (formed in the HF solution because of the electrical current) attack on them generating the pores. By using this electrochemical methodology it is also possible to create multilayer structures by alternating different current densities. For instance, if we start making the first layer with a current density J1 then the final porosity (and the refractive index) is going to be approximately determined by this current density. The electrochemical reaction time determines the thickness. By switching the current density to a different value J2, the reaction mainly continues at the crystalline silicon interface, leaving an almost intact first layer. Then the second layer will have a different refractive index and thickness (if we readjust the reaction time).

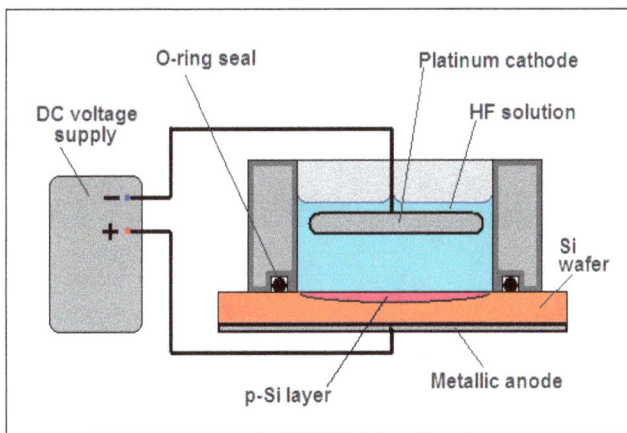

Figure 1. Experimental setup for porous silicon fabrication.

The figure 2 shows two structures that we fabricated from electrochemical etching of porous silicon. A luminescent monolayer made from p-type silicon wafers with a resistivity of 1-2 Ohm/cm (Fig. 2 a), and a multilayer prepared from p-type silicon wafers with a resistivity of 0.001-0.005 Ohm/cm (Fig. 2 b). Notice that by changing the dopant concentration, which is relate to the electrical resistivity used, the characterisitics of the p-Si structures differ. High resistivity crystalline silicon wafers give us higher porosities and small nanowires related to a given luminescent behavior. In turn, low resistivity allows us to achieve multilayers structures.

After the electrochemical etching stage, the surface of p-Si is hydrogen-terminated; this permits to immobilize large amounts of biomolecules [7]. It is possible to control several parameters of p-Si such as; pore size and consequently the refractive index, thickness, morphology, etc. by modifying the anodization conditions [6, 11]. Porosity can be measured by gravimetrical means. That is, the original crystalline silicon wafer is weighed first, then p-Si is formed and the wafer is weighed again, finally the p-Si layer is removed by adding KOH (Potassium hydroxide) and the wafer is weighed once more. With these three measurements is possible to determine the porosity. To measure the thickness, SEM (scanning electronic microscopy) techniques are normally used giving the best resolution and accuracy. Refractive index is usually determined by optical interference methods, where the refractive index can be estimated by taking adjacent maxima or minima from interference fringes coming from the p-Si sample.

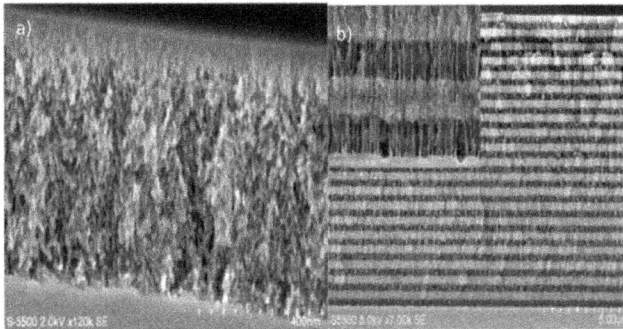

Figure 2. Crossectional SEM images of porous silicon nanostructures. A luminescent monolayer (a) and a multilayer (b). These strucutures were prepared at CIE-UNAM porous silicon laboratory.

There are other methods for obtaining p-Si such as the photoelectrochemical [5], the chemical vapour etching [8], the metal-assisted etching [9], and the 'stain etching' procedure [10]. The last two techniques mentioned do not require an electrical bias. In the stain etching procedure the power supply action is replaced by the chemical oxidant action of nitric acid. The reaction control is performed trough the addition of other additives. The results are less homogeneous than for the first process described, but they still permit to have the material quality compatible with several applications. As an example in figure 3 we show SEM images of a p-SI monolayer obtained by metal assisted etching of gold nanostructures and subsequent chemical attack of an HF/H_2O_2 electrolyte.

Figure 3. SEM images of gold nanostructures (a,b) used to fabricated a porous silicon monolayer. We show the surface (c) and the cross sectional (d) images. These structures were prepared at CIE-UNAM porous silicon laboratory.

The p-Si material can be prepared either in powder or wafer permitting to elaborate devices that can be dispersed in a given medium or reused [12]. Furthermore p-Si is a material that allows the fabrication of high quality photonic crystals [13] by applying the method described before to obtain multilayers structures. Such characteristics therefore allow several biosensing approaches usign this porous material [1].

2. Porous silicon biosensors: construction and transduction principles

The aim of a biosensor is to produce either discrete or continuous signals, which are proportional to a single analyte or a related group of analytes [14]. Because of its particular properties, the p-Si can be used as a transducer to convert this analytes into an optical or electrical signal [1]. Its large surface area enables an effective capture of the biological analytes although such a large surface area also implies high reactivity with the enviroment. This can cause the degradation of the biosensor and/or possible false positives. For this reason, stabilization of the p-Si surface via an appropriate surface chemistry is a required step for obtaining a succesful biosensor [15]. The surface chemistry should be designed in such a way as to obtain the desired effects, and yet still displaying bioactivity [16]. Also the binding affinity with the studied analytes must be taken into account [15, 17]. Some common techniques to functionalize p-Si include: oxidation [18, 19, 20], silanization [1, 15, 21, 22, 23, 24, 25], hydrosilylation of alkenes and alkynes [27, 28], radiation [29], and other chemical approaches [15, 16].

A proper pore-size distribution helps to achieve an efficient biosensor; p-Si fabricate from p+ and n+ -type silicon substrates is mesoporous, and suitable for immobilisation of biomacromolecules, while p-Si from p-type substrates, whose pore diameter is of the order of a few nm, is suitable only for very small molecules [12]. Macroporous p-Si from n-type substrates may accommodate larger molecules [12].

Once the appropriate chemical functionalization and porous distribution size is obtained, the challenge then becomes transducing the recognition of the biological analytes into a measurable signal. The requirements for efficient transduction are precision (same response to the same stimuli: repeatability) and accuracy (indicating magnitude value as close as possible to the real magnitude of the stimulus to be sensed: minimum absolute error spread) [30].

In general, the most common transduction techniques include piezoresistance, piezoelectricity, capacitive, resistive, tunneling, thermoelectricity, optical and radiation-based techniques, and electrochemical methods [31]. In the case of p-Si biosensors the most frequently used tecniques are optical and electrical/ electrochemical [1]. Here, we classified the p-Si biosensors depending on the transducing mechanism in optical and electrochemical contexts. Their characteristics and some related examples are detailed in the following sections.

3. Types of Porous silicon biosensors

3.1. Optical p-Si biosensors

Chemical or biomolecule detection can be based on changes in the optical spectral interference pattern [22, 23]. When white light passes through the p-Si an interference pattern is observed, this effect is called a Fabry–Perot fringe pattern, the binding of molecules induces changes in this pattern which are relate to a change in the refractive index of the p-Si [22]. This change is shown by a shift of the fringe pattern that can be quantified [22]. The effect depends on the refractive index value of the analized solution but also on how it penetrates into the pores [11]. The simplest kind of such p-Si biosensors is made of mono and double-layer films [1]. Some biological systems studied with these biosensors are: DNA hybridization [22, 32, 33], antibody cascading and the prototypical biotinstreptavidin interaction [22]. Using an analogous optical transduction modality it is possible to build p-Si biosensors with others complex optical structures as multilayer devices [6, 30].

These can be built up by alternating the applied current densities during the electrochemical etching generating a periodic or quasiperiodic combination of refractive indices [6]. This kind of structures offers better reflectance spectra (without side lobes) if compared with a mono or doubled-layered structure [30]. The etching parameters must be chosen to accommodate the analyte of interest whilst maximising the optical response. Some of the p-Si optical structures used in biosensing are: 1D photonic crystals [35, 36], rugate filters [37], microcavities [6, 38] and quasicrystals [39]. The use of these optical structures in biosensors allows integrability of all optical components and do not require electric contacts [11].

The photoluminescence properties of p-Si are also useful mechanisms for developing biosensors. It is possible to associate the amount of analytes studied with the changes in the photoluminescence spectra [1, 40, 41]. For example the quenching in the photoluminescence spectra after DNA deposition was used to study the transduction of DNA hybridization [42]. In this case the behavior was attributed to non-radiative recombination processes [1]. In recent years a successful implementation of this type of biosensor was obtained [6, 32, 42, 41] however until now this kind of biosensor is less accurate than its interferometric counterparts [1].

In a similar way the amount of an analyte of interest can be quantified by measuring the fluorescence signal intensity of a fluorescence molecule used as a marker fixed at a p-Si structure before and after an analyte is located into the p-Si [43].

We offer two comprehensive case examples to illustrate how the optical p-Si biosensors work. The first example is a sensor of a fluorescent molecule: fluorescein-5-maleimide (FM), by using a 1D photonic crystal or Bragg mirror [44]. The basic mirror was made by alternating layers of high (2.83) and low (1.65) refractive indexes, with a first layer that allows a good penetration of the active molecule into the porous structure. The surface of the first layer was functionalized by silanization with 3-mercaptopropyl)-trimethoxysilane (MPTS) to link the fluorescent molecule. The silicon mirror was fabricated in order to achieve a reflectance spectrum in a range that overlaps the fluorescent excitation of the molecule. The samples were analyzed by fluorescent spectrometry. The emission signal from fluorescent molecules was enhanced because of the p-Si mirror. That is, the p-Si structure provided a platform for high-sensitivity measurements. This biosensor uses two different detection platforms by using reflectance measurements as we show in the figure 4 and by analyzing the fluorescent spectrum as it can be observed in the figure 5.

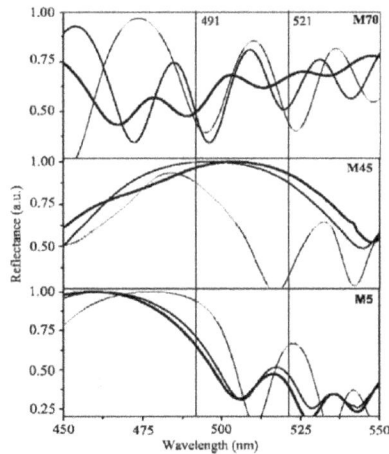

Figure 4. Reflectance spectra for freshly etched (thin line), silanized (normal line), and functionalized (thick line) mirrors. Vertical lines correspond to wavelength of excitation of 491 nm and emission of 521 nm of the FM molecule in a phosphate solution. Uncertanties in the reflectance intensity and wavelength were of ±2% and ±1 nm, respectively [44].

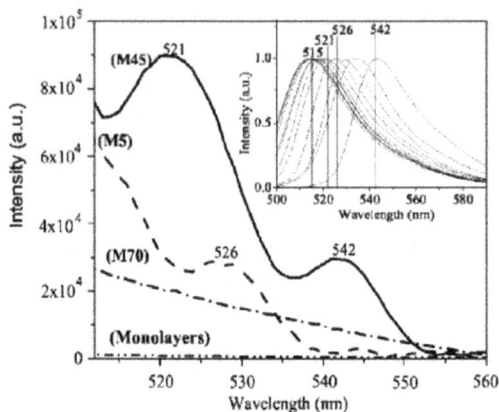

Figure 5. Fluorescence emission of FM molecules deposited on the MPTS functionalized surfaces. Monolayers correspond to sample m45 and m70 (no mirrors). Sample M45 shows the best fluorescence signal. Fluorescence emissions of FM in solution are shown in the inset for comparison; the concentrations for each spectrum from left to right, are 0.37, 0.7, 1.2, 3.77, 5.39, 7.7, and 11 mM. Uncertanties in fluorescence intensity and wavelength were of ±0.1% and ±1 nm, respectively [44].

The second example is a microcavity [45]. This microcavity is formed when a luminescent p-Si layer is inserted between two Bragg reflectors made of p-Si. The broad luminescence band is altered and very narrow peaks are detected. The position of these peaks is extremely sensitive to a small change in refractive index, such as that obtained when a biological analyte is placed in the large internal surface of p-Si. A DNA biosensor was developed by using such an oxidized microcavity [45]. After successful silanization of the p-Si surface, DNA was immobilized into the porous surface through a careful diffusion. Finally, the DNA-attached wass exposed to its complementary strand of DNA (cDNA). A red-shift in photoluminescence is observed. Full-length viral DNA molecules were also detected with the microcavity biosensor [45vis]. The advantages of optical sensing are significantly improved when this approach is used within an integrated optics context [46].

3.2. Electric and Electrochemical P-Si biosensors

The highly sensitive surface of p-Si and the possibility to measure changes in its electrical properties added to its capacity to adsorb an enormous amount of different compounds, can be used for electrical biosensor applications [47, 48]. These approaches consider the use of electrical contacts on the p-Si layer made by metal deposition to measure the changes in the electrical properties such as capacitance and conductance when an analyte is attached to the p-Si layer [49]. An example of this type of biosensor is a macroporous sensor to detect DNA hybridization by characterizing the difference between the dipolar moment in p-Si layers with and without the analyte [47]. Another DNA detector of nanoporous silicon biosensor is described in reference [50]. This biosensor is an electrochemical device that transduces the hybridization of DNA into a chemical oxidation of guanine by Ru (bpy)$^{2+}$ $_3$, the reduced form of which is then detected electrochemically.

Another effective platform to develop a p-Si biosensor is by applying electrochemical characterization. There are two main types of electrochemical transduction in biosensors: potentiometry and amperometry/voltammetry [12].

In potentiometryc biosensors the main parameter is the potential difference between the cathode and the anode in an electrochemical cell [51, 52]. This difference can be transduced as an electrical signal [12]. Amperometric and voltammetric biosensors consider the redox reaction that takes place in the anodization cell when an analyted of interest is placed. In this case the analyte is immobilised and an analyte oxidation/reduction process produces a flux of electrons measured, in terms of current intensity, cross the electrodes of the electrochemical cell [12]. These biosensors are too sensitive to pH modifications [51].

Examples of these sensors are the potentiometric and amperometric urea sensor based on nanoporous silicon technology described by Joon-Hyung Jin et al [52]. One of the electrochemical devices consists on three thin-film electrodes patterned on p-type silicon wafer by using platinum RF sputtering and silver (Ag) evaporation. The working electrode, on which the urease is inmobilized with a polymeric conductor: polypirrole (PPy) is sensitive to urea dissolved in artificially made electrolyte solution. The reference electrode is p-Si -based Ag/AgCl thin-film reference electrode (TFRE). The other is a platinum (Pt) thin-film counter electrode. In a potentiometric urea sensor, urea concentration is related to the measured potential applied between the working and reference electrode according to the Nernst equation. The other device is developed under amperometric regime. In this case the urease-catalyzed hydrolytic reaction of urea causes current flow between the working and counter electrode and the amount of current flow is proportional to the urea concentration that represents a change of pH, which is based on the Cottrell equation. In this study [50] it was found that urea sensitive electrodes (PSUE's) and Ag/AgCl TFRE's based on p-Si layers provides better adhesive strength between thin-films, and silicon-based electrodes. This reduces the leaching out of TFRE components and enhances the sensitivity of a sensing electrode. The presence of carbon, nitrogen and sulfur, which were attributed to the urease-doped PPy films were confirmed by EDX characterization. The p-Si-based Ag/AgCl TFRE can be recommended as an ideal non-polarizable reference electrode to determine the electrochemical cell potentials and currents of sensing electrodes. Amperometry for monitoring the urea concentrations caused by urease-catalyzed reactions is superior to a potentiometric method in that the amperometric urea sensors gives a longer linear range, higher sensitivity and shorter response times than the potentiometric urea sensors, especially at low urea concentrations.

Another very interesting application of porous silicon biosensors is for liver diagnosis [53]. Min-Jung Song et al presented a study of a biosensor array system consisting of cholesterol, bilirubin and glutamate sensors. The p-Si electrochemical system consisted of porous silicon layers formed on each working electrode that increased greatly the effective surface area. The electrodes in the sampling wells minimized a cross-interference effect to permit multiple sampling by immobilization of the enzymes using a silanization technique. The biosensor arrays tested used aqueous samples of the enzymes prepared in a 50 mM phosphate buffer solution (pH 8). All measurements were performed at room temperature at amperometric detection regime of each sensor was carried out using at a potential of +0.6 V vs. Ag/AgCl for the biosensors of the hydrogen peroxide generated in the silanized layer where the enzymatic reactions occur. In general, normal cholesterol concentrations in the human do not exceed 200 mg per 100 ml [53]. Higher cholesterol concentrations are considered abnormal.

In this case, the current detected is linearly proportional to cholesterol concentrations in the range of 1 mM to 50 mM; sensitivity was measured at approximately 0.2656 μA/mM. The bilirubin calibration curve covers a large concentration range between 0.002 mM and 0.020 mM, which includes normal levels (0.2 ~ 1.0 mg/dl), and levels typical of abnormal serum bilirubin. The sensitivity of the calibration curve approximated 0.15354 mA /mM. The detection of the ratio of alanine aminotransferase (ALT) and aspartate aminotransferase (AST) that in human serum indicates an abnormal symptom of the liver is also based upon electrochemical oxidation at the Pt electrode surface. Since L-glutamate is a product of both ALT and AST reactions occurring in the buffer solution, the enzyme activities can be determined from the current changes at the L-glutamate sensor. On average, the serum ALT and AST levels measured in healthy people by optimized conventional ALT and AST assays approximates 10 U/l at 25 °C and any increase in enzyme levels that exceed 100 U/l is taken to indicate liver disease. The sensitivity determined from the semi-logarithmic plot approximated 0.13698 μA/(U/l) for ALT over the range of 1.3 U/l to 250.0 U/l. For AST, the sensitivity was about 0.45439 μA/(U/l) in the same concentration ranges as for ALT.

This device offers several important advantages which include

1. readout within minutes from application of microvolumes of sample,

2. reduced physical dimensions of the device,

3. relative stability of the reagents used, and

4. simple electronics applicable to the further development of a hand held device useful for point of care biomarker liver analyses.

In the following paragraph we will describe in detail an example of an electrochemical sensor [48].

Porous silicon samples were prepared from p+-type, boron doped silicon wafers with a resistivity of 0.008-0.012 Ohm cm by standard anodization (electrolyte: 15% of HF) at a current density of 30 mA.cm^{-2}. The porosity measured by the gravimetrical method was approximately 62%. The pore size was estimated by TEM and ranged from 50 nm-75 nm in diameter. These diameters are large enough to allow the sensing molecules to penetrate and attach. For DNA, the diameter of the nucleotides is approximately 5Å, which is small enough to fit into the porous matrix. Stabilization of p-Si is necessary to passivate its surface and this was done by thermal oxidation. Thermal oxidation of p-Si requires several precautions and high temperatures (>700°C). Covering the whole internal surface with a thin SiO$_2$ layer stabilizes the structure, permits water penetration into the pores and facilitates probe and target penetration [54]. Al l p-Si samples were thermally oxidized in oxygen ambient at 900°C for 10 minutes.

The electrochemical instrumentation used for these experiments included a BAS 100B/W Electrochemical Analyzer and a BAS VC-2 voltammetry cell (model MF-1065). It is well suited for small sizes and has a special micro-cell for volumes as small as 50µL. The micro-cell, which included the working electrode, separated a small volume containing the sample from a bulk solution containing the reference and auxiliary electrode with a salt bridge. A platinum wire served as auxiliary electrode and the modified p-Si samples function as working electrodes. It is important to mention that p-Si, especially oxidized p-Si, is not conducting and it is in fact the p+ doped silicon that was conducting the electrical current. The top area of the exposed PSi samples was 0.8 cm² and all lateral areas were insulated with a commercial epoxy resin (see figure 6). The epoxy resin was deposited very carefully and dried for one hour. The samples were attached to the electrochemical system as shown in fig. 6. Potentials were measured relative to an aqueous, saturated Ag/AgCl double junction (reference electrode). The voltammetry experiments were carried out at different scan rates in an electrochemical buffer solution composed by 50 mM sodium phosphate (pH 7) with 0.7 M NaCl. A schematic representation of the electrochemical measurement set up and the electrode arrangement is shown in fig. 6.

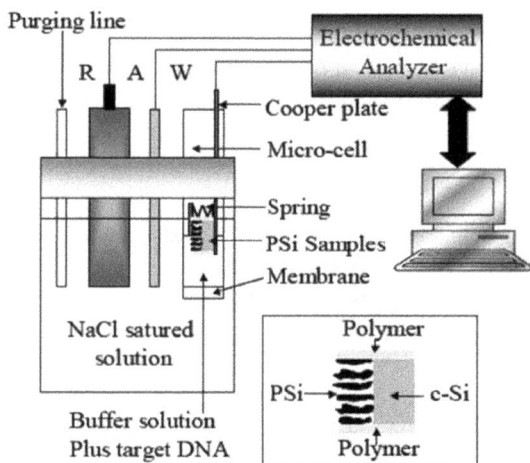

Figure 6. Measurement system: the p-Si electrode is used as working electrode. A platinum wire is the auxiliary electrode and Ag/AgCl the reference electrode. Inset: cross section showing the different parts of the working electrode [48].

Three different synthetic oligonucleotides were obtained from MWG Biotech, INC, and have the following sequences: (probe): 5'-TAI-CTA-TII-AAT-TCC-TCI-TAI-ICA-3',(target):5'-GCCTAC-GAG-GAA-TTC-CAT-AGC-T-3' and (two-base mismatch target):5'-GCC-TAC-GAG-GAA-TTG-GAT-AGC-T-3. Tris(2,2'-bypyridyl) ruthenium (II) chloride hexahydrate was purchase from Strem Chemicals. All other chemicals were of analytical grade and purchased from Aldrich and Fluka. Deionized distilled water was obtained from Millipore.

The detection of DNA consists of the following steps: p-Si silanization, probe immobilization, hybridization, and voltammetric detection.

Several methods may be employed to bind DNA to different supports [55]. One method commonly used for binding DNA involves silanization of an oxidized surface. The function of silane coupling agents is to provide stable bond between two non-bonding surfaces: for example, an inorganic surface to an organic molecule. 3-glycidoxypropyltrimethoxysilane was used to silanize the oxidized p-Si A 5% aqueous solution of silane was prepared (pH 4.0). This converts silane into a reactive silanol through hydrolysis. The p-Si samples were then immersed into the continuosly stirred solution and left overnight. 3-glycidoxypropyl-trimethoxysilane is hydrolyzed to a reactive silanol by using double distilled water (pH 4). p-Si samples were then submerged into silanol solution for approximately 17 hours. Constant stirring of the solution was necessary to continuously mix the solution.

After successful silanization, DNA was immobilized onto the surface of p-Si through diffusion. Aqueous solutions of DNA containing 150 µl of DNA (50 µM) were carefully placed directly above de p-Si layer. The DNA molecules covalently bond to the silanized surface, where they become immobilized. The samples were then placed in a steam container where they were heated in an oven at 37ºC for approximately 20 hours. The DNA attached samples were then rinsed in double distilled water and dried with nitrogen.

The DNA attached to p-Si was exposed to its complementary strand DNA (target), the mismatch sequence (mismatch probe) and itself (probe). Binding was allowed to proceed for 1 hour at room temperature into hybridization buffer containing 1 M NaCl, 10-20 mM sodium cacodylate, 0.5 mM EDTA, 150 mM KCl and 5 mM MgCl2. Throughout the steps, binding was confirmed using Fourier Transform Infrared Spectroscopy (results not shown here).

Cyclic voltammetry (CV) was carried out having the DNA modified, p-Si electrode as working electrode, an Ag/AgCl as the reference electrode, and platinum wire as the counter electrode. 6 µl of $Ru(bpy)$ $^{2+}_3$ (0.1µM) was poured into 150 µl of electrochemical buffer solution. After allowing the solution to diffuse into the samples for 15 minutes, CV was performed. Solutions were deoxygenated via purging with nitrogen for 10 minutes prior to measurements.

p-Si DNA-electrodes and $Ru(bpy)$ $^{2+}_3$ were used for specific gene detection. $Ru(bpy)$ $^{2+}_3$ exhibits a reversible redox couple at 1.05 V and oxidizes guanine in DNA at high salt concentration [56] according to:

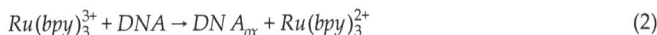

$$Ru(bpy)_3^{2+} \rightarrow Ru(bpy)_3^{3+} + e^- \tag{1}$$

$$Ru(bpy)_3^{3+} + DNA \rightarrow DNA_{ox} + Ru(bpy)_3^{2+} \tag{2}$$

where DNA_{ox} is a DNA molecule in which guanine has been oxidized by $Ru(bpy)$ $^{3+}_3$. If the DNA probe contains guanine then $Ru(bpy)$ $^{3+}_3$ will oxidize guanine in DNA, even without the presence of the target DNA. In order to prevent that the DNA probe reacts with $Ru(bpy)$ $^{2+}_3$ the guanine has been replaced for another less reactive nucleotide. Some previous results show that the addition of an oligonucleotide that does not contain guanine produces a small enhancement in the oxidation current [56]. Those results have shown that the inosine 5′-monophosphate is 3 orders of magnitude less electrochemically reactive than guanosine 5′-monophosphate and still recognizes cytidine [56]. This fact is very important to recognize all four bases in the target sequence. Nevertheless there is a drawback that can have consequences on the hybridization efficiency. Since the deaminated hypoxanthine in the ionosine can only form two of the three hydrogen bonds in a Watson-Crick base pair, it may be desirable to use a guanine derivative that is redox-inert but capable of forming all three hydrogen bonds. However some studies have shown [56] that the specificity afforded by inosine substitution was sufficient but they propose 7-deazaguanine as alternative. For this reason the DNA probe sequence does not contain the guanine base but the target does. Figure 7 (top) shows the CV obtained in solution for the hybrid DNA (probe-target) at different scan rates (target DNA concentration of 0.5 x 10^{-10}M). Figure 7 (bottom) shows that the anodic current of $Ru(bpy)^{2+}_3$ is linearly proportional to the scan rate.

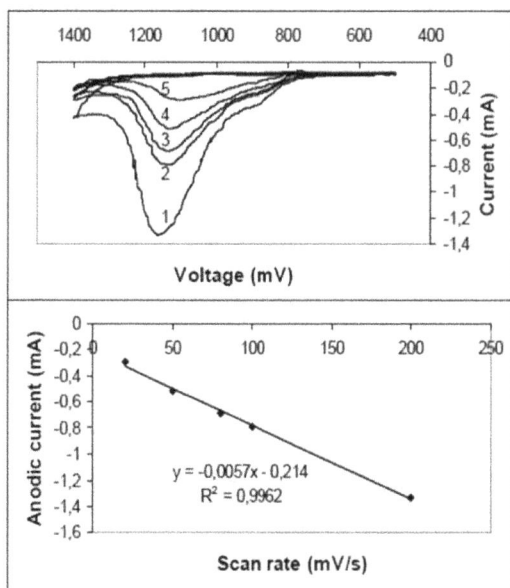

Figure 7. (top) Cyclic voltammograms of the probe-target DNA sequence in 0.1 µM $Ru(bpy)^{2+}_3$ solution at different scan rates (mV.s⁻¹): (1) 20; (2) 50; (3) 80; (4) 100; (5) 200. (bottom) The anodic current changed with the scan rate. The target concentration was 0.5x10^{-10}M [48].

This result is congruent with a process that is controlled by adsorption. Figure 8(top) shows the CV (scan rate of 50 mV.s⁻¹) of varied concentrations of target DNA (probe-target sequence, curves 2 to 5) and different targets (probe-mismatch target sequence, curve 1 and probe-probe sequence, curve 6). In curve 1, the mismatch target sequence contains two more pairs of base G than the target sequence and that is why the current in this case is bigger than the current obtained in the probe-target sequence cases (curves 2 to 5) or the probe-probe sequence (curve 6). Moreover in curve 6 the current intensity decreases as a consequence of the absence of the

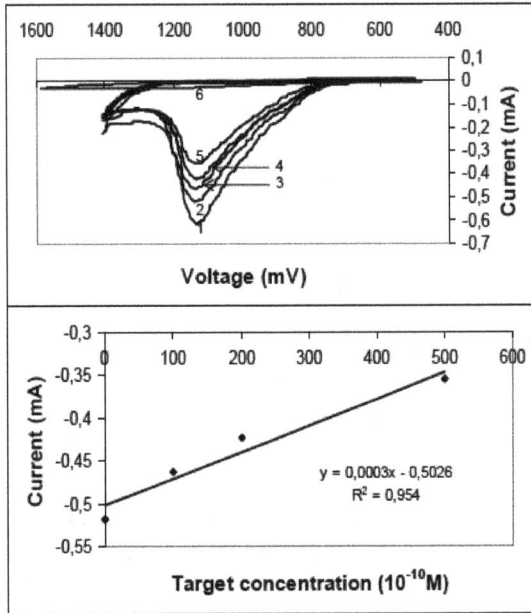

Figure 8. top) Cycled voltammograms of different concentrations of target DNA: (2) 0.5 x 10⁻¹⁰M, (3) 100 x 10⁻¹⁰M, (4) 200 x 10⁻¹⁰M, (5) 500 x 10⁻¹⁰M. Curve 1 shows the CV for the probe-mismatch target DNA sequence (0.5 x 10⁻¹⁰M) and curve 6 for the probe-probe DNA sequence (0.5 x 10⁻¹⁰M). (bottom) The anodic current changed with the concentration of the target DNA. In all cases, 0.1 µM of $Ru(bpy)^{2+}_3$ was used [48].

base G in the probe. Nevertheless a significant increase in current was observed for curves 2 to 5 where the target DNA undergoes hybridization to the complementary DNA. This current increase suggests that the hybridization was successful and that the electron transfer from the guanines of the hybridized strand to $Ru(bpy)^{2+}_3$ is responsible for the increase in the current. In comparing curves 1, 2 and 6 (same DNA concentration but different target sequence) it is observed that the sensor responds differently to each target and therefore a good selectivity is achieved. Figure 8 (bottom) shows the anodic peak currents of $Ru(bpy)^{2+}_3$ at four different concentrations (curves 2 to 5). The peaks are linearly related to the concentration of the target DNA sequence between 0.5 x 10⁻¹⁰ and 500 x 10⁻¹⁰M. The detection limit of this approach was 5 x 10⁻¹¹M. The sensitivity achieved in this work is similar to the one obtained in reference [57], where a sensitivity of 9.0 x 10⁻¹¹M was reported for a sensor that uses gold substrates instead. In summary those results clearly show that the p-Si sensor shown here has a good selectivity and sensitivity to the target compound, two very important characteristics that a sensor ought to have.

An overview of the requirements for a good performance of a p-Si biosensor was presented and the generalities of the fabrication of different kinds of these biological sensors as well. In the next section we will discuss the new materials, uses and future of porous silicon.

4. Future of porous silicon biosensors

Sensors allow our systems and devices to be in relation with the real events that we need to register or control. So, precision (same response to the same stimuli: repeatability) and accuracy (indicating magnitude value as close as possible to the real magnitude of the stimulus to be sensed: minimum absolute error spread) are two main requirements for any sensor when the industry selects a structure type for market use. However, other properties will define the success of a new kind of sensor in the market. These are: technological compatibility with the existing devices, geometric dimension requirements, low noise insertion, ease of adjustment and setup, low power consumption, performance standardization (linear if possible), low thermal or aging characteristic drifts, robustness, reliability, low obsolescence, and very wide field of applications. p-Si is a material that accomplishes all of these requirements with enough margins to think that it will become increasingly popular in the short term. For instance, integrated circuits (IC) are made of crystalline Silicon, which means it is fully compatible for associating a p-Si sensor to any electronic device. The electrochemical technology used to create a p-Si layer does not collide with the IC lithography. The geometric dimensions required to create this type of sensor are sufficiently small to be integrated in an IC. The homogeneity of the porous and its radius control (internal surface density control) as well as its layer stability is improving very fast.

An important factor to take into account in the implementation of a p-Si biosensor is its chemical stability after sample storage. Due to its high superficial area, p-Si based materials tends to be oxidized when are exposed to air ambient conditions. This oxidation plus the addition of other molecules present in the air ambient could modify the biosensor reponse to an analyte after certain amount of time. Pasivation techniques and surface functionalization described before have been proved being successful to prevent or minimize these stability issues. Work has to be done in order to improve the existing methods and assure the reproducibility of p-Si biosensors reponse over a lapse time of years.

Porous silicon has proven to be a succesful material for biosensing applications [1, 58]. In some cases even femtomolar concentrations in biomolecules was demostrated [1]. The wide range of applications in sensing biological substances include: healt applications [7, 43, 58], virus detection [43], inmunosensors [59, 1], DNA biosensors [58,60], drug delivery [16], biosecurity on food [61], biological warfare agents [62], implantable biosensor technology [58], among others [12, 15, 63]. Some new trends in the fabrication of p-Si biological sensor devices are the use of nanomaterials [54]. The sensitivity and performance of biosensors are being improved by using nanomaterials for their construction allowing simple and rapid in vivo analyses [61]. Recently, the plasmonic properties of metal nanoparticles have been used to develop a p-Si biosensor that present Ramman enhacement [64]. Another attractive method for monitoring biomolecular interactions in a highly parallel fashion is the use of microarrays. This p-Si novel porous chip was demonstrated as stable and reproducible, and the fluorescent bioassay reproducibility has been shown [65].

Lately integrated systems of p-Si has been developed [66, 67] e.g a guided mode biosensor based on grating coupled p-Si waveguide [68].

Another type of biosensensing approach that is appearing is an acoustic wave transducer that is coupled with a bioelement e.g an antibody. When the analyte molecules (antigen) get attached to a membrane, the membrane mass changes, resulting in a modification in the resonant frequency of the transducer that can be measured [66]. Rencently p-Si has been proposed as a good material for this kind of biosensors [64].

Notwithsranding the many advantages of p-Si mentioned during this work, several challenges will need to be overcome to be able to make biosensors a viable commercial product. They fall into two main areas: those concerning the fabrication of p-Si for cost effective and robust devices, and those addressing the ability to handle real-world sample matrices such as whole blood [1]. Both are presently the focus of intensive research and it is reasonable to believe that new and exciting developments will occur in a very near future.

Author details

M. B. de la Mora[1], M. Ocampo[2], R. Doti[2], J. E. Lugo[2*] and J. Faubert[2]

*Address all correspondence to: je.lugo.arce@umontreal.ca

1 Instituto de Física. Universidad Nacional Autónoma de México. Circuito de la Investigación Científica Ciudad Universitaria, México

2 Visual Psychophysics and Perception Laboratory, School of Optometry, University of Montreal, Canada

References

[1] Andrew, Jane., Roman, Dronov., Alastair, Hodges., & Nicolas, H. Voelcker. (2009). Porous silicon biosensors in the advance. *Trends in Biotechnology Review*, 27, 230-240.

[2] Bobby, Pejcic., Roland De, Marco., & Gordon, Parkinson. (2006). The role of biosensors in the detection of emerging infectious diseases. *Analyst*, 131, 1079-1090.

[3]   Industry Experts. (2012). Biosensors- A Global Market Overview. http://www.repor-tlinker.com/p0795991/Biosensors-A-Global-Market-Overview.html, accessed June).

[4]   Herino, R., Bomchil, G., Barla, K., Bertrand, C., & Ginoux, J. L. (1987). Porosity and pore size distribution of porous silicon layers. *Journal of Electrochemical Society*, 134, 1994-2000.

[5]   Cullis, A. G., Canham, L. T., & Calcott, P. D. J. (1997). The structural and lumines-cence properties of porous silicon. *Journal of Applied Physics*, 82, 909-965.

[6]   Fauchet, P. M., Tsybeskov, L., Peng, C., Duttagupta, S. P., von Behren, J., Kostoulas, Y., Vandyshev, J. M. V., & Hirschman, K. D. (1995). Light-emitting porous silicon: materials science, properties, and device applications. *IEEE Journal of Selected Topics in Quantum Electronics*, 1, 1126-1139.

[7]   Mathew , Finny P., & Alocilja , Evangelyn. C. (2005). Porous silicon-based biosensor for pathogen detection. *Biosensors and Bioelectronics*, 20, 1656-1661.

[8]   Ben, A. Jaballah, Hassen, M., Hajji, M., Saadoun, M., Bessais, B., & Ezzaouia, H. (2005). Chemical vapour etching of silicon and porous silicon: silicon solar cells and micromachining applications. *Physica Status Solidi (a)*, 202, 1606-1610.

[9]   Zhipeng, Huang., Nadine, Geyer., Peter, Werner., Johannes de, Boor., & Ulrich, Go-sele. (2011). Metal-Assisted Chemical Etching of Silicon: A Review. *Advanced Materi-als*, 23, 285-308.

[10] V´azsonyi, E., Szil´agyi, E., Petrika, P., Horv´atha, Z. E., Lohner, T., Frieda, M., & Jal-sovszky, G. (2001). Porous silicon formation by stain etching. *Thin Solid Films*, 388, 295-302.

[11] Luca, De Stefano, Rendina, Ivo., Moretti, Luigi., Tundo, Stefania., & Mario, Andrea. Rossi. (2004). Smart optical sensors for chemical substances based on porous silicon technology. *Applied Optics*, 43, 1.

[12] Andrea, Salis, Setzu, Susanna., Monduzzi, Maura., & Mula, Guido. (2011). Porous Sil-icon-based Electrochemical Biosensors. Biosensors- Emerging Materials and Applica-tions, Chapter InTech., 17, 334-352.

[13] Vincent, G. (1994). Optical properties of porous silicon superlattices. *Applied Physics Letters*, 64, 2367-2369.

[14] Anthony, P. F. Turner. (2012). Biosensors: Past, Present and Future. http://www.cran-field.ac.uk/health/researchareas/biosensorsdiagnostics/page18795.html,   accessed June).

[15] Kristopher, A., Kilian, Till., Bocking, J., & Gooding, Justin. (2009). The importance of surface chemistry in mesoporous materials: lessons from porous silicon biosensors. *Chemical Communications*, 630-640.

[16] Thesis:. (2012). Biocompatibility and biofunctionalization of mesoporous silicon par-
      ticles. *Luis Maria Bimbo. Division of Pharmaceutical Technology Faculty of Pharmacy Uni-
      versity of Helsinki Finland.*

[17] Chaniotakis, N., & Sofikiti, N. (2009). Novel semiconductor materials for the devel-
      opment of chemical sensors and biosensors: a review. *Analytica Chimica Acta,* 615, 1-9.

[18] Chen, Huajie., Hou, Xiaoyuan., Li, Gubo., Zhang, Fulong., Yu, Mingren., & Wang,
      Xun. (1996). Passivation of porous silicon by wet thermal oxidation. *Journal of Applied
      Physics,* 79, 3282-3285.

[19] Debarge, L., Stoquert, J. P., Slaoui, A., Stalmans, L., & Poortmans, J. (1998). Rapid
      thermal oxidation of porous silicon for surface passivation. *Materials Science in Semi-
      conductor Processing,* 1, 281-286.

[20] Martin-Palma, R. J., Martinez-Duart, J. M., Salonen, J., & Lehto-P, V. (2008). Effective
      passivation of porous silicon optical devices by termal carbonization. 103,
      083124-083124-4.

[21] Selena, Chan., Yi, Li., Lewis, J. Rothberg, Benjamin, L. Miller, & Fauchet, Philippe. M.
      (2001). Nanoscale silicon microcavities for biosensing. *Materials Science and Engineer-
      ing C,* 15, 277-282.

[22] Lin, V. S. Y., Motesharei, K. K., Dancil, P. S., Sailor, M. J., & Ghadiri, M. R. (1997). A
      Porous Silicon-Based Optical Interferometric Biosensor. *Science,* 278, 840-843.

[23] Keiki-Pua, S. Dancil, Douglas, P. Greiner, & Michael, J. Sailor. (1999). A Porous Sili-
      con Optical Biosensor:Detection of Reversible Binding of IgG to a Protein A-Modified
      Surface. *Journal of American Chemical Society,* 121, 7925-7930.

[24] Xuegeng, Li., Yuanqing, He., & Mark, T. Swihart. (2004). Surface Functionalization of
      Silicon Nanoparticles Produced by Laser-Driven Pyrolysis of Silane followed by HF-
      HNO_3 Etching. *Langmuir,* 20, 4720-4727.

[25] Sweetman, Martin J., Shearer, Cameron. J., Shapter, Joseph. G., & Voelcker, Nicolas.
      H. (2011). Dual silane surface functionalization for the selective attachment of human
      neuronal cells to porous silicon. *Langmuir The Acs Journal Of Surfaces And Colloids,* 27,
      9497-9503.

[26] Lorraina, N., Hiraouia, M., Guendouza, M., & Hajia, L. (2011). Functionalization con-
      trol of porous silicon optical structures using reflectance spectra modeling for bio-
      sensing applications. *Materials Science and Engineering: B,* 176, 1047-1053.

[27] Linford, M. R., & Chidsey, C. E. D. (1993). Alkyl monolayers covalently bonded to
      silicon surfaces. *Journal of American Chemical Society,* 115, 12631-12632.

[28] Schmeltzer, J. M., Lon, J., Porter, A., Stewart, M. P., & Buriak, J. M. (2002). Hydride
      Abstraction Initiated Hydrosilylation of Terminal Alkenes and Alkynes on Porous
      Silicon. *Langmuir,* 18, 2971-2974.

[29] Boukherroub, R., Petit, A., Loupy, A. J., Chazalviel, N., & Ozanam, F. (2003). Micro-wave-Assisted Chemical Functionalization of Hydrogen- Terminated Porous Silicon Surfaces. *Journal of Physica Chemical B*, 107, 13459-13462.

[30] Lugo, J. E., Ocampo, M., Doti, R., & Faubert, J. (2011). Porous Silicon Sensors- from Single Layers to Multilayer Structures . *Biosensors- Emerging Materials and Applications Chapter 15, InTech.*

[31] Jeffrey, Fortin. (2009). Chapter 2 Transduction Principles. *A. Zribi and J. Fortin (eds.), Functional Thin Films and Nanostructures for Sensors, Integrated Analytical Systems, Springer Science + Business Media.*

[32] Claudia, Steinem., Andreas, Janshoff., Victor, S., Lin, Y., Volcker, Nicolas H., & Gha-diri, Reza M. (2004). DNA hybridization-enhanced porous silicon corrosion: mecha-nistic investigations and prospect for optical interferometric biosensing. *Tetrahedron*, 60, 11259-11267.

[33] Yuri, L., Bunimovich, Young., Shik, Shin., Woon-Seok, Yeo., Michael, Amori., Gabri-el, Kwong., & Heath, James R. (2006). Quantitative Real-Time Measurements of DNA Hybridization with Alkylated Nonoxidized Silicon Nanowires in Electrolyte Solu-tion. *Journal of American Chemical Society*, 128, 16323-16331.

[34] Bisi, O., & Stefano, Ossicini. L. Pavesi. (2000). Porous silicon: a quantum sponge structure for silicon based optoelectronics. *Surface Science Reports*, 38, 1-126.

[35] Orosco, M. M., Pacholski, C., Miskelly, G. M., & Sailor, M. J. (2006). Protein- Coated Porous-Silicon Photonic Crystals for Amplified Optical Detection of Protease Activi-ty. *Advanced Materials*, 18, 1393-1396.

[36] Cheng, L., Anglin, E., Cunin, F., Kim, D., Sailor, M. J., Falkenstein, I., Tammewar, A., & Freeman, W. R. (2008). Intravitreal properties of porous silicon photonic crystals: a potential self-reporting intraocular drug-delivery vehicle. *British Journal of Ophthal-mology*, 92, 705-711.

[37] Cunin, F., Schmedake, T. A., Link, J. R., Li, Y. Y., Koh, J., Bhatia, S. N., & Sailor, M. J. (2002). Biomolecular screening with encoded porous-silicon photonic crystals. *Nature Materials*, 1, 39-41.

[38] Elizabeth, C., Wu, Jennifer. S., Andrew, Lingyun., Cheng, William. R., Freeman, Lindsey., & Pearson, Michael. J. Sailor. (2011). Real-time monitoring of sustained drug release using the optical properties of porous silicon photonic crystal particles. *Biomaterials*, 32, 1957-1966.

[39] Dal, L., Negro, C. J., Oton, Z., Gaburro, L., Pavesi, P., Johnson, A., Lagendijk, R., Righini, M. Colocci, & Wiersma, D. S. (2003). Light transport through the band-edge states of Fibonacci quasicrystals. *Physics Review Letters*, 90, 1-4.

[40] Starodub, N. F., Shulyak, L. M., Shmyryeva, O. M., Pylipenko, I. V., Pylipenko, L. N., & Melnichenko, M. M. (2011). Nanostructured Silicon and its Application as the

Transducer in Immune Biosensors. *NATO Science for Peace and Security Series A: Chemistry and Biology*, 2, 87-98.

[41] Huimin, Ouyang., Marie, Archer., & Philippe, M. Fauchet. (2007). Porous Silicon Electrical and Optical Biosensors. *Frontiers in Surface Nanophotonics, Springer Series in Optical Sciences*, 133.

[42] Girolamo Di, Francia., Vera La, Ferrara., Sonia, Manzo., & Salvatore, Chiavarini. (2005). Towards a label-free optical porous silicon DNA sensor. *Biosensors and Bioelectronics*, 21, 661-665.

[43] Andrea, M., Rossi, Lili., Wang, Vytas., & Reipa, Thomas. E. Murphy. (2007). Porous silicon biosensor for detection of viruses. *Biosensors and Bioelectronics*, 23, 741-745.

[44] Palestino, G., de la Mora, M. B., del Rio, J. A., Gergely, C., & Perez, E. (2007). Fluorescence tuning of confined molecules in porous silicon mirrors. *Applied Physics Letters*, 91, 1219091-3.

[45] Selena, Chan., Yi, Li., Lewis, J., Rothberg, Benjamin. L., & Miller, Philippe. M. Fauchet. (2001). Nanoscale silicon microcavities for biosensing. *Materials Science and Engineering C*, 15, 277-282.

[46] Kirill, Zinoviev., Laura, G., Carrascosa, Jose., Sanchez del, Rıo., Borja, Sepulveda., Carlos, Domınguez., & Lechuga , M. Laura. (2008). Silicon Photonic Biosensors for Lab-on-a-Chip Applications. *Advances in Optical Technologies*, 1-6.

[47] Marie, Archer., Marc, Christophersen., & Philippe, M. Fauchet. (2003). Porous Silicon Electrical Biosensors. 737. *Material Research Society Symposium Proceedings*.

[48] Lugo, J. E., Ocampo, M., Kirk, A. G., Plant, D. V., & Fauchet, P. M. (2007). Electrochemical Sensing of DNA with Porous Silicon Layers. *Journal of New Materials for Electrochemical Systems*, 10, 113-116.

[49] Lenward, Seals., James, L., Gole, Laam., Angela, Tse., & Hesketh, Peter J. (2002). Rapid, reversible, sensitive porous silicon gas sensor. *Journal of Applied Physics*, 91-94, 2519.

[50] Joon-Jyung, Jin., Se-Hwan, Paek., Chi-Woo, Lee., & Nam-Ki, Min. (2003). Fabrication of Amperometric Urea Sensor Based on Nano-Porous Silicon Technology. *Journal of the Korean Physical Society*, 42, S 735-S738.

[51] Marion, Thust. M. J., Schoning, S., Frohnhoff, R., Arens-Fischer, P., & Luth, Kordos H. (1996). Porous silicon as a substrate material for potentiometric biosensors. *Measurement Science and Technology*, 7, 26-29.

[52] Joon-Hyung, Jin., Evangelyn, C. Alocilja, & Grooms, Daniel L. (2010). Fabrication and electroanalytical characterization of label-free DNA sensor based on direct electropolymerization of pyrrole on p-type porous silicon substrates. *Journal of Porous Materials*, 17, 169-176.

[53] Min-Jung, Song., Dong-Hwa, Yun., Nam-Ki, Min., & Suk-In, Hong. (2007). Electrochemical biosensor array for liver diagnosis using silanization technique on nanoporous silicon electrode. Journal of bioscience and bioengineering , 103, 32-37.

[54] Archer, M., & Fauchet, P. M. (2003). Electrical sensing of DNA hybridization in porous silicon layers. *Physics. Status Solidi*, 198, 503-508.

[55] Isola, N., Stokes, D. L., & Vo-Dinh, T. (1998). Surface-Enhanced Raman Gene Probe for HIV Detection". *Analytical Chemistry*, 70, 1352-1356.

[56] Napier, M. E., Loomis, C. R., Sistare, M. F., Kim, J., Eckhardt, A. E., & Thorp, H. H. (1997). Biomolecule Recognition with Electron Transfer: Electrochemical Sensors for DNA Hybridization. *Bioconjugate Chemical*, 8, 906-913.

[57] Yan, F., Erdem, A., Meric, B., Kerman, K., Ozsoz, M., & Sadik, O. A. (2001). Electrochemical DNA Biosensors for Gene Related Microcystis species. *Electrochemical Communications*, 3, 224-228.

[58] Chen, Jianrong., Miao, Yuqing., He, Nongyue., Wu, Xiaohua., & Li, Sijiao. (2004). Nanotechnology and biosensors. *Biotechnology Advances*, 22, 505-518.

[59] Meskinia, O., Abdelghani, A., Tlili, A., Mgaieth, R., Jaffrezic-Renault, N., & Martelet, C. (2007). Porous silicon as functionalized material for immunosensor application. *Talanta*, 71, 1430-1433.

[60] Luca De, Stefano., Paolo, Arcari., Annalisa, Lamberti., Carmen, Sanges., Lucia, Rotiroti., Ilaria, Rea., & Ivo, Rendina. (2007). DNA Optical Detection Based on Porous Silicon Technology: from Biosensors to Biochip. *Sensors*, 7, 214-221.

[61] Radke, S. M. (2005). A microfabricated biosensor for detecting foodborne bioterrorism agents. *Sensors Journal IEEE*, 5, 744-750.

[62] Gooding, Justin J. (2006). Biosensor technology for detecting biological war- fare agents: Recent progress and future trends. *Analytica Chimica Acta*, 559, 137-151.

[63] Miller, Benjamin L., Amarjeet, S. Bassi, & Knopf, George K. (2006). Porous Silicon in Biosensing Applications. *Smart Biosensor Technology*, 271-290.

[64] Yang, Jiao., Dmitry, S., Koktysh, Nsoki., Phambu, Weiss., & Sharon, M. (2010). Dual-mode sensing platform based on colloidal gold functionalized porous silicon. *Applied Physics Letters*, 97, 1531251-3.

[65] Anton, Ressine., Gyorgy-Varga, Marko., & Thomas, Laurell. (2007). Porous silicon protein microarray technology and ultra-/superhydrophobic states for improved bioanalytical readout. *Biotechnology Annual Review*, 13, 149-200.

[66] Mohanty, Saraju P. (2012). Biosensors : A Survey Report. http://biocapteurs.wikispaces.com/file/view/BiosensorSurveyReport.pdfaccessed June)

[67] Iryna, V., Gavrilchenko, Arthur. I., Benilov, Yuriy. G. Shulimov, & Skryshevsky, Valeriy A. (2009). Thermally induced acoustic waves in porous silicon. *Physica Status Solidi (c)*, 1725-1728.

[68] Xing, Wei., & Sharon, M. Weiss. (2011). Guided mode biosensor based on grating coupled porous silicon waveguide. *Optics Express*, 19, 11330-11339.

[69] Jesus, Alvarez., Paolo, Bettotti., Isaac, Suarez., Neeraj, Kumar., Daniel, Hill., Vladimir, Chirvony., Lorenzo, Pavesi., & Juan-Pastor, Martinez. (2011). Birefringent porous silicon membranes for optical sensing. *Optics Express*, 19, 26106-26116.

Chapter 5

Review on the Design Art of Biosensors

Shengbo Sang, Wendong Zhang and Yuan Zhao

Additional information is available at the end of the chapter

1. Introduction

"Biological microelectromechanical systems" (BioMEMS) is a special class of Microelectro-mechanical systems (MEMS) where biological matter is manipulated for analyses and meas-ures of its activity, characterisations under any class of scientific study. The BioMEMS-based devices are an attractive area of development based on microtechnology. The techonolgy has more exciting developments in the application of MEMS technology in recent decades. For scientific analysis and measurement, various novel sensor and detection platforms in the BioMEMS and microfluidic fields are required and have been reported, in addition to basic components, such as microchannels, micropumps, microvalves, micromixers and microreac-tors for flow management at microscopic volumes [1]. Any of the most important applica-tions based on BioMEMS are: biomedical and biological analysis and measurements, micro total analysis systems (μTAS) and lab-on-a-chip systems [2-5], which will give new applica-tions in biomedicine and biology, especially the ability to perform point-of-care measure-ments. The advantages of such systems are that they can deliver and process the biological or biomedical samples in microvolumes for testing and analysis in an integrated way there-fore dramatically reducing the requirement to the manipulation steps and the samples, and improving data quality and quantitative capabilities. The BioMEMS technology also helps to reduce overall cost and time for the measurement. At the same time it improves the sensitiv-ity and specificity of the analysis.

To the BioMEMS technology and application, biosensors play a critical role in the process of information gathering with the technologically advanced development of our civilization, demand for information. With new applications in the areas – genetics, diagnostics, drug discovery, environment and industrial monitoring, quality control as well as security and threat evaluation [6], the need for high throughput label-free multiplexed sensors for biolog-ical sensing has increased in the last decade.

A biosensor is a device for the detection of an analyte that combines a biological component with a physicochemical detector component [7]. In general, one concept of biological sensors encompasses two main features in addition to the associated signal processors used for the display of the results in a user-friendly way: the sensitive biological element which is a chemically receptive or selective layer, and the transducer or the detector element (it can work in a physico-chemical way, optical, piezoelectric, electro-chemical, etc.) that transforms the signal induced by the interaction between the analytes and the biological element into another more easily measured and quantified signal, Figure 1. The chemical layer provides specific binding sites for the target analyte of interest, such as molecules, proteins and cells. To most biological and chemical sensors, sensitivity has been increased tremendously in recent years, but it still has some deficiency and needs more improvement. The selectivity of the receptive layer can be designed employing principles of molecular and biomolecular recognition; for example antigen-antibody binding (i.e. any chemicals, bacteria, viruses, or pollen binding to a specific protein)[8]. Other surface functionalizations such as self-assembled monolayer [9, 10] and polymer coatings are also employed. The selectivity is then achieved by a specific chemical reaction on the functionalized sensor's surface. However, absolute selectivity remains a major challenge. In fact, most sensing technologies are faced with the issue of non-specific interactions which can complicate the sensor response, produce false positives, and affect the reproducibility and the suitability of the sensor system for a particular application. Therefore, the chemical layer must be designed to maximize the sensor's sensitivity to the specific response.

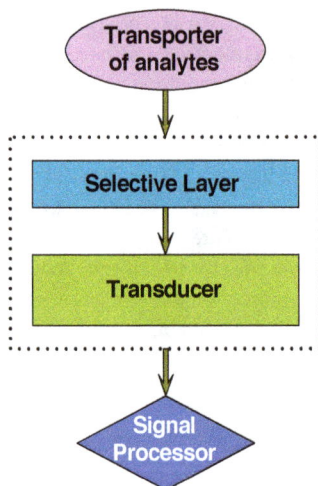

Figure 1. Generalized schematic representation of a biological sensor [11]

Once the analyte is recognized by the chemical layer, the transducer converts the chemical stimulus into a measurable output signal, as shown in Figure 1. Both the chemical layer and the physical transducer impose limitations on the performance of a certain class of sensors. Nevertheless it is often the physical transducer which determines the limits of detection attainable. The search for new transduction principles is therefore constantly stimulated [8]. In fact, each step depicted in Figure 1 has an influence on the sensor's performance. From the mechanism that drives the analyte to the sensor (e.g. microfluidic, activated diffusion, etc.), to the instrument reading the output signal of the transducer; all stages are the subject of extensive research efforts.

Besides, microfluildic technology is a frequent technology which is used in biosensors. Microfluidic devices or components have emerged in the beginning of the 1980s and were quickly used in the development of inkjet print heads, Deoxyribonucleic Acid (DNA) chips, lab-on-a-chip technology and micro-thermal devices. Microfluidics can precisely control and manipulate fluids and analytes that are geometrically constrained to a small size, typically sub-millimeter, and scale. One of the most attractive applications of microfluidics has been in biomedical and life science diagnostics [12]. µTAS applications are attractive because of the potential of such systems to allow faster analysis of biological material. Further they can reduce the requirement to the amount of reagent and the number of processing steps. In addition, miniaturization of such systems can result in higher repeatability and precision of analysis, lower power consumption, and the potential to create portable diagnostic tools for on-site analysis. These advantages result not only in time and cost savings for diagnostic tests, but can also be life saving in time-critical environments such as critical medical diagnostics or biowarfare pathogen detection.

2. Design art of biosensors

The design art of biosensors can be broadly classified into label-free and label based on the detection technology. It can also be further classified as shown in Figure 2. Label-based techniques rely on the specific properties of labels like fluorescence, chemiluminescence etc. for detecting a particular target. However, the process of labeling and purification processes is associated with sample losses, which is critical when sample quantity is limited. Labeling processes can also have a detrimental effect on the functionality and stability of molecules like proteins. Mass spectrometry, surface plasmon resonance (SPR) and other optical method are label-free techniques, which can conquer these disadvantages. The following subsections are the detailed discussion to the respective advantages and disadvantages of every design art.

2.1. Label-free biosensor

2.1.1. Surface Plasmon resonance (SPR) biosensors

Surface Plasmon Resonance (SPR): SPR-based biosensors measure the refractive index near a sensor surface through an optical method for getting some information. When a light beam impinges onto a metal film at a specific (resonance) angle, the surface plasmons can be resonated with the light. As a result, it can induce the absorption of light. For the widely used Kretschmann configuration, a beam is focused onto the metal film. There is a range of incident angles provided focused light and the reflected beam will have the same range of the angles while the projection of the beam forms a band. A dark line will appear in the reflected band if the SPR occurs within the spread angles. An intensity profile of this band can be monitored and plotted against the range of angles as shown in Figure 3.

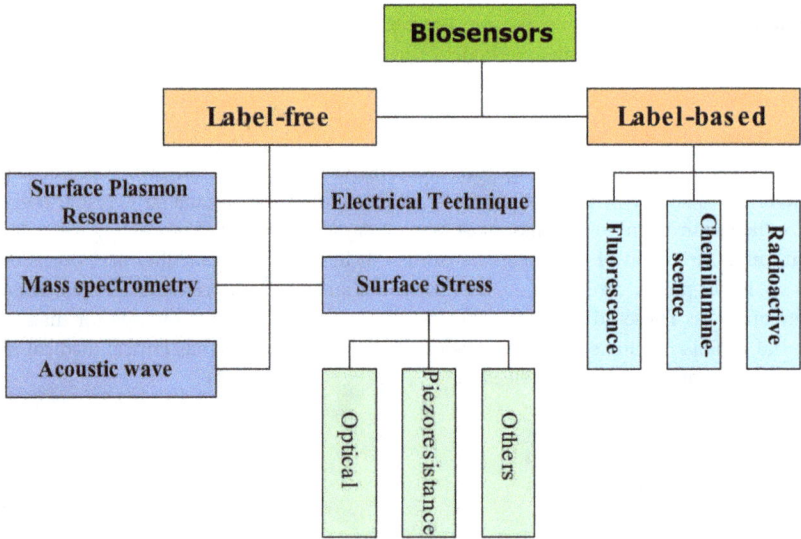

Figure 2. Classification of the design art of biosensors based on the detection method [11]

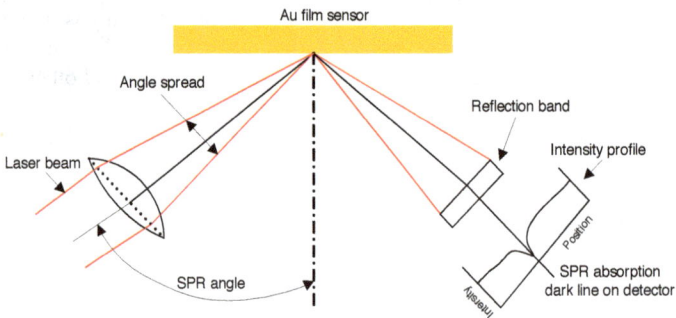

Figure 3. Principle of SPR detection in the mode of measuring the SPR angular shift[11]

Many researchers have worked on developing SPR biosensors for studying various kinds of biological reactions, and many reports have been published. The first application of SPR in biosensor was demonstrated in 1982 and the first commercial SPR sensor was introduced in 1990's [13, 14]. Biosensing Instrument Incorporated uses a different approach to detect the SPR angle change, for example, using the position-sensitive detector. Only the position shift of the dip is measured, so it offers a highly sensitive detection scheme to measure extremely small angle changes of the SPR. In the range of the SPR angle spread, the system delivers exceptionally high angular resolution in its measurement [15]. Like Figure 4 illustrates, SPR is observed as a sharp shadow in the reflected light from the surface at an angle that is dependent on the mass of material at the surface [15]. When biomolecules bind to the surface and change the mass of the surface layer, the SPR angle will shift (from I to II in the lower left-hand diagram). This change in resonant angle can be monitored non-invasively in real time as a plot of resonance signal (proportional to mass change) versus time [15].

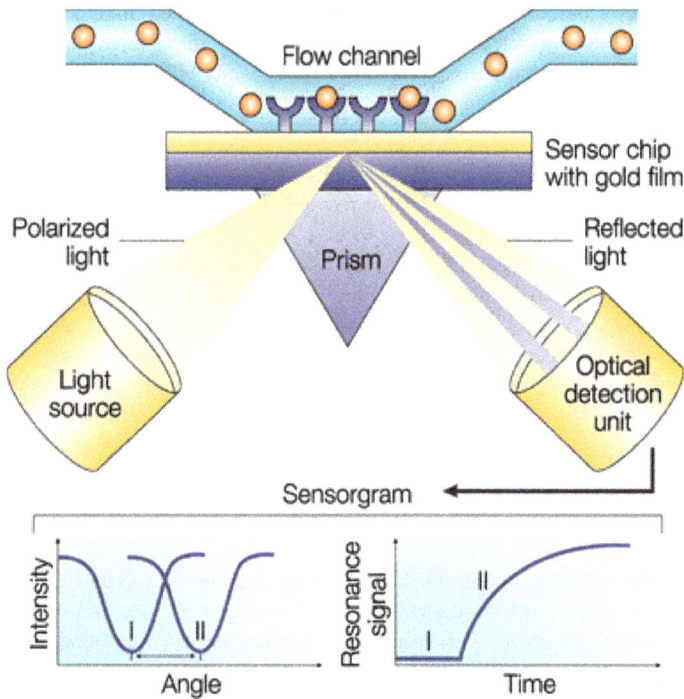

Figure 4. Schematic of SPR sensing method [15]

Earlier works were focusing mainly on antigen-antibody interactions [16], the streptatividin-biotin reaction [17], and some Immunoglobulin G (IgG) examinations, especially to test new algorithms in biospecific molecular interaction analysis, to characterize newly developed

SPR set-ups [18]. Current researches include far more advanced ways to improve the sensitivity of biosensor through functionalization layer [19] and silver mirror reaction which the biosensor based on Ag/Au film could make the resonant wavelength move to longer wavelength following with the sensitivity enhancement of the SPR biosensor [20]. One of the new areas is the examination of protein–protein or protein–DNA interactions [21], even detecting conformational changes in an immobilized protein [22]. A domain within the tumor suppressor protein adenomatous polyposis coli(APC) has been examined regarding its biochemical properties [23], as well as the binding kinetics of human glycoprotein with monoclonal antibodies [24]. Work has been done on the activator target in the Ribonucleic Acid(RNA) polymerase II holoenzyme [25].

SPR biosensor technology exhibits various advantageous features. Versatility generic SPR sensor platforms can be tailored for detection of any analyte, providing a biomolecular recognition element; Analyte does not have to exhibit any special properties such as fluorescence or characteristic absorption and scattering bands because label-free binding between the biomolecular recognition element and analyte can be observed directly without the use of radioactive or fluorescent labels; The speed of analysis binding event can be observed in real-time providing potentially rapid response and flexibility SPR sensors can perform continuous monitoring as well as one time analyses [26].

However, SPR biosensors exhibit two inherent limitations: On the one hand, specificity of detection specificity is solely based on the ability of biomolecular recognition elements to recognize and capture analyte. Biomolecular recognition elements may exhibit crosssensitivity to structurally similar but nontarget molecules. If the nontarget molecules are present in a sample in a high concentration, sensor response due to the nontarget analyte molecules may conceal specific response produced by low levels of target analyte. On the other hand, sensitivity to interfering effects similar to other affinity biosensors relying on measurement of refractive index changes, SPR biosensor measurements can be compromised by interfering effects which produce refractive index variations. These include nonspecific interaction between the sensor surface and sample (adsorption of nontarget molecules by the sensor surface), and background refractive index variations (due to sample temperature and composition fluctuations) [26].

2.1.2. Mass spectrometry biosensors

Mass spectrometry (MS) is an analytical technique that can be used to measure the mass-to-charge ratio of charged particles, the mass of particles, the elemental composition of a sample and the chemical structures of molecules by ionizing chemical compounds.

In order to measure the characteristics of individual molecules, a mass spectrometer converts molecules to ions so that they can be moved about and manipulated by external electric and magnetic fields. Mass spectrometers are generally composed of three fundamental parts, namely the ionization source (a small sample is ionized, usually to cations by loss of an electron), the mass analyzer (the ions are sorted and separated according to their mass and charge) and the detector (that registers the number of ions at each m/z value) [27].A typical procedure usually contains five steps: A sample is loaded onto the MS instrument and undergoes vaporization. The components of the sample are ionized by one of a variety of methods (e.g., by impacting them with an electron beam), which results in the formation of charged particles (ions).The ions are separated according to their mass-to-charge ratio in an analyzer by electromagnetic fields. The ions are detected, usually by a quantitative method. Finally, the ionsignal is processed into mass spectra.

The mass spectrometry can be applied to identify and, increasingly, to precisely quantify thousands of proteins from complex samples, which is believed to have a broad impact on biology and medicine [28]. But, the size of these equipments is large in general, which makes them unfeasible for field applications which require portable devices, especially for biosensors.

2.1.3. Acoustic wave biosensors

The detection mechanism of acoustic wave sensors is an acoustic (mechanical) wave. The velocity and/or amplitude of the acoustic wave can be affected by the changes of the characteristics of the propagation path when the acoustic wave propagates through or on the surface of the material [39]. Based on the sensor, the changes of velocity can be monitored by measuring the frequency or phase characteristics, and then it can be analyze based on the corresponding physical quantity being measured [29]. Acoustic wave based biosensors offer a promising technology platform for the development of label-free, sensitive and cost-effective detection of biomolecules in real time.

Emerging applications for acoustic wave devices as sensors include as torque and tire pressure sensors [30~33], gas sensors [34~37], biosensors for medical application [38~41], and industrial and commercial applications such as: vapor, humidity, temperature, and mass sensors [42~44]. Additional capabilities of acoustic wave sensors include remote operation and passive interrogation [44].

Surface acoustic wave sensors, as a class of MEMS, are widely used recently. The sensor can transform an input electrical signal into a mechanical wave which can be easily influenced by physical phenomena. Then, the changed mechanical wave is transduced back into an electrical signal. The presence of the desired phenomenon can be detected through the difference between the input and output electrical signal (amplitude, phase, frequency, or time delay). The basic surface acoustic wave device consists of a piezoelectric substrate, an input interdigitated transducer (IDT) on one side of the surface of the substrate, and a second output interdigitated transducer on the other side of the substrate.

As shown in Figure 5, Self-assembled monolayer (SAW) devices have the interdigitated transducers (IDTs) excitation electrodes fabricated on the one side of the piezoelectric film. As a result, the SAW devices have the acoustic waves propagating along the surface of the piezoelectric substrate. The SAW device could be resonator or delay line depending of the design of the IDTs. For SAW resonators the IDTs are fabricated in a central position and reflectors are added on both sides of the input and output IDTs to trap the acoustic energy within a cavity. The surface between the IDTs is coated with antibodies sensitive to the analyte to be detected. The analyte molecules binding to the immobilized antibodies on the sensor surface influence the velocity of the SAW and hence the output signal generated by the driving electronics.

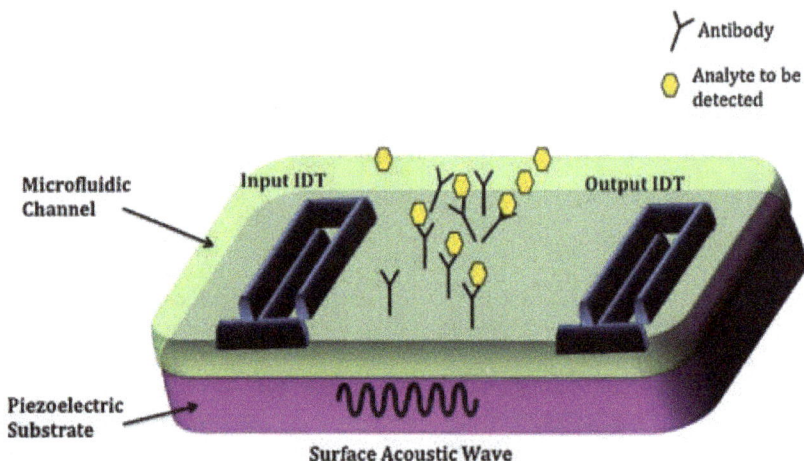

Figure 5. SAW delay line biosensor integrated in a microfluidic channel[45]

Virtually all acoustic wave devices and sensors use a piezoelectric material to generate the acoustic wave. The technology has been utilized in the commercial range for more than 60 years. And the high mass sensitivities of acoustic wave devices make them an attractive platform for monitoring immunochemical and other biomolecular recognition events. However, not all acoustic wave devices are suitable for liquid operation. If the sensor has surface normal deformations and the velocity of acoustic wave is greater than the compressional wave velocity of sound in liquid, then they can couple to compressive waves in the liquid and cause severe attenuation of the sensor signal. In contrast, devices in which surface particle motion is parallel to the sensor surface dissipate energy into the liquid primarily by viscous coupling, which does not produce severe losses and therefore are suited for liquid phase sensing.

2.1.4. Electrochemical biosensors

Biosensors based on electrochemistry provide an attractive means to analyze the content of a biological sample due to the direct conversion of a biological event to an electronic signal. Electrochemical biosensors build a bridge between the powerful analytical methods and the recognition process of the biological specificity.

Electrochemical biosensors are normally based on enzymatic catalysis of a reaction that produces or consumes electrons (such enzymes are rightly called redox enzymes).Electrochemical sensor consists of biological materials as sensitive components, electrode (solid electrodes, ion selective electrode, gas sensor electrode etc) as a conversion components, electric potential or current as the detection signal. The sensor substrate usually contains three electrodes: a reference electrode, a working electrode and a counter electrode. The target analyte is involved in the reaction that takes place on the active electrode surface, and the reaction may cause either electron transfer across the double layer (producing a current) or can contribute to the double layer potential (producing a voltage). We can either measure the current (rate of flow of electrons is now proportional to the analyte concentration) at a fixed potential or the potential can be measured at zero current (this gives a logarithmic response). Note that potential of the working or active electrode is space charge sensitive and this is often used. Further, the label-free and direct electrical detection of small peptides and proteins is possible by their intrinsic charges using biofunctionalized ion-sensitive field-effect transistors [46].

Electrochemical sensors can be classified into amperometric, potentiometric or conductometric sensors based on whether current, potential or resistance is being measured during an electrochemical reaction (oxidation or reduction) between the analyte of interest and the electrode surface.

Macro scale electrochemical sensors have been used in chemical and biological sensing for a very long time. Recently, due to many electrochemical sensor researches, sensors utilizing nanoscience and nanomaterials exploit unique properties (i.e. nanoporous electrodes, nanoparticles, nanotubes, etc.). A microbiochip that is based on an electrical detection system has been used for the detection of alpha-fetopro-tei (AFP) antigen [47].

The inherent advantages of electrochemical biosensors are their robustness, easy miniaturization, excellent detection limits, also with small analyte volumes, and ability to be used in turbid biofluids with optically absorbing and fluorescing compounds [48, 49]. The main adverse problems is long term stability and reliability associated with incorporation of liquid electrolytes, life time and cycle time issues due to small amount of reactants (consumable electrodes like Ag/AgCl). Secondly, non specific reactions taking place between the electro active impurities on the surface and the sample also limit the sensitivity of these sensors [50].

2.1.5. Surface stress sensors

Surface stress [51, 52] is a macroscopic quantity that is governed by microscopic processes. Although being a macroscopic quantity, the measurement of the surface stress involved in a system can lead to insight into the microscopic mechanisms basic for the generation of surface stress without detailed knowledge of the atomistic processes involved. Recent investigations of surface reconstruction, interfacial mixing, and self-organization at solid surfaces have renewed interest in the study of surface stress [53-57]. Now the surface stress existing between biological molecules, cells and some special functional materials has been used in the biosensors for biological and medical research based on the surface stress analysis.

Biological sensing of numerous analytes based on surface stress can be achieved using cantilever or membrane as the sensitive element of sensors. It is possible to sensitize one surface

of the sensititive element differently than the opposing surface. When the analytes of interest interact with the sensitized surface, a surface stress is induced, and the cantilever or
membrane bends due to the different surface stresses acting on both sides of the cantilever
or membrane. The sensor's specificity, i.e. the sensitivity of the sensor to a specific analyte, is
determined by the chemical functionalization of the sensitized surface of biosensors. Very
specific surface functionalizations can be achieved using molecular self-assembled monolayers (SAMs) as sensing layers assembled on the surface of biosensor sensitive element. Thiol-
chemistry has been favored as a versatile method of sensitizing a surface.

Microcantilever and micromembrane surface stress sensors have lots of applications in numerous flieds. Integrated microfluidics enabled individual cantilever/membrane addressing
in the array for selective functionalization. Hansen *et al.* were able to show that microcantilevers are sensitive enough to detect single base-pair mismatch in DNA hybridization [58].
Wu *et al.* developed a Prostate-specific antigen (PSA) detection assay using a single microcantilever at clinically relevant levels in a large background concentration of human serum
proteins – albumin and plasminogen [59]. Yue*et al.* [60] showed that passivation of the back-
side with inert coatings like polyethylene glycol (PEG) is absolutely necessary in order to
make reliable protein interaction measurement in cantilever detection system. Si-Hyung
"Shawn" Lim developed a 2-D multiplexed cantilever sensor array plat-form for high-
throughput target specific coating material search, which performed chemical sensing experiments using toluene and water vapour [61]. In addition, misun Cha demonstrated the
capability of the thin membrane transducer (TMT) for detecting biomolecular reactions such
as hybridization, single nucleotide polymorphism (SNP), and aptamer–protein binding [62].
Vasiliki Tsouti reveals the structure characteristics that should be considered in the design
of the biosensors in the case of flat Si membranes based on the simulation results [63]. Srinath Satyanarayana presents the design and fabrication of a novel parylene micro membrane surface stress sensor that exploits the low mechanical stiffness of polymers, measuring
the sensor response to organic vapors like isopropyl alcohol and toluene [64]. But all these
kinds of biosensor must associate with some detection methods to measure and analyse the
usable information gotten from the analyte, for example, optical method, piezoresistance,
and capacitive method.

2.1.5.1. Optical detection methods

The invention of the atomic force microscope (AFM) in 1986 [65] and its impact on the fields
of biotechnology and nanotechnology has created a new modality of sensing: the cantilever.
The most simple way of measuring cantilever deflection resulting from surface stress between the analytes with cantilever to get some measured information is by optical beam deflection as in most AFM instruments [66].

In the optical beam deflection technique, a laser diode is focused on the end of the free cantilever and the reflected laser beam is monitored using a position sensitive photodetector, as
shown in Figure 6 [9]. The typical displacement sensitivity achieved using this technique is
on the order of 10^{-9} m [8]. Fritz *et al.* demonstrated DNA immobilization and hybridization
using microcantilever measured by optical deflection detection [67]. And Yue *et al.* demon-

strated a 2D cantilever array with integrated microfluidics using a single laser source and a Charge-coupled Device (CCD) camera for simultaneous interrogation of several hundred cantilevers for DNA and protein sensing [68]. The advantages are its simplicity, linear response, and lack of electrical connections. However it suffers some limitations. A calibration is needed in order to obtain the recorded signal in terms of the actual cantilever deflection. Index of refraction changes of the surrounding medium of the cantilever can produce artificial deflection and the technique cannot be used in opaque media such as blood.

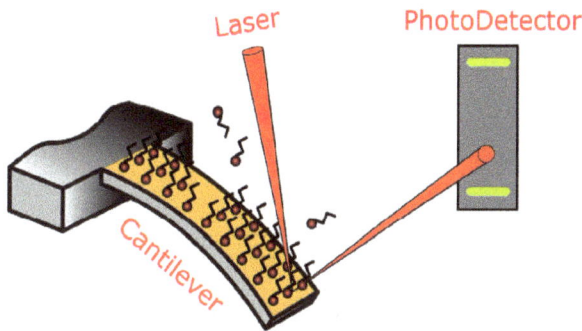

Figure 6. The optical beam deflection technique is used to monitor the deflection of the cantilever [9]

Another optical method which can attain better performance is interferometry [69]. When using a fiber optic interferometer [69, 70], the interference signal from the reflected light of the cleaved end of the fiber optic and of the cantilever surface is a direct measure of the average cantilever displacement in the field of view. Deflection in the range of 10^{-11} m to 10^{-13} m can be measured [71]. Fiber optic interferometer is a mature technology and has many advantages, good performance, low loss, high bandwidth, safety and relatively low cost, for example, which is suitable for biosensors. The principle is as schematically shown in diagram Figure 7, the interference is formed inside an optical fiber. When the laser diode light passes at the fiber end-face, a portion is reflected off at the fiber/air interface (R_1) and the remaining light still passed through the air gap (L) with a second reflection occurring at the air/membrane interface (R_2). R_1 is the reference reflection named the reference signal (I_1) and the sensing reflection is R_2 called sensing signal (I_2). These reflective signals interfere constructively or destructively in the fiber due to the difference of the optical path length between the reference and sensing signals, which is called the interference signal [70]. Therefore, small deflection of the membrane causes a change in the air gap (L), which changes the phase difference between the sensing and reference signals producing fringes.

However, optical detection systems for cantilever arrays are still typically large and are more suited for bench-top applications than for portable handheld use. Nonspecific adsorption on the back side (non functionalized side) of the cantilever because of sensor immersion in liquid sample during measurement is a significant source of noise in these sensors.

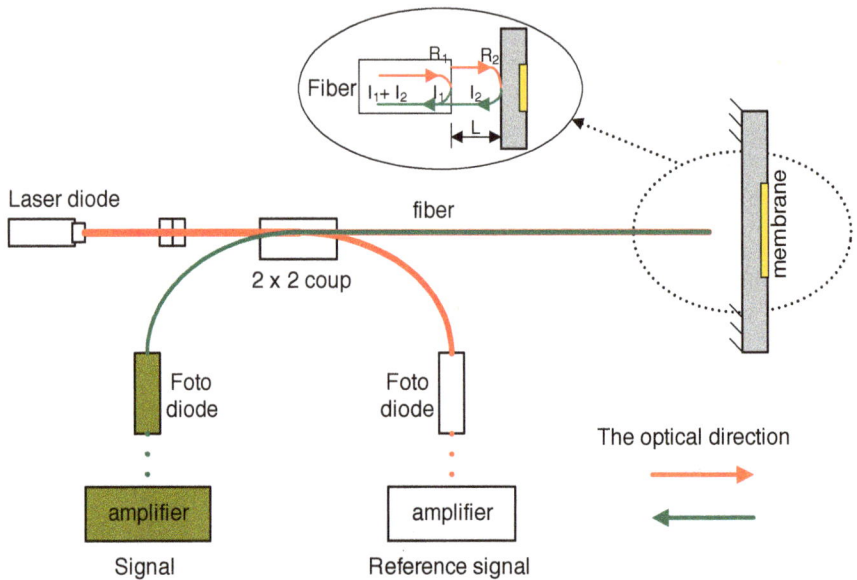

Figure 7. Schematic diagram of a fiber interferometer [72]

2.1.5.2. Piezoresistance detection methods

Piezoresistivity is the variation of the bulk resistivity under applied stress. When a silicon cantilever is stressed because of its bending caused by surface stress, a highly doped region will change resistance in a sensitive way. The variation of cantilever resistance is typically measured using a Direct Current (DC)-biased Wheatstone bridge. The advantage of piezoresistivity technique is that the sensor and the detection scheme can be easily integrated into lab-on-a-chip type devices. In addition it is more compatible with large array formats. Marie at al developed a cantilever system using piezoresistive detection instead of optical deflection method for sensing DNA hybridization [73].

Nevertheless, this method possesses electrical connections which need to be protected for experiments performed in liquids and requires current to flow through the cantilever. This results in heat dissipation and thermal drifts which causes parasitic cantilever deflections.

2.1.5.3. Others

There are some less widely used and readout schemes existed methods, such as the capacitive method, piezoelectric method and electron tunneling. More recently, displacement detection methods for nanoscale cantilevers were implemented Cleland et al. developed a scheme based on capacitively coupling a nanobeam to a single electron transistor achieving sensitivity down to 10^{-14} m[73].

2.2. Label based techniques

Label based techniques use 'tags' or labels to detect a particular analyte in a background of other materials. Fluorescence, chemiluminescence and radioactive are three popular label based techniques in biosensors.

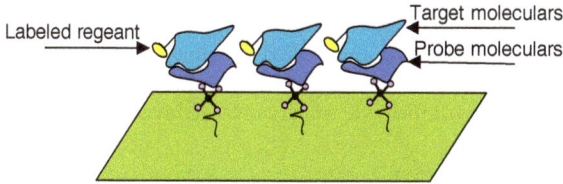

Figure 8. Schematic diagram of fluorescence: Surface with different probe molecules is exposed to solution with pre-labeled target molecules, presence of fluorophores on the surface indicates a specific binding reaction and the presence of a target molecule [11]

2.2.1. Fluorescence

Fluorescence is the short-time (< 1 μs) category of luminescence, which is mostly exploited as an optical phenomenon in cold bodies. For doing this, the molecular component used absorbs a photon and can conesecutively emit a photon with a longer (less energetic) wavelength. Molecular rotations, vibrations or heat can be produced because the absorbed photons have different energy; with the emitted photons, for example, the emitted light can be in the visible range even if the absorbed photon is in the ultraviolet range. The phenomenon depends on the absorbance and Stokes shift of the particular fluorophore.The fluorescence sensor's principle can be shown via an example in Figure.8: the probe molecules, such as antibodies, are immobilized onto the surface of probing microchamber using cross-linkers or covalent methods; the target molecular with labelled reagent, such as antigens with fluorophore are loaded on the microchamber and they will bind to probes; and then the bound targets can be detected. Fluorescence is generally preferred and the most widely used detection method for reasons of sensitivity, stability, and availability of fluorescent scanners tailored for microarray use [74].

The disadvantage of this technique is that most fluorophores are bleached quickly upon exposure to light and are very sensitive to environment conditions such as solution's hydrogen ion concentration(pH) value. And both direct and indirect labeling methods have also their disadvantages. Indirect labeling is more complicated and time consuming, while fluorescence tags in direct labeling may be less stable and more disruptive to the labelled proteins as compared to small molecule tags in indirect labeling.

2.2.2. Chemiluminescence

Chemiluminescence or chemoluminescence is the phenomenon of light emission as the result of a chemical reaction, which leads to limited emission of heat (luminescence).

Chemiluminescence (CL) analysis promises high sensitivity with simple instruments and without any light source, so chemiluminescence has become an attractive detection method in μTAS in recent years [75-78]. On the other hand, many flow sensors based on CL reaction and molecular recognition using enzymes, which have great sensitivity in environmental, biomedical and chemical analysis, have been developed [79-84]. However, limited feature resolution because of signal bleeding and limited dynamic range is the reported drawbacks of this method [85]. Furthermore, sensing with chemiluminescence can be performed only once, unlike fluorescence-based methods which can be archived for future imaging.

2.2.3. Radioactivity

Radioactivity-based detection method is comparable to fluorescence method except that the labels are radioisotopes instead of fluorophores. Techniques using radioactive labels offer robust and reproducible protocols in applications that require ultimate sensitivity and/or resolution. Quantification of results is possible by the following exposure of signal to autoradiography film or reusable storage phosphor screens in automated imaging systems. However, automated liquid handling of radioactivity is difficult due to the need for safe handling and disposal of radioactivity, and is generally limited to manual, low-throughput applications.

3. A novel surface stress-based biosensor

Surface stress-based membrane biosensors, whose bottom surface can be sealed by the biological solutions, are ideal for the development of novel surface stress based biosensors with capacitive readout. Capacitive readout has the advantage of easy and accurate detection as well as it is suitable for device miniaturization. However, it is not applicable in the cantilever biosensors due to faradaic currents in the electrolytic solutions.

Due to the advantage of membrane biosensors, a new biosensor is proposed by Micro Nano System Research Center (MNSRC), which can be used to detect cells. It consists of micro-fluidics, sensitive membrane and capacitive readout unit. The sensitive element of this biosensor — membrane (Figure 9) is formed by the Polydimethyl siloxane (PDMS) thin film with part gold coated on its surface and the trapezoidal structure in the silicon substrate. The surface stress changes when analyte species adhere to the probes which are immobilized on the membrane surface. The change of surface stress causes an out of plane deflection that alters capacitance.

The parameters of membrane (Table 1) have been decided based on the FE(finite element) analysis software ANSYS just like reference [86] and the fabrication techniques. The dimension of one whole bio-sensor chip is $13 \times 7 mm^2$(length×width).

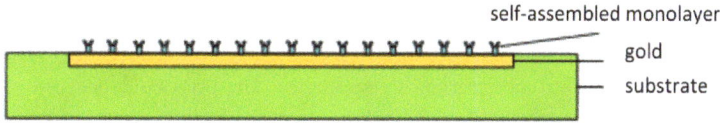

Figure 9. Schematic diagram of membrane[86].

Material	Length(μm)	Width(μm)	Thickness(μm)
PDMS	400	400	1
Gold	390	390	0.02

Table 1. The membrane parameter [87]

Based on the parameters, the biosensors were fabricated [87].One biosensorchip contains two micro-membranes, one acting as the active membrane and the other as reference, as shown in Figure 10. The active membrane is sensitized to react with specific analytes. The selective biochemical reactions between the analytes and the membrane will induce a surface stress change that causes the membrane to deflect. The analytes will not present on the reference membrane, it is only used to remain sensitive to other environmental factors that can also result in a deflection of the active membrane, such as temperature variations (bimetallic effect), laminar or turbulent flow around the membrane, vibrational (including acoustic) noise, non-specific binding (including the swell induced by solution) and so on. The differential signal is solely due to the interaction of the analytes with the membrane. This design has very sensitive surface stress measurements for analytes detection, which is one main novelty of the biosensor.

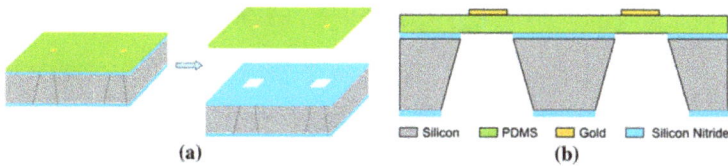

Figure 10. The schematic structure of biosensor(a: 3D schematic diagram of biosensor, b: Cross-section view of biosensor) [87]

One biosensor test systems was set up based on the fiber optic interferometer (FOI) debugged for the test [88]. Figure 11 shows the comparison of active signals and reference signals obtained from FOI-based biosensor test system. Lines 1 and 2represent the basic reference signals when 20μl pure medium (no E.coli) was loaded on the membranes. The voltage had a big change at the loading point because the membrane deflections were first mainly caused by the medium weight. With the evaporation of medium, the influence factors of weight and reflex became smaller and the signals increased again. The active signals shown by lines 3 and 4 were achieved when20μl medium with living and with dead E. coli (1.7 × 10³cells/μl)was loaded into the reservoirs, respectively. The primary deflections also came from the weight and the reflex of E. coli medium, so at the loading point the active signals became smaller as well.

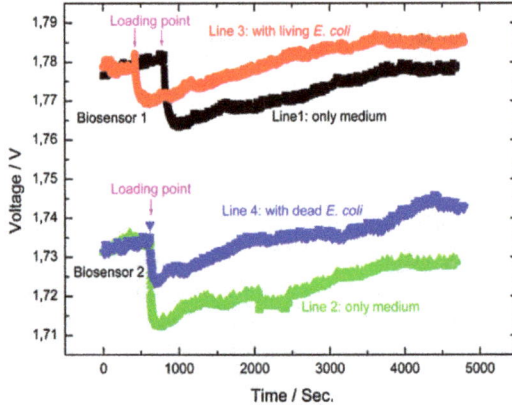

Figure 11. The comparative signals between only medium and the medium containing living or dead E. coli[88]

For E.coli detection, the experimental results show that the conditions of E. coli can be detected based on the deflection tests. For the biological and medical applications, other kinds of cells or molecules can also be detected based on the biosensor test systems. The analysis of some pathological changes of cells or molecules can be carried out based on the alteration of membrane deflection. Furthermore, the deflection signal can be translated into electric signal though adding an electrode on the substrate, which is still promoting for portable sensitive biosensor by us.

4. Conclusion

Labeling of biomolecules with fluorescent or other similar tags for detection can result in sample losses during the labeling and purification process and occasional loss of functionality, especially in proteins and other small molecules. Label-free detection technique can conquer the disadvantage, and the optical interference method is easy and simply to operate and high precision to detect the deflection of membrane for getting the information/properties of the biological analytes. Label-free detection measurements of biosensors are becoming a promising alternative approach to traditional label-based methods. Surface stress-based micro membrane biosensor is a relatively new class of sensor, this kind of biosensor allows the detection in opaque media and the miniaturization of the sensor to incorporate it in portable devices for point-of-use sensing. The technique is advantageous since the process does not require any labeling of target molecules or the addition of redox probes.

PDMS attracts extensive attention worldwide as a membrane of sensors. One kind of PDMS micro membrane biosensors was successfully fabricated through conquering many challenges, e.g., integration of PDMS processing with conventional micro-fabrication processes and the fabrication of perfect PDMS thin film.

Acknowledgements

This work was financially supported by the National Natural Science Foundation of China (Grant 51105267 and Grant No. 91123036) and the National Research Foundation for the Doctoral Program of Higher Education of China (Grant No.20111402120007).

Author details

Shengbo Sang[1,2*], Wendong Zhang[1,2] and Yuan Zhao[1,2]

1 MicroNano System Research Center, Taiyuan University of Technology, Shanxi, China

2 Key Lab of Advanced Transducers and Intelligent Control System of the Ministry of Education,Taiyuan University of Technology, Shanxi, China

References

[1] Wang, Wangjun and Soper, Steven A (2007). Bio-MEMS: Technologies and Applications. United States of America: CRC Press. ISBN 978-0-8493-3532-7.

[2] Lee, Abraham P. and Lee, L. James (2006). BioMEMS and Biomedical Nanotechnology Vol. 1 Biological and Biomedical Nanotechnology. United States of America: Springer. ISBN 978-0387-25563-7.

[3] Ozkan, Mihrimah and Heller, Michael J (2006). BioMEMS and Biomedical Nanotechnology Vol. 2 Micro Nano Technology for Genomics and Proteomics. United States of America: Springer. ISBN 978-0387-25843-0.

[4] Desai, Tejal and Bhatia, Sangeeta (2006). BioMEMS and Biomedical Nanotechnology Vol. 3 Therapautic Micro Nano Technology. United States of America: Springer. ISBN 978-0387-25565-1.

[5] Bashir, Rashid and Wereley, Steve (2006). BioMEMS and Biomedical Nanotechnology Vol. 4 BiomolecularSenging, Processing and Analysis. Springer. ISBN 978-0387-25845-4.

[6] Srinath Satyanarayana (2005). Surface Stress and Capacitive MEMS Sensor Arrays for Chemical and Biological Sensing, University of California, Berkelery, Ph.D. dissertation.

[7] International Union of Pure and Applied Chemistry. "Biosensor". Compendium of Chemical Terminology, Compiled by A. D. McNaught and A. Wilkinson

[8]

[9] Michel Godin (2004). Surface Stress, Kinetics, and Structure of Alkanethiol Self-Assembled Monolayers, Department of Physics McGill University, Canada, Ph.D dissertation.

[10] A.S. Widge et al., (2007). Self-assembled monolayers of polythiophene conductive polymers improve biocompatibility and electrical impendance of neural electrodes, Biosensors and Bioelectronics, 22, 1723-1732.

[11] Shengbo Sang (2011).A novel PDMS micro membrane biosensor based on the analysis of surface stress, VDM Verlag, Jun. 5th.

[12] A.Manz, N.Graber, H.M .Widmer (1990). Miniaturized Total Chemical Analysis Systems: A Novel Concept for Chemical Sensing, Sensors and Actuators B, 1, 244–248.

[13] B.Liedberg, C.Nylander, I.Lundstrom (1983). Surface-Plasmon Resonance for Gas-Detection and Biosensing, Sensors and Actuators, 4 (2), 299-304.

[14] C.Nylander, B.Liedberg, T.Lind (1982). Gas-Detection by Means of Surface- Plasmon Resonance, Sensors and Actuators, 3 (1), 79-88.

[15] Matthew A. Cooper (2002). Optical biosensors in drug discovery, Nature Reviews Drug Discovery 1, 515-528.

[16] Chengjun Huang, Kristien Bonroy, Gunter Reekman, Kris Verstreken, Liesbet Lagae, Gustaaf Borghs(2009). An on-chip localized surface plasmon resonance-based biosensor for label-free monitoring of antigen–antibody reaction, Microelectronic Engineering 86, 2437–2441

[17] Chang Duk Kang, Sang Wook Lee, Tai Hyun Park, Sang Jun Sim (2006). Performance enhancement of real-time detection of protozoan parasite, Cryptosporidium oocyst by a modified surface, Enzyme and Microbial Technology 39, 387–390.

[18] Christina Boozer, Qiuming Yu, Shengfu Chen, Chi-Ying Lee, Jir ˇı´ Homola, Sinclair S. Yee, Shaoyi Jiang(2003). Surface functionalization for self-referencing surface Plasmon resonance (SPR) biosensors by multi-step self-assembly, Sensors and Actuators B 90, 22–30.

[19] Dominique Barchiesi, Nathalie Lidgi-Guigui, Marc Lamy de la Chapelle (2012). Functionalization layer influence on the sensitivity of surface plasmon resonance (SPR) biosensor, Optics Communications 285, 1619 –1623.

[20] Liying Wang, Ying Sun, Jian Wang, Xiaonan Zhu, FeiJia, Yanbo Cao, Xinghua Wang, Hanqi Zhang, Daqian Song (2009). Sensitivity enhancement of SPR biosensor with silver mirror reaction on the AgAu film, Talanta 78,265–269.

[21] D.R. Mernagh, P. Janscak, K. Firman, G.G. Kneale (1998). Protein –protein and protein – DNA interactions in the Type I restriction endonuclease R.EcoR124I, Biol. Chem. 379, 497 – 503.

[22] H. Sota, Y. Hasegawa, M. Iwakura (1998). Detection of conformational changes in an immobilized protein using surface plasmon resonance, Anal. Chem. 70, 2019 – 2024.

[23] J. Deka, J. Kuhlmann, O. Muller(1998). A domain within the tumor suppressor protein APC shows very similar biochemical proper-ties as the microtubule-associated protein Tau, Eur. J. Biochem. 253,591 – 597.

[24] V. Regnault, J. Arvieux, L. Vallar, T. Lecompte (1998).Immunopurification of human Beta(2)-glycoprotein I with a monoclonal antibody selected for its binding kinetics using surface plasmon resonance biosensor, J. Immun. Methods 211,191 – 197.

[25] S.S. Koh, A.Z. Ansari, M. Ptashne, R.A. Young (1998). An activator target in the RNA polymerase II holoenzyme, Mol. Cell 1,895 – 904.

[26] Jiří Homola (2003).Present and future of surface plasmon resonance biosensors, Anal BioanalChem377,528–539.

[27] Xiaojun Feng, Xin Liu, QingmingLuo, and Bi-Feng Liu (2008). Mass Spectrometryin Systems Biology Anoverview, Mass Spectrometry Reviews,27, 635 – 660

[28] R.Aebersold, and M.Mann (2003). Mass spectrometry-based proteomics, Nature, 422 (6928), 198-207.

[29] H. Wohltjen et al (1997). Acoustic Wave Sensor—Theory, Design, and Physico-Chemical Applications, Academic Press, San Diego39.

[30] D. Cullen, T .Montress(1980). IEEE Ultrasonics Symposium, pp. 519–522.

[31] D.Cullen, T .Reeder (1975). IEEE Ultrasonics Symposium, pp. 519–522.

[32] P.G.Ivanov, V.M.Makarov, V.S.Orlov, V.B.Shvetts (1996).IEEE Ultrasonics Symposyum, 61–64.

[33] A.Pohl, G.Ostermayer, L.Reindl, F.Seifert (1997). IEEE Ultrasonics Symposium, pp. 471–474.

[34] N.Levit, D.Pestov, G.Tepper (2002). Sensors and Actuators B: Chemical 82 (2–3), 241–249.

[35] T.Nakamoto, K.,Nakamura, T.Moriizumi (1996). IEEE Ultrasonics Symposium, pp. 351–354.

[36] E.Staples (1999). IEEE Ultrasonics Symposium, pp. 417–423.

[37] H.Wohltjen, R.Dessy (1979). Analytical Chemistry 51(9), 1458–1475.

[38] J.C.Andle, J.F.Vetelino (1995). IEEE Ultrasonics Symposium, pp. 452–453.

[39] D.S.Ballantine, R.M.White, S.J.Martin, A.J.Ricco,E.T.Zellers, G.C.Frye, H.Wohlt-jen (1997).Acoustic Wave Sensors:Theory, Design and Physico-Chemical Applications. Academic Press, San Diego, CA.

[40] B.A.Cavic, G.I.Hayward, ,M.Thompson (1999). The Analyst, 1405–1420.

[41] A.Janshoff, H.-J.Galla, C.Steinem (2000). Angewandte Chemie International Edition 39, 4004–4032.

[42] W.Bowers, R.Chuan, T.Duong (1991). Review of Scientific Instruments 62 (6), 1624–1629.

[43] J.Cheeke, N.Tashtoush, N.Eddy (1996). IEEE Ultrasonics Symposium,pp. 449–452.

[44] A.L.Smith, R.Mulligan, J.Tian, H.M.Shirazi, J.Riggs (2003). IEEE International Frequency Control Symposium and PDA Exhibition, St Philadelphia, pp.1062–1065.

[45] Ioana Voiculescu, Anis Nurashikin Nordin (2012), Acoustic wave based MEMS devices for biosensing applications, Biosensors and Bioelectronics ,33, 1–9.

[46] Simon Q. Lud, Michael G. Nikolaides, Ilka Haase et al,2006, Field Effect of Screened Charges: Electrical Detection of Peptides and Proteins by a Thin-Film Resistor, ChemPhysChem,7, 379–384

[47] Joon-Ho Maeng, Byung-Chul Lee, Yong-Jun Ko, Woong Cho, YoominAhn,Nahm-Gyoo Cho, Seoung-Hwan Lee, Seung Yong Hwang(2008),A novel microfluidic biosensor based on an electrical detection system for alpha-fetoprotein, Biosensors and Bioelectronics, 23, 1319–1325

[48] M. S.Wilson (2005).Electrochemical immunosensors for the simultaneous detection of two tumor mark-ers. Analytical Chemistry77(5), 1496–1502.

[49] P.D'Orazio (2003).Biosensors in clinical chemistry. Clinica Chimica Acta 334(1-2), 41–69.

[50] H.Suzuki (2000). Advances in the microfabrication of electrochemical sensors and systems, Electroanalysis, 12 (9), 703-715.

[51] H. Ibach (1997). Surface Science Reports 29, 193

[52] W. Haiss (2001). Reports on Progress in Physics 64, 591.

[53] R. Berger, E. Delamarche, H. P. Lang, Ch. Gerber, J. K. Gimzewski, E. Meyer, and H.-J. Güntherodt (1997). Science 276, 2021.

[54] C. E. Bach, M. Giesen, and H. Ibach(1997). Physical Review Letters 78, 4225.

[55] U. Tartaglino, E. Tosatti, D. Passerone, and F. Ercolessi(2002). Physical Review B 65, 241406.

[56] R. M. Tromp, A. W. Denier van der Gon, and M. C. Reuter(1992). Physical Review Letters 68, 2313.

[57] K. Pohl, M. C. Bartelt, J. de la Figuera, N. C. Bartelt, J. Hrbek, and R. Q. Hwang(1999). Nature 397, 238.

[58] K. M.Hansen, H. F.Ji, G. H.Wu, R.Datar, R.Cote, A. Majumdar, T.Thundat (2001). Cantilever-based optical deflection assay for discrimination of DNA single-nucleotide mismatches, Analytical Chemistry, 73 (7), 1567-1571.

[59] G. H.Wu, R. H.Datar, K. M.Hansen, T.Thundat, R. J. Cote, A.Majumdar (2001). Bioassay of prostate-specific antigen (PSA) using microcantilevers, Nature Biotechnology, 19 (9), 856-860.

[60] M.Yue, H.Lin, D. E.Dedrick, S.Satyanarayana, A.Majumdar, A. S.Bedekar, J. W. Jenkins, S. Sundaram (2004). A 2-D microcantilever array for multiplexed biomolecular analysis, Journal of Microelectromechanical Systems, 13 (2), 290-299.

[61] Si-Hyung "Shawn" Lim, Digvijay Raorane, Srinath Satyanarayana, Arunava Majumdar (2006).Nano-chemo-mechanical sensor array platform for high-throughput chemical analysis ,Sensors and Actuators B 119,466–474.

[62] Misun Cha, Jaeha Shin, June-Hyung Kim, Ilchaek Kim, Junbo Choi, Nahum Lee, Byung-Gee Kim,Junghoon Lee (2008). Biomolecular detection with a thin membrane transducer.Lab Chip8, 932–937.

[63] R.Marie, H.Jensenius, J.Thaysen, C. B.Christensen, A.Boisen, (2002). Adsorption kinetics and mechanical properties of thiol-modified DNA-oligos on gold investigated by microcantilever sensors, Ultramicroscopy, 91 (1-4), 29-36.

[64] R.G. Knobel and A.N. Cleland (2003). Nature 424, 291.

[65] G.Binning, C. F.Quate, Ch. Gerver (1986). Phys. Rev. Lett. 930-935.

[66] M. Godin, O. Laroche, V. Tabard-Cossa, L. Y. Beaulieu, P. Grütter, and P. J. Williams(2003). Review of Scientific Instruments 74, 4902.

[67] J.Fritz, M. K.Baller, H. P.Lang, H.Rothuizen, P.Vettiger, E.Meyer, H. J.Guntherodt, C.Gerber, and J.K.Gimzewski (2000). Translating biomolecular recognition into nanomechanics, Science, 288 (5464), 316-318.

[68] D. R.Baselt, G. U.Lee, R. J.Colton (1996). J. Vac. Sci. Technol. B 14, 798-803.

[69] D. Rugar, H. J. Mamin, P. Güthner (1989). Applied Physics Letters 55, 2588.

[70] A. Moser, H. J. Hug, Th. Jung, U. D. Schwarz, H.-J. Güntherodt (1993). Meas. Sci. Technol. 4, 769-775.

[71]   R. Raiteri, H.-J. Butt and M. Grattarola (2000). Electrochem. Acta 46, 157–163.

[72]   Vasiliki Tsouti, Stavros Chatzandroulis (2012). Sensitivity study of surface stress bio-sensors based on ultrathin Si membranes. Microelectronic Engineering 90, 29–32.

[73]   Srinath Satyanarayanaa, Daniel T. McCormickb, Arun Majumdara (2006), Parylene micro membrane capacitive sensor array for chemical and biological sensing, Sensors and Actuators B 115 ,494–502.

[74]   G.Macbeath, and S. L.Schreiber (2000). Printing proteins as microarrays for high through put function determination, Science, 289 (5485), 1760-1763.

[75]   H. Masahiko, T. Kazuhiko, N. Riichiro, F. S. Norman and J. K. Gregory (2000). J. Chromatogr., A, 867, 271–279.

[76]   Y. Xu, F. G. Bessoth, J. C. T. Eijkel and A. Manz (2000). Analyst, 125, 677–683.

[77]   X. J. Huang, Q. Pu and Z. L. Fang(2001). Analyst, 126, 281–284.

[78]   K. Tsukagoshi, M. Hashimoto, R. Nakajima and A. Arai (2000). Anal. Sci,16, 1111–1113.

[79]   Z. Zhang and W. Qin(1996). Talanta, 43, 119–124.

[80]   B. Li and Z. Zhang(2000). Sens. Actuators, B, 69, 70–74.

[81]   G. J. Zhou, G. Wang, J. J. Xu, H. Y. Chen(2002). Sens. Actuators, B, 81, 334–339.

[82]   M. C. Ramos, M. C. Torijas and A. Navas Diaz(2001). Sens. Actuators, B, 73, 71–75.

[83]   Y. Huang, C. Zhang and Z. Zhang(1999). Anal. Sci, 15, 867–870.

[84]   B. Li, Z. Zhang and Y. Jin(2001). Anal. Chem., 73, 1203–1206.

[85]   B.Schweitzer, P.Predki, M,Snyder (2003). Microarrays to characterize protein interac-tions on a whole-proteome scale, Proteomics, 3 (11), 2190-2199.

[86]   Shengbo Sang, Hartmut Witteb (2009). Finite Element Analysis of the Membrane Used in a Novel BioMEMS, Journal of Biomimetics, Biomaterials and Tissue Engi-neering Vol.3 pp 51-57.

[87]   Shengbo Sang, Hartmut Witte(2010). Fabrication of a surface stress-based PDMS mi-cro-membrane biosensor, MicrosystTechnol 16, 1001–1008.

[88]   Shengbo Sang, Hartmut Witte (2010). A novel PDMS micro membrane biosensor based on the analysis of surface stress, Biosensors and Bioelectronics25, 2420–2424.

Chapter 6

New Insights on Optical Biosensors: Techniques, Construction and Application

Tatiana Duque Martins,
Antonio Carlos Chaves Ribeiro,
Henrique Santiago de Camargo,
Paulo Alves da Costa Filho,
Hannah Paula Mesquita Cavalcante and
Diogo Lopes Dias

Additional information is available at the end of the chapter

1. Introduction

Since Clark's enzymatic electrode in 1962s [1], biosensors have been proposed for a range of application and, aiming clinical analysis application, the amperometrical, potentiometrical and optical are the biosensors which have achieved most significant development. In concept, optical biosensors are those based on the detection of changes on absorption of UV/ visible/Infrared light when chemical reactions occur or on the quantity of light emitted by some luminescent process. Regarding to supramolecular nanostructures and their ability of enhancing the sensing activity when applied to biosensors construction, a very instigating work, presented by Jin Shi et al. [2] showed a way of turning carbon nanotubes into more water-soluble compounds and, consequently, more biocompatible by modifying their surface with a synthetic DNA sequence. In this way, the carbon nanotubes can overlay the biosensor electrode more efficiently, enhancing the biosensing activity. Lieden et al. [3] also took advantage of the properties of nanotubes in biosensing. In their work, they showed that a biosensor can present a rate of detection tree times faster when carbon nanotubes are used in the nanobiosensor construction, preventing the attachment of protein to the device components. The change in electric resistance of carbon nanotubes when proteins touch them is immediate, which confer to the device a fast recognition ability, and leads to an increased efficiency of the biosensor. Yet, other fascinating works are the one presented by

Chen et al. [4] and the work presented by Park et al.,[5] in which a piezoelectric nanogenerator was developed to feed several devices, including implantable biosensors, which make of this research field very promising, since nowadays, implantable biosensors present the disadvantage of the need to be recharged or replaced when discharged.

In a common sense, biosensor is a device constructed to inform about a system, requiring as less human action as possible. It is formed of sample holder, a biological recognition element which must be selective, a physical transducer to generate a measurable signal proportional to the concentration of the analytes and the signal processing unit, which gives to the analysts graphical, numerical or comparative information that they shall interpret (fig.1). The recognition element can be of almost any type of biological system, from antibodies, proteins and peptides to viruses, microbes, cells and tissues.The selection of the appropriate recognition element considers not only what is the information to be obtained, but also the ease of construction of the devices employing such element and, of course, their durability. As an example, some microbes have been employed instead of antibodies and proteins mostly due to their facile production via cell culture and in vitro stability. Nevertheless, it is evident that microbes might lead to a lack of selectiveness, due to their non-specific metabolisms. Recent research had proposed that highly selective microbial biosensors can be constructed by inducing a desired microbe metabolic pathway and by adapting them to the substrate of interest, using selected conditions of cell-culturing. Also, as alternative to manipulate the selectivity and sensitivity of microbial biosensors at the DNA level, the genetically engineered microorganisms (GEMs) had been proposed. [6]

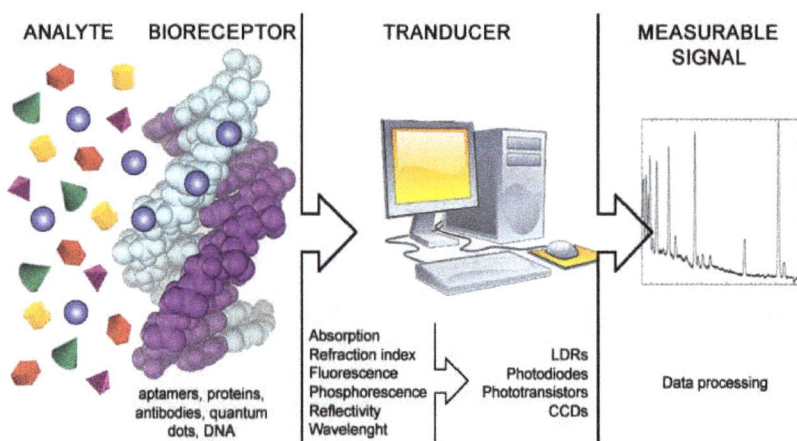

Figure 1. General scheme of a biosensor.

Although it is of extreme importance to accurately choose a recognition element to propose a suitable biosensor, it is of great importance to define a way to turn the recognition event into a signal that can be detected and interpreted. The transducer figures now as the element which is able to transform the biochemical response into a recognizable physical signal. For example, techniques of immobilizing microorganisms on transducers had played important roles in the fabrication of microbial biosensors. [7] Some traditional methods of immobilization include adsorption, encapsulation, entrapment, covalent binding, and cross-linking, mainly via sol-gel processes and, in special, immobilization of microorganisms in conducting polymers are of great interest due to their unique electrochemical properties, which can be exploited on the transduction function. [8, 9]

A number of possible transducers can be proposed, once the interaction between analyte and recognition element is defined. Available techniques are of a large number and to elect the proper one very often is not a trivial task. Among various sensing techniques, there are piezoelectrical, calorimetrical, enthalpimetrical, DNA microarray, Surface Plasmon Resonance (SPR), Impedance Spectroscopy, Scanning Probe Microscopy (SPM), Atomic Force Microscopy (AFM), Quartz Crystal Microbalance (QCM), Surface Enhanced Raman Spectroscopy (SERS), but electrochemical and optical techniques are the widest used in the development of microbial biosensors, due to their numerous possibilities, which turn possible the construction of a number of selective sensors.Electrochemical biosensors are classified as amperometric, potentiometric, conductometric, voltammetric, depending on which detection principle is employed in the biosensor. In a quick overview on these important devices, an amperometric biosensor is the one that operates at a given applied potential between the working and the reference electrodes. A current signal, related to analyte's concentration in the sample, is then generated, due to the reduction or oxidation process suffered by an electroactive metabolic product. A conductometric biosensor is that in which a conductivity change is observed upon production or consumption of ionic species involved in the metabolic process. It became a very attractive device due to its enhanced sensitivity and fastness brought about through sophisticated modern analytical techniques. Additionally, they are suitable for miniaturization once it requires no reference electrode in the system. [10] Its disadvantage lies on that all charge carriers lead to a change of conductivity, which directly affects the device selectivity and is known as relatively poor.

The potentiometric biosensor is based on the potential difference between working and reference electrodes. In these biosensors, the measured species is not consumed, as it is in the amperometric biosensor. Its response is on the activity of the species in comparison to the reference electrode, with the output signal recorded in voltage units and, independently of the sensor size, the signal is proportional to natural the analyte concentration. Its great advantage lies on sensitivity and selectivity, if the working electrode is species-selective. However, a highly stable and accurate reference electrode is always a requirement.

The most versatile electrochemical technique applied to biosensors is voltammetry, since both current and potential difference combined, consist on a reasonable system response. Its major advantage its successfully application as a multi-component detector.

2. Optical biosensors

Optical biosensors are those which can sense phenomena related to the interaction of micro-organisms with the analytes and correlate the observed optical signal to the concentration of target compounds, based on the measurement of photons involved in the process, rather than electrons, as in the afore mentioned techniques. More specifically, optical detection is based on the measurement of luminescence, fluorescence, colour changes, by the measurement of absorbance, reflectance or fluorescence emissions that occur in the ultraviolet (UV), visible, or near-infrared (NIR) spectral regions.

Even though the first biosensor reported, the Clark's amperometric enzyme electrode for glucose [1] is dated of 1962, the first fiber-optic biosensor was described only in 1975, by Lubbers and Opitz [11, 12] and, since then, with the incredible rapidity of new detection techniques development, various optical biosensors have been demonstrated. Among optical properties, fluorescence is by far the most often exploited one by a significant number of techniques, mostly based on parameters such as intensity, lifetime, anisotropy, quenching efficiency, non-radiative and luminescence energy transfer, and so on.

Although fluorescent biosensors have been reported since the late nineties, [13- 15] very often there is a confusion between labelling and the description of a biosensor, which turns the exact moment of the fluorescent biosensor proposal not very clear to point. Nevertheless, amazing development ondetection methods in biosensors, their construction, even at nano-scale, and applications have been reported in the last decade. [16- 22]

2.1. Classes of optical biosensors

Due to the enormous quantity of biosensors already proposed, it is a very hard task to classify them all. Nevertheless, for our objective, it is convenient to classify them into two classes, not based on the detection method, which, of course, must be an optical method, but on the recognition one. With it in mind, optical biosensors can be sub-classified as:

- probing biosensors: this class consists on biosensors that have based their activity in differences of interactions between analyte and recognition element, ruled by affinities. This behavior leads to changes in the optical response that can be measured by several optical methods.

- reacting biosensors: in this class, their optical responses are related to chemical processes, such as chemisorptions, catalytic reactions of any kind, formation of new chemical bonds and so on. These chemical (and definitive changes) can also generate optical changes which are detectable by countless optical methods.

With respect to the detection mechanism, biosensors can be of fluorescence, phosphorescence, reflection, UV/Vis/IR absorbance, lifetime, which is the characteristic measured by Förster Resonant Energy Transfer (FRET) based biosensors, responsible for the wide range of applications in which biosensors had acted lately. Also, refractive indexes changes can be detected by several and modern techniques, such as interferometry and Surface Plasmon

Resonance, which had opened a new and promising field for biosensor research. Further information about them will be presented later in this chapter.

Among all optical properties exploited in biosensors construction, the most common method of detecting and quantifying biological compounds is still based on the fluorescence activity due to the fact that fluorescent properties of most organic fluorophores are susceptible to environment changes, which is indispensable to sensing applications. The most important advantage of these biosensors is that they are proposed for general use, thus, they provide the possibility of multiple compounds detection within a single device, they are able to perform remote sensing and they are ease to build. Nevertheless, there are some requirements for the construction of the fluorescent recognition unit that must be attended, regarded to structure of the fluorophore and to its photophysical activity.

When proposing a biosensor, the hardest task is to address the right fluorophore to compose it, the one that will provide the needed answer. This is because there is a number of fluorescent probes which can be applied on biosensing and a number of device designs that can be proposed within the same objective.

Since in sensing applications, detection is based on changes of fluorescent responses of a particular fluoroprobe, when it is inserted into the analyte environment and interacts with it by a variety of mechanisms, it is of extremely importance to profoundly understand the photophysical behavior of the system and the fluorescence parameters that will be determinant in detection. Once this step is achieved, a biosensor can be built aiming to countless applications, which include immunoassays, nucleic acid detection, cellular and sub-cellular labeling, resonance energy transfer studies, diagnostic assays and disease monitoring and treatment.

Most of fluoroprobes used in biosensors present an environment-dependent luminescent behavior. Therefore, attention must be given to the environment characteristics that may interfere with the photophysics of fluoroprobe-analyte, such as pH dependence of the luminescent response; self-quenching at high concentrations of fluoroprobes; susceptibility to photo-bleaching; short excited state fluorescent lifetimes, which may confer low sensibility to salvation relaxation; small Stokes shifts, which favors self-absorption effects and undesired luminescent emission shifts and short-term stability when in presence of water or in aqueous medium [23, 24].

2.2. Fluorescent biosensors

In fluorescent biosensors, a particularly interesting characteristic that have been more and more exploited is the effect that local interactions can cause on fluoroprobe's electronic excited states, even when no chemical changes are included. With this respect, mechanisms of controlling and orienting these effects are proposed and studied. For example, non-radiative energy transfers leading to a more selective system or a more efficient luminescent response, achieved by combining common organic fluorophores with metallic nanoparticules, metallic complexes or with nanostructures of carbon or peptides have presented interesting results. In our previous work, with leadership of Prof. Alves, [25] diphenylalanine peptide nano-

tubes were physically modified with a fluoroprobe containing a polar head, 1-pyrenyl-car-
boxylic acid, and their steady-state and dynamic fluorescence responses were determined
(fig.2). Also, the mechanism of interaction between peptides to form the supramolecular
structure and their interaction with the fluorophore were studied and revealed. By computational simulations, it was shown that the nanotube formation and the fluorophore adsorption are governed by π-stacking interactions, which makes of electrostatic interactions
essential. Moreover, since the connection of nanostructured materials, especially biomaterials such as peptides, with solid well-structured surfaces make the manipulation of such
structures an interesting alternative for device fabrication. Although the afore mentioned
work focused on the supramolecular chemistry of nanostructure obtainment and the role of
pH conditions to ensure the right nanotube dimensions, it rose the perspective of constructing an electronic device for biological recognition. Aiming at the application of this nanostructured peptides to a biosensor construction, it is of great importance to determine
whether the interaction of the peptide nanostructure with the probable substrates can influence the optical response measured. The first step is to determine the nature of the interaction between the fluorescent peptide nanotubes with Indium-Tin Oxide (ITO) electrode,
elected as a good candidate to act as anode in a future biosensor device.The modified Phe-
Phe nanotube was then deposited on ITO electrode and the time-resolved fluorescence of
this system was detailed studied.

Since the amphoteric ITO is a transparent conductive oxide of low electrical resistivity (10^{-6}
to $10^{-4}\Omega$.cm) and small bandgap (of 3.3 eV), high physical-chemical stability and good surface morphology, it is widely used as electrode in a variety of devices, which includes sensors. When applied to biosensors, ITO can interact with protein residues due to –OH groups
on its surface, which enables the adsorption of carboxylic and amine residues via hydrogen
bonds.

By combining the final fluorescent nanotubes with ITO electrodes, a non-radiative resonant
energy transfer (FRET) was detected between the thin layer of mixed oxide and the fluorophore doping in nanotube surface. In fact, the results show that when pyrenyl-doped nanotubes are formed at neutral and higher pH ranges, FRET occurs, leading to a small excited-
state lifetime of pyrenyl moieties, but, at lower pH ranges, the excited states of pyrenyl
moieties are stabilized and the lifetime rises. This energy transfer process is favored, in this
case, by the strong electrostatic interaction between the charged nanotube and the charge
transporter ITO of dipole-dipole induced type. When nanotubes are formed in low pH ranges, the final structure presents an overall superficial positive charge, due to protonation of
carboxylic and amine groups of peptide residues and carboxylic groups of the fluorophore,
which is responsible for a less effective electrostatic interaction between combined peptide
structure and 1- pyrenyl-carboxylic acid moieties and the ITO surface. These results contribution laid in terms of the supramolecular control of the structures, showing that they can
be designed and actually obtained as desired and many aspects of biosensing activity can be
exploited in the same device conception. In this example, both a fluorescent-based biosensor
for local environment monitoring and a FRET-based fluorescent biosensor can be designed,
depending only on the approach.

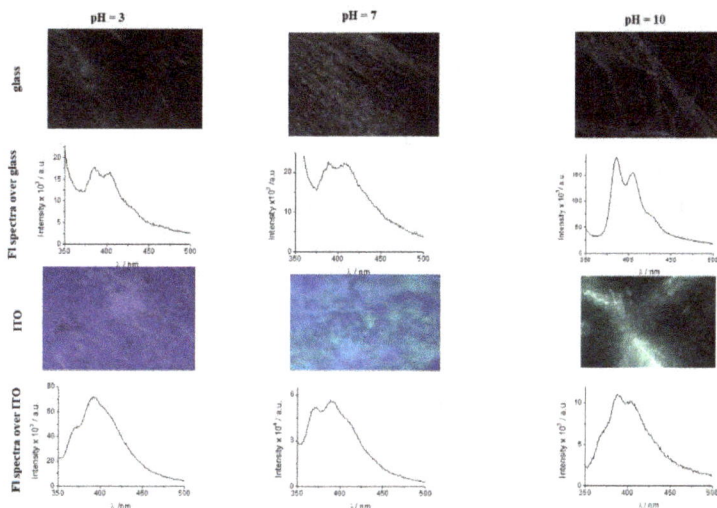

Figure 2. Epifluorescence images (200 times increased) and respective fluorescence spectra of Phe-Phe nanotubes samples obtained at of 0.07% m/m of 1-pyrenyl-carboxilic acid and at distinct pH ranges, deposited over glass and glass covered by ITO substrates.

2.3. FRET-based fluorescent biosensors

FRET is often exploited in biosensors due to the variety of systems that can present this effect and to the quantity of sensing elements that can be employed in the biosensor construction. An example is the use of colloidal luminescent semiconductor nanocrystals, the quantum dots (QDs), to biosensors. These fluorescent compounds are very well known as fluorescent labels for a series of studies, in particular, for imaging of biological and non-biological systems. [26-28] Nevertheless, they also give rise to more robust biosensors, since their unique fluorescent properties can overcome the organic-based fluorophores liabilities.

The major property that inform about the occurrence of non-radiative energy transfer process is changes in the fluorophoreexcited state lifetimes. Due to the variety of compounds that can have their excited state disturbed, several detection techniques can be applied, several device architectures only dependent on the application, rather than in the detection technique, generating biosensors with significant sensitivity and selectivity enhancement, which can be even at the one single molecule limit. FRET is also essential to the conductive polymers, DNA, aptamer or protein-based biosensors and all these possible architectures make use of the interpretations of luminescent signals that reveals several mechanisms for FRET. Indeed, FRET is a resonant process that depends on several conditions of the environment where it takes place. In a short description, this non-radiative excitation energy transfer occurs always that a donor chromophore and an acceptor become close to each other and their electronic energy levels interact, with some pre-requisites. As described by Förster [29],

FRET is determined by a long range dipole–dipole interaction between the donor and the acceptor and his formulations for these events are widely applied from solutions to solid systems, which contain the chromophores of biochemical interest. As an advantage, FRET offers an experimental approach to determine molecular distances through luminescent spectral measurements, which correspond to the efficiency of energy transfer between a donor and an acceptor located at two distinct specific sites, with separation limited to a range of 10–80 Å. Because of the sensitiveness of this technique corresponds to the inverse sixth power dependence of the transfer efficiency to the donor-acceptor distance (equation 1), FRET is assumed to consist of a sensitive technique for detection of global structural alterations. Förster formalism assumes that donor and acceptor are stationary in the timescale of their electronic excited-states lifetimes and, as a consequence, the donor-acceptor separation is static, giving a single distance between them. Nevertheless, the dynamic nature of large systems such as proteins and polymers cannot be ignored and the distances between them are expressed as a distribution. Förster mechanism involves an inductive resonance transfer in which the excitation process creates an electric field around the donor, due its charge transport. As a second oscillator, the acceptor, come closer to the donor, it inductively oscilates and, if it occurs with the adequate frequency, the energy of the donor is transferred to the acceptor. The energy transfer is maximum when both oscillators are similar. To observe this phenomenon, it is necessary that electronic transitions of both donor and acceptor are permitted and then, that coulombic interactions, such as dipole-dipole, which are distance dependent by a factor of R^{-3}, occur. This leads to a probability of occurrence of FRET proportional to R^{-6}. Förster predicts that the energy transfer occurs if there is a coupling between transitions and radiation field, at a rate constant (k_{DA}) given by:

$$k_{DA} = \frac{9000 k^2 ln 10}{128 \pi^5 n^4 N_A \tau_{DA} r^6} \int \frac{F_D \varepsilon_A(v)}{v^4} dv \qquad (1)$$

In which k^2 describes donor-acceptor dipole relative orientation; N_A is the Avogadro Constant; F_D stands for the donor corrected fluorescence intensity; ε_A is the acceptor molar extinction coefficient; τ_{DA} is the donor fluorescence lifetime; r is the distance between donor and acceptor; n, the refractive index and v the wavenumber.

When the probability of FRET occurrence is 50%, the distance in which it takes place is a reference distance, called Förster Distance (R_0), defined as the distance in which the FRET rate K_{DA} is equivalent to the fluorescence rate of the donor in the absence of the acceptor τ_D^{-1}. They are related by:

$$k_{DA} = \frac{1}{\tau_D} \left(\frac{R_0}{r} \right)^6 \qquad (2)$$

When $R = R_0$, $K_{DA} = 1/\tau_D$

2.4. DNA – based fluorescent biosensors

Despite specific requirements of the systems of interest, these assumptions can be applied to any measurements that can identify the energies of the electronic transitions involved and the excited states lifetimes. In DNA-based biosensors, high sensitivity detection and real time information are crucial. In their work, Liu and Bazan [30] proposed homogeneous biosensor assays, which were based on the detection of distinct luminescent responses of a water-soluble conjugated polymer and took advantage of its characteristic of self-assemble to improve the biosensor capability over those employing small molecules. In this biosensor, the interaction between the oligonucleotide hybridized with a cationic polythiophene and a single-stranded DNA or a double-stranded DNA in the presence of cationic poly(fluorene-co-phenylene) leads to conformational changes on polymer backbone and to changes in the fluorescent response, as the cationic poly(fluorene-co-phenylene) acts as donor in the fluorescence energy-transfer assay and, hence, to a signal amplification(fig.3). In the presence of the single-stranded DNA, the positively charged polymer interacts with it, but without energy transfer and only emission from the poly(fluorene-co-phenylene) is detected. When interacting with the double-stranded DNA, emission from the poly(fluorene-co-phenylene) decreases and emission from the hybridized oligonucleotide is observed. The signal transduction is then controlled by specific electrostatic interactions.

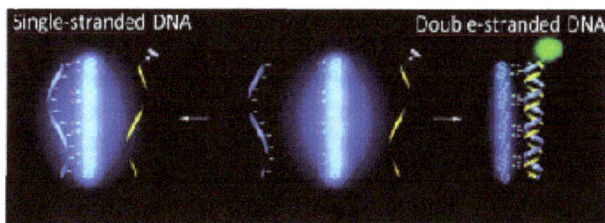

Figure 3. Scheme of the transduction mechanism on Liu´s conductive polymer FRET-based biosensor for DNA detection. Reprinted (adapted) with permission from [30]. Copyright (2004) American Chemical Society.

The approach of employing oligonucleotides in the transduction process of biosensors has become very attractive and popular in such a way that a new class of biosensor has arose: the *aptamer-based* ones.

2.5. Aptamer-based biosensors

Although most of the aptamer-based biosensor utilizes optical methods for detection, it is not exclusive. In fact, they consist of a versatile tool for biosensors, since they behave as efficiently as antibodies, selectively interact with the target and consist of innovative approaches for biosensor construction. There are numbers of aptamers that can be selected from the Systematic Evolution of Ligands by Exponential (SELEX) enrichment, which consists of an *in vitro* iterative process of adsorption, recovery and re-amplification of single-stranded DNA combinatorial lists. This routine of select an aptamer is necessary due to the specificity of their interactions and the variety of biosensors that can be proposed.

In a recent work, Yildirim et al. [31] showed an environmental application for aptamer-based optical biosensors. In their approach, β-estradiol 6-(O-carboxy-methyl)oxime-BSA was covalently immobilized on an optical fiber surface to develop an aptamer-based biosensor for rapid, sensitive and highly selective detection of 17β-estradiol, a endocrine disrupting compound that is a common water pollutant. In an indirect competitive detection approach, samples of 17β-estradiol were premixed with a fluorescence-labeled DNA aptamer. In the sensor surface, a higher concentration of 17β-estradiol led to a less intense fluorescence of the labeled aptamer, by creating a dose-response curve of 17β-estradiol, with a detection limit as low as 0.6 ng mL^{-1}.

2.6. Quantum dots-based fluorescent biosensors

Some of the quantum dot's photophysical properties overcome by several orders of magnitude those of common fluorophores. For example, they present very broad absorption spectra, from UV region towards blue-visible region, corresponding to a large wavelength range; their molar extinction coefficients are of hundred times larger than those of small organic fluorophores and can reach values of several millions. [32] Also, they present the ability of tuning their photoluminescence as a function of the core size, which turn possible assign a determined quantum dot for an application.[33, 34] This possibility becomes a great advantage of quantum dots in comparison to organic luminescent polymers, which cannot have their behavior well predicted only by their chain size: it is important to know their solubility and the interaction forces that act in a given system, since their final photophysical properties are intimately related to their chain conformation and, so, to inter and intrachain energy transfer processes.

Regarding to the photoluminescent properties of quantum dots, their tunable fluorescence combined to the very broad absorption lead to a large effective stokes shifts and to the probability of an efficient excitation of a mixed population of quantum dots, at a single wavelength, several nanometers delocalized from their fluorescent maximum. The characteristics of size-tunable luminescence and of broad absorption spectra make of quantum dots suitable for multi-color (or, as usually called multiplexed) immunoassays. Their photostability and sensitivity also make them good options for a number of imunoassays, especially because they provide flexibility to the analytical techniques [35, 36].

In Pinwattana et al. [37] work, quantum dots were conjugated to a secondary anti-phosphoserine antibody in a heterogeneous sandwich immunoassay, acting as labels and generated amplified electrochemical signals, analyzed by square-wave voltammetry, which is not an optical technique, but it demonstrates the amplitude of analytical methods that the choice of quantum dots as actives in biosensors permit. Their experiments consisted of the addition of the model phosphorylated protein, bovine serum albumin, to a primary bovine serum albumin antibody-coated polystyrene microwells, followed by the addition of a quantum dot labeled anti-phosphoserine antibody. This quantum dot label was then removed by acid attack and the free label was detected, leading to current responses that were proportional to the concentration of the phosphorylated bovine serum albumin.

Since quantum dots are usually obtained from organometallic precursors, they are poorly water-soluble and this is, in some cases, an issue to biosensing, since it seems improbable that, in these conditions, a quantum dot will interact with a biosystem. Nevertheless, there are several methods available to efficiently exchange or functionalize their native organic ligands with a desired ligand that can better both solubility and bioconjugation potential, by either chemical or physical processes. One of the most common modifications is to attach a biomolecule to a functional group on the quantum dot surface, which can be amines, carboxyls or even thiols. In this matter, amines andcarboxyls can be easily modified by 1-Ethyl-3-(3-dimethylaminopropyl) carbodiimide, a common reactant. Thiol groups are usual sites for maleimide.

The quantum dot surface modification can also be conducted by direct interaction with the biomolecule. In this case, interaction forces balance may govern the stability and the yield of the modified quantum dot. Examples are metal-affinity between polyhistidine appended proteins and Zinc atoms present on quantum dot's structure, which leads to coordination of the biomolecule to the metallic center [38, 39] or dipole interactions between thiol groups of cysteine residues with sulfur atoms present in the surface.[40, 41] These modifications can transform quantum dots into efficient elements to immunoassays applications. Examples were presented by Puchades et al. In their review, [42] they showed that by electing a quantum dot, taking into account its size and by coating it with a variety of substances, from antibodies to silica, the quantum dot can by assigned to a specific immunoassay. For example, immunoassays which employ fluorescence spectroscopy as analytical technique had used distinct quantum dots as labels, such as quantum dots conjugated to antibodies, covered by biotin or bare quantum dots, while lanthanide-quantum dots were assigned to time-resolved fluorescence experiments and streptavidin-modified quantum dots were used in either in fluorescent or chemiluminescent experiments.

Such a variety of methods to adapt the inorganic nanoparticles to immunoassays leads to another classification, with respect to the exploited mechanism. In this sense, immunoassays are classified as non-competitive or sandwich assays, when involving an analytical path in which the antigen in the sample is bound to the antibody site and a second labeled antibody is bound to the antigen, resulting in a response that is directly proportional to the concentration of the analyte; and as competitive immunoassays, when involving the competition of the antigen in the sample and the labeled antigen to bind to specific antibodies, resulting in a response that is inversely related to concentration. In a brief comparison, direct assays are faster, once they employ only one antibody, eliminating the secondary antibody cross-reactivity. In contrast, indirect immunoassays are much more sensitive.

As a competitive immunoassay example, Ding et al. [43] coated a microtiter platewith ovalbumin haptenand observed the decrease of the fluorescenceat higheranalyte concentrations. This indirect immunoassay was used to determine sulfamethazine residuesin chicken-muscle tissues with alimit of detection of 1 ng/mL for the immunoassay. Competitive assays also contemplate those using fluorescent probes as internal standards, such as quantitative assays.

As an example of non-competitive detection strategies is the new indirect immunoassay proposed by Li et al., [44] in which quantum dot fluorescent labels were combined to enzymaticchemiluminescentlabels. This system was used to simultaneously detect three cancer markers in human serum and thelimits of detection were the same for all markers, in the ng/mL range. Seeking for a new strategy that to be applied to the construction of a biosensor for most distinct application, they coupled the quantum dot to other two chemiluminescent enzymes, creating a hybrid multiplexed detection system for lung cancer.

2.6.1. The Surface Enhanced Resonance-Raman Scattering applied to quantum dots biosensors

It is noteworthy that quantum dots can be applied to surface-enhanced resonance Raman scattering (SERS) measurements as a powerfultool for ultrasensitive analysis, since unlike fluorescence techniques, SERS-active groups do not self-quench. An example is Han et al. work, [45] where fluorescein was used as Raman probe for a microtiter plate covered by antigen (human IgG) samples of unknown concentration. In their experiments, a solution of fluorescein-conjugated antibody was added to the sample and fluorescein SERS spectrum was recorded with a limit of detection of 0.2 ng mL^{-1} in a several-days stable device.

2.7. Labeled and labeled-free optical biosensors

Among optical biosensors that can be constructed based on this spectroscopic technique, there are those classified as labeled-fluorescent biosensor, as the above mentioned examples, but there is also a evolving rinsing class of optical label-free biosensors [46]. The most crucial distinction between these techniques is that label-free biosensors can directly evaluate the properties of the system, instead of the fluorescent response of a labeled material to the effect of its local environment. In their review, Fan et al. [46] make a clear distinction between labeled and label-free optical biosensors, with respect to techniques of detection, sample preparation, sensibility and versatility. They mentioned that although there are some great differences between fluorescence-based and label-free detection techniques, both are widely used in optical sensors construction.These distinct characteristics of the label-free devices make of these optical biosensors the most versatile among all types of sensing technologies that are only able of label-free detection, as in the case of surface acoustic wave and quartz crystal microbalance technologies. It also discusses Raman, refractive index and absorbance as detection methods. These approaches enable the optical detection to be yet more versatile, enabling the construction of a series of other biosensors, only by specifying even more de detection technique. It is not unusual that an optical label-free biosensor mixes more optical structures to enhance its sensing performance. For example, among the refractive index optical biosensors, they can be:

• Surface plasmon resonance

The first work that employed the plasmon resonance phenomenon to sensing was developed by Liedberg et al. in 1983.[47] Since then, this remarkable technique has been widely employed in many biochemical and biotechnological fields and greatly developed. Several types of biorecognition elements are currently used, depending on the application. This technique is so versatile that have been employed in a wide range of processes, including food and environmental monitoring and clinical analyses.

In simple words, surface plasmon is a charge density wave over a metallic surface. In the case that a thin layer of a metal is deposited on glass, there must be distinct dielectric constants on both faces of the film: the one in contact with glass surface and the other in contact with air. Then, a charge density oscillation occurs at these interfaces, leading to the phenomenon occurrence. It is observed as a sharp minimum of the light reflectance when the incident angle is changed, leading to a very important sensitivity to refractive indexes variations. These systems can be excited by some methods, as the waveguide coupling, where a dielectric and a metal are positioned over the substrate, generating a waveguiding layer, which creates an interface between metal and the waveguide. Light, therefore, propagates in this waveguide through totalinternal reflection, giving rise to an evanescent field at the interface, which excites the surface plasmon wave. [48, 49] Although this is a very popular technique, there are other methods to excite the surface plasmon wave that can lead to better detection limits, such as prism coupling. Nevertheless, waveguide coupling consists of an alternative due to easily combine to other optical components.

Some advances of these techniques had been presented in the last few years. In their work, Sacarano et al.[50] presented the SPR imaging technique, or *"SPR microscopy"* as the "most attractive and powerful advancement of SPR-based optical detection", which presents the advantage of coupling the sensitivity of the SPR measurements with the spatial capabilities of imaging. In this approach, the entire biochip surface is visualizedin real time, which, as a perspective, might enable experiments based on the continuous monitoring of immobilized spot arrays, with controlled size and shape and with no need of labeling.

• Interferometry:

Based on the interferometry technique of improving analytical signals, some types of interferometer-based biosensors were developed, such as Mach-Zehnder, Young's multi-channel and Hartman, among the most commonly used. They are based on the concept that a guided wave suffers a phase change when its evanescent field interacts with the sample. This interaction produces an optical phase change that is quantitatively related to the sample. In these constructions, a sensitive biosensor must present a long interaction length between guided wave and sample. Although many are the interferometric components that can be employed in a biosensor construction, these are by far the most commonly found:

• Mach-Zehnder interferometer:

It is composed of beam splitter that divides a coherent, polarized single frequency of a laser beam into two branches. The first one is passed through a window that leads to the reference branch of the interferometer. The second one passes through the detection branch window, in which the evanescent field interacts with the sample. Both beams are kept apart by a thick coating layer. They recombine at the output, resulting in an interference pattern that is detected at the photodetector. This type of interferometer gives rise to excellent bulk refractive index detection capability, but until now, there had not been much development of devices based on this interferometer.

• Young's interferometer:

Similar to Mach-Zehnder interferometer, this is also based on the passage of a laser beam into a slit, reaching a splitter. The laser beam is divided into reference and sensing branches, but instead of recombining at the output, the optical output of the two branches combine to form interference fringes on a CCD detector. This improves the signal, giving information about spatial intensity distribution along the CCD.

• Hartman interferometer:

In this configuration, optical elements are placed over a planar waveguide, organized in strips. A laser beam enters the device by an input grating, reaching the optical elements composed of functionalized molecules and leaving the device by the output grating. [51] Integrated optics is positioned after the output, creating interference between pairs of functionalized strips.

• Backscattering interferometer:

The most common backscattering interferometers employ in their construction a simple optical arrangement composed of a coherent light source (ususaly a low-power He-Ne or red-diode laser), a microfluidic path, and, of course, a phototransducer. The backscattering interferometry technique presents some advantages when compared to the above mentioned techniques. Since it is based on microfluidic concepts, it shows comparable performance as former interferometers, but using a much smaller sensing area,which permits a wide range of configuration. In this approach, the coherent laser beam is focused on a small sensing area, the interaction of the laser beam with the fluid-filled microchannelleads toan interference pattern that is registered in the photodetector, which is sensible to the laser reflected intensity. When a biological sample is placed over the illuminated surface, a laser of distinct intensity is detected due to a phase change caused by the light reflection over this surface.

• Photonic Technologies

The photonic technology has been object of many research fields and of a very rapid improvement, compared to other technologies. In this sense, there are many methods to employ photonicprinciples, and a wide range of scientific and technological issues to apply it, resulting in several equipment proposals. The broad range of applications of photonic devices permits to glimpse the importance of this emerging field. It is possible to incorporate different types of lasers, dielectric waveguide structures and photodetectors in a variety of possible equipments, that enables the perspective of explore from ultraviolet to far infrared, extending the fundamental research approach and the application possibilities, that can explain why photonics application, in other potential technologies, has grown in such impressive way.

In the biosensor perspective, many materials and concepts of application have been developed and a wide range of the proposed devices employ optical fiber and waveguides, as well as photonic crystals. Here some characteristics of such devices are pointed.

• Optical fiber

The two basic concepts of optical fiber based biosensors are the *Fiber Bragg's grating* and the *long-term grating*. They differ from each other not in principle, but in construction. While the fiber Bragg's grating concept requires the etching of the fiber (or grating) surface, followed by the physical pattern of the surface, the long-term grating is a configuration based on periodic grating of 100 μm to 1 mm, which make them much larger than the common Fiber Bragg's gratings and confer them the advantage of an increased sensing to refractive index changes. Moreover, they are easier to build and can be customized by chemically removal of the coating. Either fiber Bragg's or long term grating designs can lead to very high refractive index sensitivity and low detection limits, which consist of the most desired characteristics of promising biosensors.

• Optical waveguide

This elegant technique has been applied for biosensing in the last decade with a considerable success. Due to that, many structures of construction had been proposed usually directed by the analyte of interest. In this concept, some popular structures are:

• *Resonant mirror*, in which a low refractive index spacer composed of a metal or a dielectric layer, is sandwiched between a high refractive index substrate, usually a prism from where light reaches the biosensor and is refracted, and a high refractive index waveguide layer. There, the incident light at the resonant angle is coupled into the high-index waveguide layer and has a strong reflection at the output side of the system. Then, it is conducted through the waveguide, creating the evanescent field that leaves the waveguide. This approach turns the resonant angle sensitive to any change of refractive index and the signal is produced.

• *Metal coat waveguide*, which differs from the former structure by employing a low refractive index waveguide layer, separated from the high refractive index substrate by a metal layer. In this configuration, light is guided through the low refractive index and the light intensity is increased by the metal spacer, leading to an increased light-sample interaction and, therefore, to an increased sensitivity.

• Optical ring resonator

In this configuration, light suffers a total internal reflection in the boundaries of a curved interface between a high and a low refractive media. This process leads light to propagate in the circulating waveguide form or in the whispering gallery modes, as illustrated by Fan and cited by some other works therein.[46]Devices based on this technique can be constructed in a much smaller scale than those based on the former techniques with similar sensing capability, which is their great advantage. They can be constructed in a number of configurations, in which the microfabricated ring shaped, disk shaped or microtoroid shaped resonators on a chip, the stand-alone dielectric microspheres and the so called capillary-based opto-fluidic ring resonators are the most common examples. The chip-based ring resonators present some advantages which include the capability of opto-electronic integration, but, apart from the microtoroid configuration, they usually present problems of low quality factors (Q-factor), which are designated as all intrinsic and extrinsic losses occurred in the optical resonant cavity system [52] and in this case, these problems are related to their surface roughness. These types of ring resonators are very well presented and discussed in Fan's review. [46]

• Photonic crystal

This class of biosensors is, in fact, an evolution of the optical fiber based ones. They are formed by photonic crystal microcavities, obtained by introducing a defect in periodically organized microstructured holes, usually of silica, by altering their dimensions. Some can be embedded with molecules, which are responsible for the occurrence of a change in the refractive index of the biosensor, leading to a detectable signal in the form of a spectral shift of the resonant wavelength of the photonic crystal cavity. Also, polymers can be used as a coating layer for the photonic crystal cavities, as showed by Chakravarty et al., [53] which doped the photonic crystal microcavities with a quantum dot and coated it with anion-selective polymer. With this procedure, they were able to construct a sensor with good properties, such as a very specific and accurate detection for changes of perchlorate anions and calcium cations at submicro concentrations in solution, while Lee et al. [54] presented a photonic crystal suitable for protein and single particle detection. In their experimental and theoretical work, they claimed their device achieved a sensing volume of $0.15 \ \mu m^3$, and that it presented a limit of detection as small as 1 fg. They also determined its performance for particles in the size range of a variety of viruses, using latex spheres as models.

In a recent work, Aroua et al.[55] have studied, also by experimental and theoretical means, a label-free biosensor in order to determine it characteristics, the field intensity and the resonant wavelength shift when the nanocavities of the photonic crystal are filled with blood plasma, water or dried air. With this protocol, they showed that the enhancement on sensitivity is related to the photonic crystal design parameters.

2.8. Carbon nanotubes and graphene-based biosensors

Some new materials had also found a great deal of applications, especially in biosensing, such as carbon nanotubes and lately, graphene. As for single-walled carbon nanotubes, (SWNTs), they are known to exhibit unique intrinsic properties, which include a semiconductive behavior and photophysical propertiesdependent of their structure. For example, nanotubes with some chirality, band gap fluorescence is observed, as well as strong resonance Raman scattering.In this way, hybrid materials of SWCNTs and biomolecules is a way to obtain good materials for biosensing applications, since the fluorescence band-gap of SWNTs is highly sensitive to its environment and show shifts when the nanotube is in contact with other molecules.

In their work, Jin et al. [56] proposed the construction of a platform for selectively determine the hydrogen peroxide efflux from living cells, in order to biosensing human carcinoma, in an array of fluorescent single-walled carbon nanotubes. In this biosensor, the carbon nanotubes have their fluorescence quenched when H_2O_2 is liberated by A431 human epidermal carcinoma cells, in response to the epidermal growth factor. They show that this array is able to distinguish between peroxides originated on the cell membrane from other contributions.

Also to show the versatility of carbon nanotubes, Chen et al. [57]presented a sensitive method for multiplexed protein detection by using functionalized single-walled carbon nanotubes (SWNTs) as multicolor Raman labels. They claim that this method is a good alternative for standard fluorescence-based techniques since, unlike fluorescence, Raman detection benefits from the sharp scattering peaks of SWNTs with minimal background interference. Also, it can be combined to surface-enhanced Raman scattering substrates, allowing protein detection sensitivity down to 1×10^{-15}Mol L^{-1}, which is three orders of magnitude minor then the detection limit of fluorescence-based methods. They used these modified SWNT to Raman detection of human autoantibodies against proteinase 3, a biomarker for the Wegener's granulomatosisautoimmune disease, and by conjugating different antibodies to pure (12)C and (13)C SWNT isotopes, they had demonstrated the multicolor Raman protein detection.

In their work,Morales-Narvaéz and Merkoçi[58] took advantage of graphene´sinnovative mechanical, structural (several graphenes present lattice-like nanostructures), electrical, thermal and opticalproperties. They employed graphene oxide (GO) as a biosensing platform due to its ability of nanoassemble in wire form when in presence of biomolecules, its processability in solution and due to its heterogeneous chemical and electronic structure, which confers to GO the ability to be used as insulator, semiconductor or semi-metal. Also, they presented graphene oxide as a universal highly efficient long-range quencher, with the perspective of been applied to several novel biosensing strategies.

Phan and Viet also worked on testing graphene application on biosensors by replacing carbon nanotubes for graphene ribbons in biosensors,[59] which were able to sense the transition of DNA secondary structure from the native right-handed form to the alternate left-handed form.

Although studies ongraphene´s propertiesare still preliminary, it is thought as a promising platform for biosensing. In their review, Yang et al. [60] discuss all aspects of functionality, performance, properties, fabrication, handling and challenges of these carbon-based materials as part of biosensors. Ina critical analysis, they present the great opportunities yet to come with the use of these materials and point us what is necessary to have in mind when proposing a new architecture for biosensors and new materials to be employed. Nevertheless, graphene application on bioelectronics is still controversial.

3. Construction basics

The devices architectures that have been found in the literature and in the market are as numerous as the applications that they find. As well resumed by Reardon et al., [61] optical sensing opens a wide range of methodologies that can be based in either one of the electromagnetic wave parameters analysis, as their phase, polarization and amplitude. Analysis and detection methods applied to the amplitude parameters are the most common, due to the direct information that they can give about the system and due to the possibility of coupling this information to those obtained by polarized systems or phase optics.

There are no limits to explore regarding to optical properties or to instrumentation to be applied, alone or combined. Nevertheless, the most common optical approach is to promote the interaction of the analyte with light to produce light as response. Conventional methods permit the detection of absorption, reflection or scattering of light, through which it is possible to infer about the interaction between light and the analyte through signal changes.

Most elegant methods employ the detection (and promotion) of emitted light from the analyte, which in general relies on larger wavelengths that those used to illuminate the sample. The wavelength shift can be caused by either nonlinear interaction processes between light and analyte, as harmonic or inharmonic oscillations, such as vibrational modes that give rise to Raman shifts or by the electronic excitation of a system that result in the loss of a portion of the excitation energy as photons through luminescent phenomena that occur at shifted wavelength range. These luminescent processes include fluorescence and phosphorescence, which differ from each other with respect to the quantum levels involved in each electronic transition and that are affected by the interaction with the environment and, consequently, to analyte concentration.

There are diverse new methods of optical detection involving combined techniques to deliver real-time imaging of the sample. In this sense, fluorescence or Raman confocal microscopies are coupled to time-resolved fluorescence or to Raman spectrophotometers and images obtained can be evaluated in terms of concentration, energetic processes, charge transfers and localized events.Once detection methods and finality of the biosensor are defined, the size scale must be determined and, then, convenient materials for the construction of the biosensor must be selected and tested.

With respect to materials, it is important to consider environmental, economical and sustainable issues when choosing the ideal material combination for a biosensor. It is important to elect safe and ease of processing materials that can lead to a cheap method of fabrication and, in addition, can be recycled. Also, it needs to be chemically stable and permits a good selectivity and sensibility to the biosensor. Thus, the functional nanostructures can be composed of metals, semiconductors, magnetic materials, quantum dots, molecular or polymeric dyes and some hybrid materials, proposed to achieve any specific property that is an issue in other materials. [62]

Electrodes must be also adequately elected, with respect to work function as well as to processing. Usually, gold surface is preferred due to its possibility of modification with sulfur containing biological elements, such as SH-protein and antibodies, but also due to its optical and low potential properties, which enables refractive index measurements by several techniques, as mentioned in previous sections. Nevertheless, there are some alternative electrodes, such as metal-modified carbon electrodes, as those described by Wang et al. in which carbon electrodes were modified with rhodium [63], ruthenium [64] and platinum [65] to achieve lower potentials. More recently, some works presented Indium Tin Oxide thin layers as alternative electrodes for optoelectrical biosensors, with transmittance similar to the glass substrate. In their work, Choi et al. [66] showed that a 100 nm thick ITO layer as electrode permits simultaneous optoelectric measurements to record optical images and microimpedance to examine time-dependent cellular growth.

Transduction methods are responsible for the identity of the biosensor and, thus for address their applicability. Since the first step on the optical transduction in a biosensor is the chemical interaction between the analyte and the indicator phase (the recognition element) that will produce the optically detectable signal, it is critical because it will determine the biosensor stability, selectivity and sensitivity as well as the optical region in which the effect will be observed and the best detection means. Also, it is important to take into account whatever the biosensor will be used in continuous measurements or in simple detection. It will inform if strong interactions will be needed and in what conditions the biosensor can be employed. If chemical reactions are the signal origin, such it is in catalysis-based biosensors, an immobilized enzyme is preferred, once it can permits steady-state measurements to explore the sample. In this case, high selectivity can be achieved using antibodies as reagents and basing sensing on competitive binding. Also, methods to immobilize the recognition element include adsorption on solid substrates, covalent bonding to a substrate and confinement by membranes with selective permeability. [67]

In optical fiber-based biosensors, immobilized recognition elements are also needed, as well as the optical fiber to enable remote measurements. The immobilization technique and the reagent must be well selected to avoid undesired effects such attenuation on fiber's transmission efficiency, coupling of light into the fiber and attenuation characteristics of the fiber itself. Nevertheless, this is a good choice for transduction since it enables a range of photophysical processes to be detected, including light absorption, absorption followed by luminescence, light reflection and scattering, among others. Optical instrumentation is then selected, based on the most important photophysical effect. [68]

Finally, when proposing a biosensor, it is always good to make two considerations: first, nanoscaled materials are not really small to living bodies, they can be sensed as intrudersand be attacked and second, biosensors, in some way, will be employed in a living body. Issues related to toxicity, lifetime, stability, durability, mechanical properties, body adjustment, coherence and adaptability must be evaluated by a scientific and systematic method. It is not unusual to find information that was mistakenly collected, and conflicting results can be found on the same subject, leading to confusion. For example, carbon nanotubes are said to be health safe, nevertheless it is known that their structure is very similar to that of asbestos, great villain to miner's lungs. Actually, aspired nanosize fibers of asbestos are recognized by the lung cells as an aggressive agent, which can cause lung cancer, pleural disorders, pleural plaques, pleural thickening, and pleural effusions. Another aspect is that asbestos present some properties such as electricity insulate and is a chemically resistant material and it is quickly incorporated in many materials.

4. Applications of biosensors

Biosensors find various applications, limited only by the creator's imagination. Since they can be constructed on unlimited configurations, based on the most diverse detection and recognition methods, employing a whole world of polymeric materials, biological elements, luminescent organic and inorganic molecules, in film form or solution, their application can be as numerous as the elements combinations on their construction.Some of them can be numbered:

• Medicine development: for example, cellular biosensors are engineered to optically report specific biological activity. Since they employ living cells in their construction, they can be used to form a "data basis"of cell behaviour when in the presence of several actives, which can be used for design new drugs and medicines.

• Drug Abuse or addiction: detection (and quantification) at nanoscale are possible due to the recent biosensors development. This feature is essential in the determination of the limits at which a person can be susceptible to addiction reactions. In this way, biosensor can limit whether a drug can be used in therapeutics, avoiding the addiction. Furthermore, DNA biosensors act as potential detection devices for investigation of DNA–drug interactions, which can be used to elucidate the addiction mechanisms.

• Clinical diagnostics: this must be the most moving topic for the biosensors development. They are capable of detecting single-nucleotide polymorphisms, which are caused by gene mutations. Faster detection methods are developed every day, aiming the early detection of breast cancer, prostate cancer, AIDS, genetic diseases, bacterial and viral infections. Although clinically relevant point mutations are detected by piezoelectrical biosensors, optical biosensors also had a large contribution, especially the FRET-based biosensors, which are able to identify minor changes related to these mutations and permit the construction of nanodevices, a feature that has been attractive due to the possibilities offered by the nanoscaled biosensors perspective.

• Genome analysis: biosensors are able to inform about the interaction nature between proteins, peptides and DNA sequences present in the gene and identify it.They can also be used in genetic-engineered products proposaland in newdrugs development.

• Food Quality control: Toxins, microbes, bacteria, fungal contamination are all detectable by optical biosensors, as presented in former sections. In this way, they provide a mechanism of controlling the quality of food and industrial processes involved, contributing with the industrial development of preventing processes. Biosensors are also useful in monitoring fermentation processes, being applied on the food and beverage fabrication processes.

• Industrial microbiology monitoring: these analyses are indispensible in cosmetic and pharmaceutical industries, among others. By using biosensors, they can be accomplished in real time, avoiding product contamination and possible destruction.

- Good fabrication practices: nanobiosensors in the strategic steps of a product fabrication are able to detect minimum defects that could lead to the final product mischaracterization.

- Environmental safety: within this focus, biosensors can be useful for determining the type and concentration of contaminants present in an environment in which non specific studies or determinations had been previously made. Also, they find application on monitoring aspects changes of the considered environment and, in this case, the objective is to track the concentration of known contaminants over time at one or more defined locations. Employing biosensors to do this task, confers them one more important advantage, which consists of using their ability to provide continuous measurements of the environment, enabling for instance, the full characterization of contaminated sites. The environmental monitoring can be extended to systems such soil, water sources, such as ground water wells, rivers, lakes, and can be extrapolated to industrial factories on monitoring water treatment stations. [61]

- Agriculture and cattle monitoring: biosensors can be used to monitor the cattle, in a way to protect it from mad cow disease or foot and mouth disease (FMD) guarantying the quality of the meat, as well as fungal contamination of vegetables, enabling prevention and the health growth of the plants. Also, it enables the correct application of pesticides and fertilizers, at controlled amounts per field size, by recognizing the effect of quantities of these compounds in different types of soil, places with distinct sun, rain or wind exposure, natural factors that cause a variation in the concentration limits or bring about new compounds to the site. In this approach, in situ biosensors would provide feedback to farmers on how to maintain their culture healthy.

- Military defense: hazards biomolecules, such as salmonella, anthrax, the H1N1 influenza virus, in the military defense point of view can represent a threat, since they can be considered powerful weapons in terrorists hands. The Salmonella bacteria fiber optic biosensor, developed by Zhou et al. [69] in the late nineties, for example, is able to detect this pathogen in food and water, with a sensitivity as low as 10^4 cell colony forming unit/ml.

- Microfluidic sensing and implantable biosensors: microfluidic technique is based on the fact that all biological molecular reaction takes place in liquid environments. The microfluidic channels work as a guide for fluids and biofluids carrying the target or the sample to be tested by the biosensor. The microfluidic technique provides a controlled flow of quantities of fluids, in microscale, containing desired compounds. As they intend to provide information on biological living systems, they usually employ as fluidic carriers buffers, aqueous suspensions of analytes (proteins or antibodies), bacterial cell suspensions, and even whole blood samples. Microfluidic devices and implantable biosensors fabrication are intimately related by the fact that both need to be fabricated using biocompatible materials. Many are the options to achieve the biocompatibility suggested and usually choices are between mimetic or inert materials. The most common inert materials employed are polydimethysyloxane, thin laminated plastic components made of polyimide or poly methylmethacrylate and conductive electroactive polymers combined with hydrogels. These are interesting due to the fact that they give rise to a class of multifunctional, bioactive polymers which are proposed aiming several biotechnological applications. As showed by Guiseppi-Elie, [70] when polymers such as polypyrrole and polyaniline are combined to hydrogels, they generate bioactive polymers that act as bioreceptor hosting membranes of enzyme-based implantable biosensors.

• Aging research: optical biosensors based in fluorescence and bioluminescence emissions are proposed as devices for monitoring certain hormones dosage in the living body, as well as the release and the action of drugs administrated via dietary paths in these hormone levels. [71] The proposal is that an implantable biosensor could, in periods of days or even in the lifespan of an individual, inform about hormone fluctuations and valuable insights into the mode of action of the intervention could result from this monitoring. It is well known that in dietary restriction and other interventions, hormones such as insulin show profound changes. To be able to continuously monitor blood levels of these hormones, drugs or other factors by less invasive methods, could allow, for example, the adjustment of the hormone levels, retarding aging processes and preventing aging diseases.

5. Insights into the future

The first insight that is important to consider must be whether there are or not advantages in developing such a variety of biosensors prior to conventional detection systems. The most important advantage that one can think of refers to quick and in vivo early detection of severe conditions, which will lead to mature diagnostics at the earlier stages of diseases that today are challenging.With that and searching for utilizing all information on materials and techniques earned from scientific issues of other natures, such as utilizing the supramolecular chemistry to design new materials and coupled methods of detection that confer richer and reliable information, biosensor development contributed to the development of a number of research fields with some common interests. For example, more accurate information on microfluidic electronics to build sensors; new organic, hybrid and inorganic materials with interesting physical, chemical and morphological properties needed to be proposed; better methods to prepare homogeneous and thin films; new supports and electrodes are constantly demanded; new techniques for characterization and properties evaluation, which ask for improvements and new approaches to exploit all features of biosensors; new methods of modification and immobilization of biological elements, to cite some of these contributions.

Now, even greater perspectives are lying in the biosensor's development into the nanotechnology approach. It had been demonstrated that the use of nanobiosensors can provide several advantages not yet considered for conventional systems. They can be built under the nanoscale limit, they need to use similar materials to those used in the large scale devices, but in much smaller sizes. We know for sure that nanoscaled materials provide quite distinct properties, which could generate distinct biosensors, maybe even more sensitive and selective biosensors. For example, metallic nanoparticles have their bandgap energies widen at nanoscale, which turns to be a very important optical effect, since it generates a quantum confinement expressed by the observed spectral blue-shifts; magnetic materials show super magnetic behavior at nanoscale [72, 73, 74] other semiconducting nanoparticles present tunneling and Coulomb blockade effects [62]and all these impressive properties are not observed for the same thick materials. Some properties are clearly added to the biosensors processed at nanoscale, along with some thought advantages. The first ones that come directly to our minds are the possibility of using a reduced amount of materials in the device construction; reduced power is needed to make biosensors to work and, with it, they can be thought as portable, once they do not require a power source ratter then a small battery; they can be more stable with time; easier to recycle or dispose and, of course, since in nanoscale, we are leading with some new properties, it is expected that the nanobiosensors produced might posses new properties and so, new capabilities. With respect to detection and operation of these devices, detection processes thought to be simpler and faster, in constructions that tend to be more friendly, since it do not need bulky detection systems, conferring also the direct advantage of low cost of construction.

Nanobiosensors are thought to be as simple as possible, so, in the most recent proposed devices, the more acceptable approach is the label-free nanobiosensor. In this way, in a biosensor, the recognition element does not need to interact with other molecules or labels. There is no need to target the analyte or activate the biosensor by processes as conjugation with an enzyme, resonant energy transfer to generate the desired fluorescence, or chemical processes generated by the incorporation of molecules in order to functionalize the recognition element to produce a luminescent response, and others. In these biosensors, the probe and target-binding or substrate reactions are expected to be recorded by the transducer in the absence of any labelrequirement. Nevertheless, applications of such nanodimensioned devices are of a great variety, it is evident that the major target is the diagnostic and medical ones. Since these biosensors consist of a nanoscale detection method, they can sense target molecules in very low concentration into the body, which consist of a key factor in early detection of diseases such as breast cancer and AIDS. In fact, this is the idea behind the home method to detect AIDS, the OraQuick® HIV test presented by Orasure Technologies Inc. Their portable sensor comprises a visually read, qualitative flow immunoassay for the detection of antibodies to HIV-1 and HIV-2. In the device, HIV-1 and HIV-2 antigens are immobilized on a nitrocellulose strip and reactive antibodies are visualized by colloidal gold labeled with protein-A. The oral fluid are collected directly on the device and the positive test appears within 20 minutes as a purple line at the visor.[75]

Nevertheless, a point that needs to be made is, clearly miniaturization is the focus of many research that have been conducted in this theme, but it also can bring about some undesired properties, which need to be taken into account when engineering a biosensor. It means that not everything in nanoscale is adequate. One must have in mind which problem needs an answer, otherwise there is no scientific method, only the old fashioned "accidental discovery".

Author details

Tatiana Duque Martins, Antonio Carlos Chaves Ribeiro, Henrique Santiago de Camargo, Paulo Alves da Costa Filho, Hannah Paula Mesquita Cavalcante and Diogo Lopes Dias

Chemistry Institute, Campus II, Federal University of Goias, Goiania, Brazil

References

[1] Clark LCJr, Lions C. Electrode Systems For Continuous Monitoring In Cardiovascu-
    lar Surgery. Annals of the New York Academy of Sciences 1968; 102 29-45.

[2] Shi J, Cha TG, Claussen JC, Diggs AR, Choi JH, Porterfield DM. Microbiosensors
    Based on DNA Modified Single-Walled Carbon Nanotube and Pt BlackNanocompo-
    sites. Analyst 2010; 136 4916-4924.

[3] Leyden MR, Messinger RJ, SchumanC, Sharf T, Remcho VT, Squires TM, Minot
    ED.Increasing the Detection Speed of an All-Electronic Real-Time Biosensor. Lab
    Chip 2012; 12 954-959.

[4] Chen X, Xu S, Yao N, Shi Y. 1.6 V Nanogenerator for Mechanical Energy Harvesting
    Using PZT Nanofibers. Nano Letters 2010; 10(6) 2133-2137.

[5] Park K, Xu S, Liu Y, Hwang GT, Kang SJL, Wang ZL, Lee KJ. Piezoelectric BaTiO3
    Thin Film Nanogenerator on Plastic Substrates. Nano Letters, 2010; 10 4939–4943.

[6] Urgun-Demirtas M, Stark B, Pagilla K. Use Of Genetically Engineered Microorgan-
    isms (Gems) For The Bioremediation Of Contaminants. Critical Reviews in Biotech-
    nolog 2006; 26(3) 145– 164.

[7] D'Souza SF. Microbial Biosensors. Biosensors and Bioelectronics 2001; 16, 337–353.

[8] Ahuja T, Mir IA, Kumar D, Rajesh K. Biomolecular Immobilization On Conducting
    Polymers For Biosensing Applications. Biomaterials 2007; 28(5) 791–805.

[9] Malhotra BD, Chaubey A, Singh SP. Prospects of Conducting Polymers In Biosen-
    sors.AnalyticaChimicaActa 2006; 578(1) 59–74.

[10] Shulga AA, Soldatkin AP, Elskaya AV, Dzyadevich SV, Patskovsky SV, Strikha VI.
    Thin-Film Conductometric Biosensors for Glucose And Urea Determination. Biosen-
    sors and Bioelectronics 1994; 9 217–223.

[11] Opitz N, Lubbers DW.New Fast-Responding Optical Method to Measure PCO2 In
    Gases and Solutions, PflugersArchiv-European Journal of Physiology 1975;355(S)
    R120.

[12] Lubbers DW, Opitz N. PO2-Optode, A New Tool To Measure PO2 Of Biological Gases And Fluids By Quantitative Fluorescence Photometry.PflugersArchiv-European Journal of Physiology 1975; 359(S) R145.

[13] de Silva, A.P.; Gunaratne, H.Q.; Gunnlaugsson, T.; Huxley, A.J.; McCoy, C.P.; Rademacher, J.T.; Rice, T.E. Signaling Recognition Events with Fluorescent Sensors and Switches. Chem. Rev. 1997, 97, 1515-1566.

[14] Johnson I. Fluorescent Probes for Living Cells. Histochemistry Journal 1998; 30 123-140.

[15] Terai T, Nagano T. Fluorescent Probes for Bioimaging Applications. Current Opinion in Chemical Biology 2008; 12 515-521.

[16] Chen JH, Fang ZY, Liu J, Zeng LW. A Simple And Rapid Biosensor For Ochratoxin A Based On A Structure-Switching Signaling Aptamer, Food Control 2012; 25(2) 555-560.

[17] Zargoosh K,Chaichi MJ,Shamsipur M,Hossienkhani S,Asghari S,Qandalee M. Highly Sensitive Glucose Biosensor Based On The Effective Immobilization Of Glucose Oxidase/Carbon-Nanotube And Gold Nanoparticle In Nafion Film And Peroxyoxalate-Chemiluminescence Reaction of a New Fluorophore. Talanta 2012; 9337-43.

[18] Hu P, Zhu CZ, Jin LH, Dong SJ. An Ultrasensitive Fluorescent Aptasensor for Adenosine Detection Based on Exonuclease III Assisted Signal Amplification. Biosensors & Bioelectronics 2012; 34(1) 83-87.

[19] Reed B, Blazeck J, Alper H. Evolution of an Alkane-Inducible Biosensor for Increased Responsiveness to Short-Chain Alkanes. Journal of Biotechnology 2012; 158(3) 75-79.

[20] Pavan S,Berti F. Short Peptides As Biosensor Transducers, Analytical And Bioanalytical Chemistry 2012; 402(10) 3055-3070.

[21] Gonzalez-Andrade M, Benito-Pena E, Mata R, Moreno-Bondi MC. Biosensor For On-Line Fluorescent Detection Of Trifluoroperazine Based On Genetically Modified Calmodulin, Analytical And Bioanalytical Chemistry 2012; 402(10) 3211-3218.

[22] Hinde E,Digman MA, Welch C, Hahn KM,Gratton E. Biosensor Forster Resonance Energy Transfer Detection By The Phasor Approach To Fluorescence Lifetime Imaging Microscopy, Microscopy Research and Technique, 2012; 75(3)271-281

[23] Lakowicz JR. Principles of Fluorescence Spectroscopy; Second Edition Kluwer Academic/Plenum Publishers: NY. 1999.

[24] Giepmans BNG, Adams SR, Ellisman MH, Tsien RY. The Fluorescent Toolbox for Assessing Protein Location and Function. Science 2006; 312 217-224.

[25] Martins TD, de Souza MI, Cunha BB, Takahashi PM, Ferreira FF, Souza JA, Fileti EE, Alves WA. Influence of pH and Pyrenyl on the Structural and Morphological Control of Peptide Nanotubes. Journal of Physical Chemistry C 2011; 115 7906–7913.

[26] Parak WJ, Pellegrino T, Plank C. Labelling of Cells with Quantum Dots. Nanotech 2005; 16, R9-R25.

[27] Michalet X, Pinaud FF, Bentolila LA, Tsay JM, Doose S, Li JJ, Sundaresan G, Wu AM, Gambhir SS, Weiss S. Quantum Dots for Live Cells, In Vivo Imaging, and diagnostics. Science 2005; 307 538-544.

[28] Alivisatos AP, Gu W, Larabell CA. Quantum Dots as Cellular Probes. Annual Reviews on Biomedicine and Engineering 2005; 7 55-76.

[29] Cheung HC, Resonance Energy Transfer InLakowicz J.R. (ed): Topics In Fluorescence Spectroscopy: vol 2: Principles, Kluwer Academic Publishers, NY 2002.p128-176.

[30] Liu B, Bazan CG. Homogeneous Fluorescence-Based DNA Detection with Water-Soluble Conjugated Polymers. Chemistry of Materials 2004; 16(23) 4467–4476.

[31] Yildirim N, Long F, Gao C, He M, Shi HC, Gu AZ. Aptamer-Based Optical Biosensor For Rapid And Sensitive Detection Of 17β-Estradiol In Water Samples. Environmental Science and Technology 2012;46(6) 3288-3294.

[32] Lin HY, Tsai WH, Tsao YC, Sheu BC. Side-Polished Multimode Fiber Biosensor Based On Surface Plasmon Resonance With Halogen Light, Applied Optics 2007; 46 800-806.

[33] Jorgenson RC, Yee SS. A Fiber-Optic Chemical Sensor Based on Surface Plasmon Resonance, Sensors AndActuators B: Chemical 1993; 12(3) 213-220.

[34] Suter JD, White IM, Zhu H, Fan X. Thermal Characterization Of Liquid Core Optical Ring Resonator Sensors, Applied Optics 2007; 46 389-396.

[35] Goldman ER, Medintz IL, Mattoussi H. Luminescent Quantum Dots in Immunoassays. Analytical and Bioanalytical Chemistry 2006; 384 560-563.

[36] Bustos ARM, Trapiella-Alfonso L, Encinar JR, Costa-Fernandez JM, Pereiro R, Sanz-Medel A. Elemental and Molecular Detection For Quantum Dots-Based Immunoassays: A Critical Appraisal. BIosensors and Bioelectronics 2012; 33(1) 165-171.

[37] Pinwattana K, Wang J, Lin CT, Wu H, Du D, Lin Y, Chailapakul O. CdSe/ZnS Quantum Dots Based Electrochemical Immunoassay for the Detection of Phosphorylated Bovine Serum Albumin, Biosensors & Bioelectronics 2012; 26 1109-1113.

[38] Zhang P. Investigation of Novel Quantum Dots/Proteins/Cellulose Bioconjugate using NSOM and Fluorescence. Journao o. Fluorescence 2006; 16 349-353.

[39] Sandros MG, Shete V, Benson DE. Selective, Reversible, Reagentless Maltose Biosensing with Core-Shell Semiconducting Nanoparticles. The Analyst 2006; 131 229-235.

[40] Tsay JM, Doose S, Weiss S. Rotational and Translational Diffusion of Peptide-Coated CdSe/CdS/ZnSNanorods Studied by Fluorescence Correlation Spectroscopy. Journal of American Chemical Society 2006; 128 1639-1647.

[41] Tsay, JM. Doose S, Pinaud F, Weiss S. Enhancing the Photoluminescence of Peptide-Coated Nanocrystals with Shell Composition and UV Irradiation. Journal of Physical Chemistry B 2005; 109 1669-1674.

[42] Chafer-Pericas C, Maquieira A, Puchades R. Functionalized Inorganic Nanoparticles Used As Labels In Solid-Phase Immunoassays, Trends in Analytical Chemistry 2012; 31 144-156.

[43] Ding S, Chen J, Jiang H, He J, Shi W, Zhao W, Shen J. Application Of Quantum Dot-Antibody Conjugates For Detection Of Sulfamethazine Residue In Chicken Muscle Tissue, Journal Of Agricultural And Food Chemistry 2006; 54 6139-6142.

[44] Li H, Cao Z, Zhang Y, Lau C, Lu J. Combination Of Quantum Dot Fluorescence With Enzyme Chemiluminescence For Multiplexed Detection Of Lung Cancer Biomarkers, Analytical Methods 2010; 2 1236-1242.

[45] Han XX, Cai LJ, Guo J, Wang CX, Ruan WD, Han WY, Xu WQ, Zhao B, Ozaki Y. Fluorescein Isothiocyanate Linked Immunoabsorbent Assay Based On Surface-Enhanced Resonance Raman Scattering, Analytical Chemistry, 2008; 80 3020-3024.

[46] Fan X, White IM, Shopova SI, Zhu H, Suter JD, Sun Y. Sensitive Optical Biosensors For Unlabeled Targets: A Review, AnalyticaChimicaActa2008; 620 8-26.

[47] Liedberg B, Nylander C, Lunstrom I. Surface Plasmon Resonance For Gas Detection And Biosensing* Sensors & Actuators1983; 4299- 304.

[48] LiedbergB,LundstromI, Stenberg E. Principles of Biosensing with an Extended Coupling Matrixand Surface Plasmon resonance. Sensors and Actuators 1993; 11 63-72.

[49] Petryayeva E, Krull UJ. Localized Surface Plasmon Resonance: Nanostructures, Bioassays AndBiosensing — A review. AnalyticaChimicaActa 2011; 706 8– 24.

[50] Scarano S, Scuffi C, Mascini M, Minunni M. Surface Plasmon Resonance Imaging for Affinity-Based Biosensors, in P. Malcovati (eds), Sensors and Microsystems: AISEM proceedings 2009, Lecture Notes in Electrical Engineering. 2010; 54(part 5). p425-428.

[51] Schneider BH, Edwards JG, Hartman NF. Hartman Interferometer: Versatile Integrated Optic Sensor For Label-Free, Real-Time Quantification Of Nucleic Acids, Proteins, And Pathogens. Clinical Chemistry 1997; 43(9) 1757-1763.

[52] Armani AM. Single Molecule Detection Using Optical Microcavities In:, Chapter 11 (), In: Chremmos I (eds.), Photonic Microresonator Research and Applications Springer Series in Optical Sciences: Springer Science+Business Media, LLC. 2010; 156.p253-273.

[53] Chakravarty S, Topol'ancik J, Bhattacharya P, Chakrabarti S, Kang Y, Meyerhoff ME. Ion Detection with Photonic Crystal Microcavities. Optical Letters 2005; 30 2578-2580.

[54] Lee MR, Fauchet PM. NanoscaleMicrocavity Sensor For Single Particle Detection Optical Letters 2007; 32 3284-3286.

[55] Aroua W, Haxha S, AbdelMalek F. Nano-Optic Label-Free Biosensors Based On Photonic Crystal Platform With Negative Refraction Optics Communications 2012; 285 1970–1975.

[56] Jin H, Heller D A, Kalbacova M, Kim JH, Zhang JQ, Boghossian AA, Maheshri, StranetMS. Detection of single-molecule H_2O_2 Signalling From Epidermal Growth Factor Receptor Using Fluorescent Single-Walled Carbon Nanotubes. Nature Nanotechnology 2010; 5 302–309.

[57] Chen Z, Tabakman SM, Goodwin AP, Protein microarrays with carbon nanotubes as multicolor Raman labels. Nature Biotechnology 2008; 26 1285–1292.

[58] Morales-Narváez E, Merkoçi A.Graphene Oxide: Graphene Oxide as an Optical Biosensing Platform. Advanced Materials, 2012; 24(25) 3298–3308.

[59] Phan AD, Viet NA. A new type of optical biosensor from DNA wrapped semiconductor graphene ribbons. Journal of Applied Physics 2012; 11, 114703 (5 pages)

[60] Yang W, Ratinac KR, Ringer SP, Thordarson P, Gooding JJ, Braet F. Carbon nanomaterials in biosensors: should you use nanotubes or graphene?, AngewandteChemie International Edition in English, 2010; 49(12) 2114-38

[61] Reardon KF, Zhong Z, LearKL. Environmental Applications of Photoluminescence-Based Biosensors. Advances in Biochemistry Engineering/Biotechnology 2009; 116 99–123.

[62] Janagama DG, Tummala RR. Nanobiosensing Electronics and Nanochemistry for Biosensor Packaging C.P. In: Wong (eds.) Nano-Bio- Electronic, Photonic and MEMS Packaging, C Springer Science+Business Media 2010. p613-663 DOI 10.1007/978-1-4419-0040-1_17

[63] Wang J, Chen Q, Pedreno M. Highly Selective BiosensingOf Lactate At Lactate Oxidase Containing Rhodium Dispersed Carbon Paste Electrodes.AnalyticaChimicaActa 1995; 304 41-46.

[64] Wang J, Fang L, Lopez D, Tobias H. Highly selective and sensitive amperometric biosensor of glucose at ruthenium-dispersed carbon paste enzyme electrodes. Analytical Letters 1993; 26(9) 1819-1830.

[65] Wang J, Naser N, Angnes L, Wu H, Chen L. Metal-dispersed carbon paste electrodes. Analytical Chemistry 1992; 64 1285-1288.

[66] Choi CK, Kihm KD, English AE. Optoelectric biosensor using indium-tin-oxide electrodes, Optics Letters 2007; 32 (11) 1405-1407.

[67] Seitz WR. Transducer mechanisms for optical biosensors. Part 1: The chemistry of transduction, Comput Methods Programs Biomedicine 1989; 30(1) 9-19.

[68] Ashworth DC, Narayanaswamy R. Transducer mechanisms for optical biosensors. Part 2: Transducer design, Computer Methods Programs in Biomedicine 1989; 30(1) 21-31.

[69] Zhou C, Pivarnik P, Rand A G, Letcher SV. Acoustic standing-wave enhancement of a fiber-optic Salmonella biosensor. Biosensors and Bioelectronics 1998; 13(5) 495–500.

[70] Guiseppi-Elie A. An Implantable Biochip to Influence Patient Outcomes Following Trauma-induced Hemorrhage, Journal Analytical and Bioanalytical Chemistry 2011, 399 403-419

[71] He Y, Wu Y, Mishra A, Acha V, Andrews T, HornsbyPJ. Biosensor technology in ag-ing research and age-related diseases, Ageing Research Reviews 2012;11 1– 9.

[72] Woods S. I., Kirtley J. R., Sun S., Koch S. H., Direct investigation of superparamagnet-ism in Co nanoparticle films. Physical Reviews Letters 2001; 87(13) 137205(4 pages).

[73] Min C, Shao H, Liong M, Yoon TJ, Weissleder R, Lee H. Mechanism of Magnetic Re-laxation Switching Sensing, ACS Nano, 2012, article ASAP DOI: 10.1021/nn301615b.

[74] Bogart L K, Taylor A, Cesbron Y, Murray P, Lévy R.Photothermal Microscopy of the Core of Dextran-Coated Iron Oxide Nanoparticles During Cell Uptake, ACS Nano, 2012, Article ASAP, DOI: 10.1021/nn300868z.

[75] Lee SR, Guillon G, Ferko G, DeTurk C, Hershberger M, Marshall T, Kardos K. Per-formance of a Rapid Point of Care Test for HIV Antibodies, OraSure Technologies, Inc.   http://www.orasure.com/docs/pdfs/products/infectious/Performance-Rapid-Point-Care-Test-HIV-Antibodies.pdf (accessed 10 July 2012)

Chapter 7

Potentiometric, Amperometric, and Impedimetric CMOS Biosensor Array

Kazuo Nakazato

Additional information is available at the end of the chapter

1. Introduction

In view of the growing concerns about such issues as food security, health care, evidence-based care, infectious disease, and tailor-made medicine, a portable gene-based point-of-care testing (POCT) system is needed. For a system that anyone can operate anywhere and obtain immediate results, a new biosensor chip must be developed. Electrical detection using complementary metal-oxide semiconductor (CMOS) integrated circuits has great potential since it eliminates the labeling process, achieves high accuracy and real-time detection, and offers the important advantages of low-cost, compact equipment.

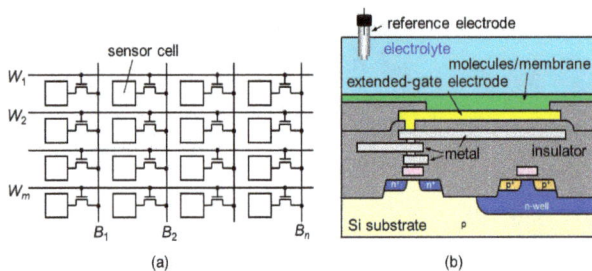

Figure 1. Integrated sensor array. (a) Matrix array arrangement where W and B are word and bit lines, respectively. (b) Schematic cross section of a sensor cell where n+ and p+ are heavily doped n-type and p-type semiconductors, respectively, and n-well is the n-type semiconductor region.

Our target is a monolithically integrated sensor array, as shown in Figure 1(a), which detects all possible biomolecular interactions simultaneously. In each sensor cell, different kinds of probes can be formed for parallel detection. In addition, the same kind of probe can be used to observe the time evolution of the spatial distribution of biomolecular interactions as well as to improve the detection accuracy since biomolecular interactions are a stochastic process. In this paper, several biosensor arrays are described based on the detection of electric potential, current, capacitance, and impedance.

2. Potentiometric sensor array

The detection of electric potential change based on a field-effect transistor (FET) [1] has shown excellent sensitivity such as for ion concentration and specific DNA sequences including single-nucleotide polymorphisms (SNPs). There are two detection principles.

One principle is the detection of electronic charge around an electrode and there is no electron transfer to the electrode. The gate potential is determined by Poisson's equation. First, a probe layer is formed on an FET. Then, target molecules are supplied. Specific molecules are selectively taken into the probe layer on the FET channel, which detects the molecular charge in the probe layer. In the case of DNA detection, the probe is single-stranded (ss) DNA with a known sequence, immobilized on the substrate. When the target ssDNA is supplied, specific hybridization occurs if the target DNA is complementary to the probe DNA. Occurrence or nonoccurrence of specific hybridization can be detected by the difference in charge since a nucleotide has a negative charge on the phosphate group.

The other principle is the detection of chemical equilibrium potential, i.e., redox potential, accomplished by electron exchange between the electrolyte/molecule and the electrode. Ferrocenyl-alkanethiol immobilized gold electrode is used to detect an enzyme reaction through a redox reaction. In this case, the gate potential is determined by the Nernst equation.

2.1. CMOS Source-Drain Follower

For the integrated sensor array, the structure must be compatible with CMOS integrated circuits. Employment of extended-gate electrodes is one solution, as shown in Figure 1(b). Molecules and/or membrane are formed on the extended-gate electrodes. Our goal is the realization of a million-sensor array on a single chip. One sensor must occupy a small area, and consume low power. Since the detection signal is in the order of 1 mV, high accuracy is essential. To meet these targets, we proposed a new integrated sensor circuit, a CMOS source-drain follower, where both the gate-source and gate-drain voltages of the sensor transistor are maintained at constant values [2, 3]. The source-drain follower has the merit of not influencing the sensing system since the input impedance is infinite for both DC and AC signals.

The basic circuitry of the CMOS source-drain follower is shown in Figure 2(a). The sensor transistor N detects the extended-gate electrode voltage V_{IN}. This circuit works as a voltage follower ($V_{OUT} = V_{IN}$) with high input and low output impedances. A benefit of the voltage follower is that the output voltage is independent of device parameters such as threshold voltage and environmental conditions such as temperature. This circuit also works as a source-drain follower for sensor transistor N when current I is kept constant.

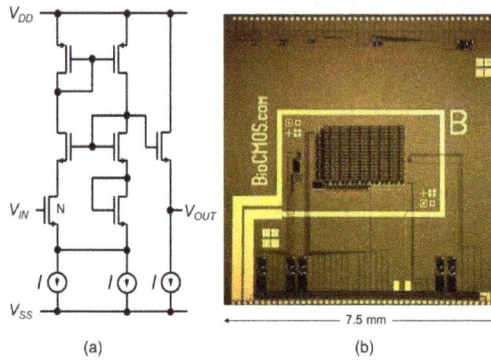

(a)   (b)

Figure 2. (a) Basic CMOS source-drain follower. V_{DD} and V_{SS} are high and low power supply voltages, respectively. (b) 16×16 integrated sensor array with CMOS source-drain followers and peripheral circuits. A heater and thermometer are also integrated on the chip.

2.2. pH Detection

Many different biosensors have been developed based on pH sensors since various biomolecular interactions produce protons. Rothberg et al. recently demonstrated a genome sequencing chip that contains 13 million pH sensors on a 17.5×17.5 mm² die [4].

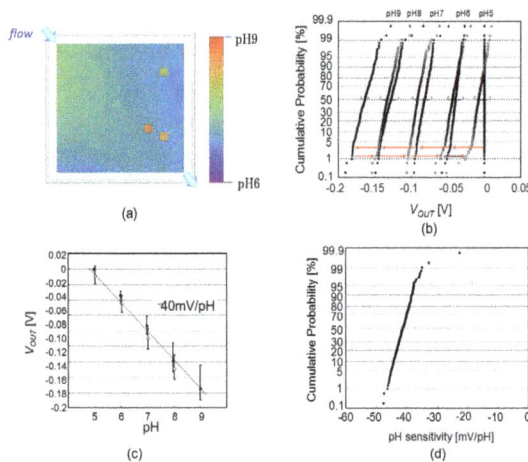

Figure 3. (a) Two-dimensional image of pH change. There are three faulty sensor units. (b) Cumulative probability of output voltage, and (c) median ±3σ plot as a function of pH from pH 5 to 9 (black) and pH 8 to 5 (white). The output voltage is the result after subtracting the initial values in order to eliminate the charge effect from the floating gate. (d) Cumulative probability of pH sensitivity of individual sensor cells.

The cumulative probability of pH sensitivity of 16×16 sensor cells with a 100-nm catalytic chemical vapor deposition (Cat-CVD) silicon nitride layer is plotted in Figure 3. Cat-CVD is a low-temperature (350°C) process and the deposited silicon nitride is of high quality, similar to that obtained by low-pressure CVD [5]. The median pH sensitivity is −41 mV/pH, which is lower than the theoretical value of −57 mV/pH. The reason for this lower value may be explained by the oxygen-rich layer on the Si_3N_4 surface [5].

2.3. DNA Detection

Using the integrated potentiometric sensor array, preliminary experiments on DNA detection were performed. Gold extended-gate electrodes were used to immobilize the probe DNA. Immobilization of a 5'-thiol-modified 20-mer oligonucleotide, and hybridization with the complementary oligonucleotide were detected in a 1 mM phosphate buffer (pH 7.0), as shown in Figure 4. Biomolecular interactions were observed as the time evolution of two-dimensional distribution. Maximum voltage change was 80 mV for immobilization and 40 mV for hybridization. In this experiment, the uniformity of biomolecular interactions was not good. Long-term drift of the sensed voltage was observed as 30 mV/h.

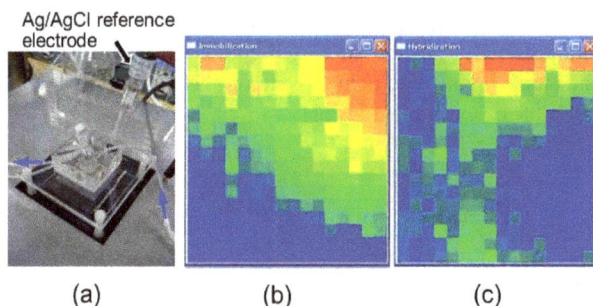

Figure 4. Preliminary experiment on DNA detection using a 16×16 potentiometric sensor array. (a) Experimental set-up. (b) Output voltage change before/after immobilization and (c) before/after hybridization.

The drift was reduced to 2 mV/h when Cat-CVD silicon nitride was deposited on the extended-gate electrode. DNA detection was also performed using the silane-coupling method for probe immobilization on Cat-CVD silicon nitride. The results show voltage changes of around 100 mV for probe immobilization, 12 mV for hybridization of complementary target DNA, and less than 1 mV for reverse-complementary target DNA.

2.4. Redox Potential Detection

The direct charge detection method using FET had a number of serious problems, as explained in Figure 5. First, the molecular charge is screened by ions in solution. Screening length is around 3 nm in the case of ion concentration of 10 mM. This can be extended if low ion concentration is used; however, in this case, a very high impedance environment is produced, and the electric potential becomes unstable. Second, the charge distribution is influenced by the shape of the molecule. It is generally understood that single-stranded DNA takes a Gaussian shape, and double-stranded DNA takes a rod-like shape. It is unclear whether it is a change in charge or change in structure that is detected. Especially in a flow system, the molecular shape fluctuates, which leads to unstable electric potential. Third, the electrode enters a floating state. Embedded charge causes a large threshold voltage variation.

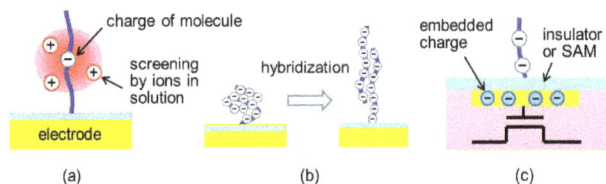

Figure 5. Problems with direct charge detection method. (a) Molecular charge is screened by ions in solution, (b) charge distribution is influenced by the shape of the molecules, and (c) electrode enters a floating state.

Instead of using the direct charge detection method, a redox potential detection method was developed using ferrocenyl-alkanethiol modified gold electrode [6, 7]. This redox potential sensor detects the ratio of oxidizer to reducer concentration, as shown in Figure 6, and is not affected by the absolute concentration and pH.

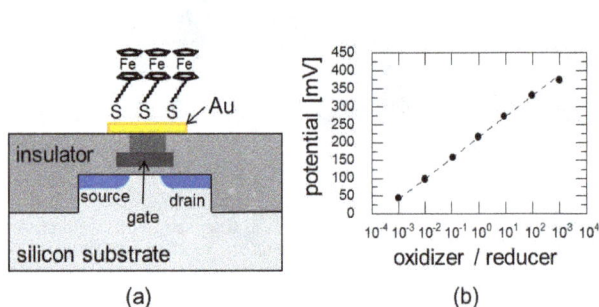

Figure 6. a) Schematic cross section of redox potential sensor. (b) Potential versus ratio of oxidizer (ferricyanide) to reducer (ferrocyanide) concentration.

We fabricated a chip that integrates 32×32 redox potential sensors, as shown in Figure 7 [8]. The sensor chip was dipped in 500 µM 11-ferrocenyl-1-undecanethiol (11-FUT) in ethanol for 24 h. Hexacyanoferrate mixture totaling 10 mM was used for the oxidizer and reducer. Six orders of concentration ratio of oxidizer and reducer were detected by this sensor array, as shown in Figure 6(b). The sensitivity was 57.9 mV/decade, which is very close to the theoretical value of 59 mV/decade at 25°C. Stability, i.e., long-term drift and fluctuation, of electric potential was examined using a bare electrode and an 11-FUT modified electrode, and 10 mM PBS solution (pH 7.4) and redox PBS solution in which the 10 mM hexacyanoferrate was additionally added to 10 mM PBS solution, as shown in Figure 7. By using redox PBS solution, the drift of electric potential was reduced by nearly one order. Furthermore, 11-FUT modification reduced the drift to nearly one fourth. This experiment showed that the drift can be drastically reduced by the redox potential detection method compared to the direct charge detection method. Of all 32×32 sensor cells, each potential of 92% was within ±1mV from the median. For the 8% abnormal output sensor cells, microscopic observation showed that an Au electrode had peeled off.

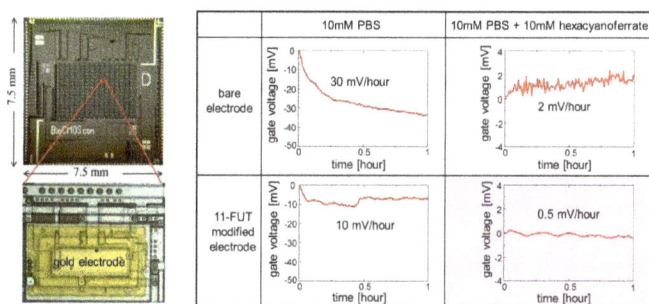

Figure 7. Redox potential sensor array (32×32) and stability of electric potential.

This redox potential sensor array successfully detected the glucose level with an accuracy of 2 mg/dL, using the following enzyme-catalyzed redox reaction:

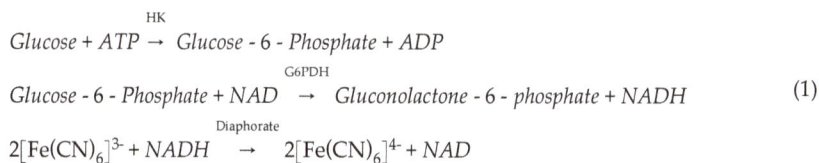

$$Glucose + ATP \xrightarrow{\text{HK}} Glucose\text{-}6\text{-}Phosphate + ADP$$

$$Glucose\text{-}6\text{-}Phosphate + NAD \xrightarrow{\text{G6PDH}} Gluconolactone\text{-}6\text{-}phosphate + NADH \qquad (1)$$

$$2[Fe(CN)_6]^{3-} + NADH \xrightarrow{\text{Diaphorate}} 2[Fe(CN)_6]^{4-} + NAD$$

where HK is hexokinase and G6PDH is glucose-6-phosphate dehydrogenase.

Continuous sample measurement was performed using a flow measurement system with a flow speed of 1 μl/s, as shown in Figure 8. We used two types of solutions: PBS solution (pH 7.4) and glucose sample solution (glucose, 9.9 mM potassium ferricyanide, 0.1 mM potassium ferrocyanide, 0.6 mM NAD, 2 mM ATP, 10 mM MgCl₂). PBS solution was used to wash out the glucose sample. As shown in Figure 9(a), the gate voltage settled in the glucose sample very rapidly. On the other hand, in the PBS solution, a long settling time was observed. Figure 9(b) shows the relationship between given and detected glucose concentrations, indicating fairly good linearity.

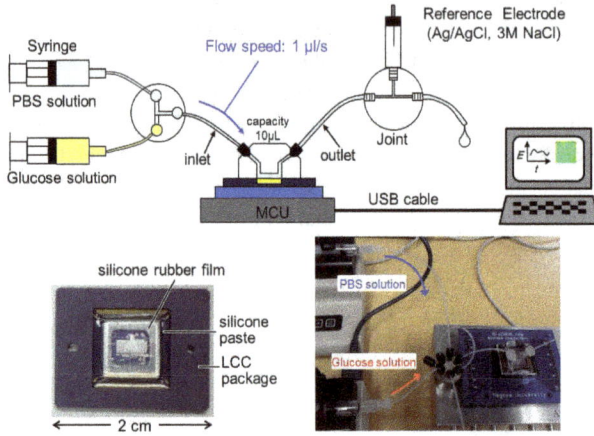

Figure 8. Setup of measurement. The chip is controlled by a microcontroller unit (MCU).

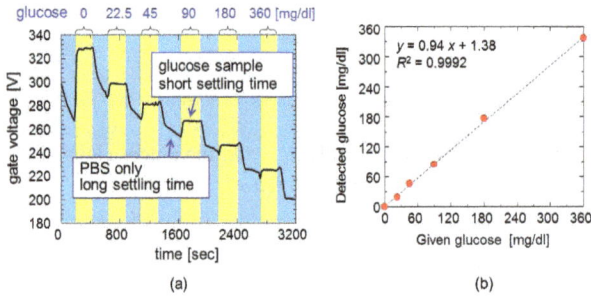

Figure 9. a) Flow measurement of glucose. Blue areas indicate the flow of PBS solution, and yellow areas indicate the flow of glucose sample solution. (b) Detected glucose vs. given glucose.

This sensor array could be applied to genome sequencing by incorporating a primer extension reaction, which produces pyrophosphate (PPi).

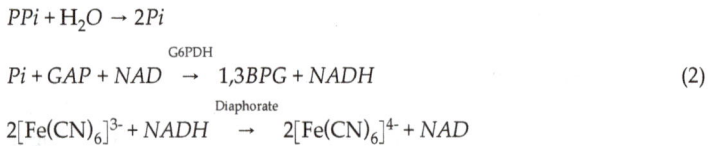

$$PPi + H_2O \rightarrow 2Pi$$

$$Pi + GAP + NAD \xrightarrow{\text{G6PDH}} 1,3BPG + NADH \tag{2}$$

$$2[Fe(CN)_6]^{3-} + NADH \xrightarrow{\text{Diaphorate}} 2[Fe(CN)_6]^{4-} + NAD$$

where GAP is glyceraldehyde 3-phosphate and BPG is bisphosphoglycerate.

3. Amperometric sensor array

Amperometric imaging offers great potential for multipoint rapid detection and the analysis of diffusion processes of target molecules. The microelectrode is one of the most versatile and powerful tools in amperometry. Although the current passing through a microelectrode is very small, it has the advantages of high mass transport density, small double-layer capacitance, and small ohmic drop. Moreover, a microelectrode has a steady-state current response in unstirred solutions. Such steady-state currents are easy to analyze and interpret. However, as it takes a few or tens of seconds before reaching a steady state, rapid multipoint measurement cannot be achieved with a simple switching scheme. Furthermore, when the inter-electrode distance is not sufficiently large, the diffusion layers begin to overlap and eventually merge to form a single planar diffusion layer. This overlapping of diffusion layers is commonly referred to as "cross talk" or the "shielding" effect. When cross talk occurs, the microelectrode array loses its unique features and becomes similar to a large-area "macro" electrode, which makes local and quantitative analysis extremely difficult. We proposed a switching circuit that measures multiple microelectrode currents at high speed, and a microelectrode structure to suppress diffusion layer expansion over the microelectrode array [9].

3.1. Switching Circuit and Microelectrode Array Structure

Figure 10(a) shows the proposed amperometric electrochemical sensor circuit. Multiple electrodes placed in an array are connected with one amperometric sensor circuit through the switches. Each electrode is connected to two switches. The electrode being measured is connected to the readout circuit via switch SWA, and on stand-by, the potential is fixed via switch SWB to maintain the steady-state current. When the reading electrode is switched, either switch of the two is kept closed. Therefore, the switching is carried out while the steady-state current is maintained. In this way, it is not necessary to wait for a steady-state current, thus realizing ultra-fast readout from each electrode.

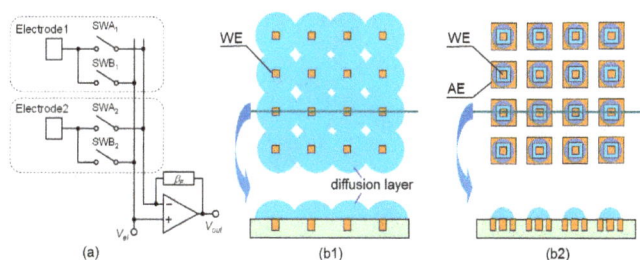

Figure 10. a) Amperometric electrochemical sensor circuit. Each electrode is connected to two switches. (b) Conventional and proposed electrode geometry.

Our working electrode (WE) structure shown in Figure 10(b2) is surrounded by a grid auxiliary electrode (AE), and the redox reaction opposite the working electrode (WE) occurs in the AE. Therefore, the diffusion layer is confined around the WE, and the overlap is decreased. The steady-state current is amplified by redox cycling, and the time to reach the steady-state is reduced.

3.2. Fabricated Amperometric Sensor Array

A 16×16 amperometric sensor array was fabricated as shown in Figure 11. Ag/AgCl and Pt wire were used as reference and counter electrodes, respectively. The solution was composed of 100 mM sodium sulfate and 1 mM potassium ferrocyanide. The WE and the AE potential were fixed at 0.65 V and 0 V (vs Ag/AgCl), respectively. Figure 11 shows the current responses of 1 mM $[Fe(CN)_6]^{4-}$ observed at the single microelectrode, conventional microelectrode array, and proposed microelectrode array. This amperometric sensor array could be applied to genome sequencing by using allele-specific primers and electrochemical reaction [10].

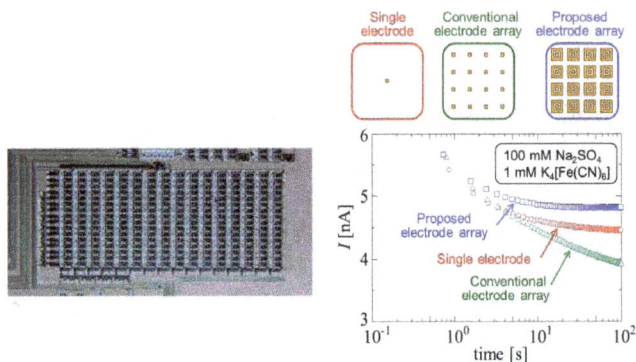

Figure 11. Amperometric sensor array (16×16) and current responses of fabricated microelectrodes. The size of the working electrode is 25×25 μm².

4. Impedimetric sensor array

4.1. Capacitance Sensor Array

We applied nonfaradaic impedimetric measurement by implementing charge-based capacitance measurement (CBCM) to realize a label-free, fully integrated capacitance biosensor. The proposed sensor exploits the capacitance changes of electrical double-layer properties as a result of biorecognition events at the sensing electrode/solution interface. Figure 12(a) shows a schematic of the proposed circuit [11]. To overcome the trade-offs between sensor area and performance, we employed a fully differential measurement circuit that would compensate for parasitic capacitances and reduce the effect of electronic noise, leading to improvement of the detection limit. A photomicrograph of the fabricated chip is shown in Figure 12(b).

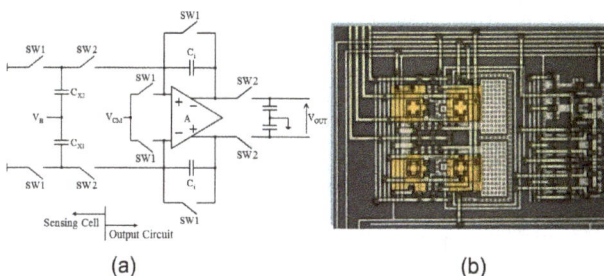

(a)   (b)

Figure 12. a) Schematic of a fully differential capacitance sensor. C_{x1} is the capacitance due to molecules to be detected. (b) Photomicrograph of a sensor chip with 4×4 μm² planar electrodes.

When probe oligonucleotides were immobilized on the electrode surface, a self-assembled monolayer serving as an insulator was formed in conjunction with the electrical double layer. The resulting interfacial capacitance is a total of these series capacitances. When complementary oligonucleotides were introduced to the probes, hybridization occurred and this interface property, i.e., double-layer thickness due to ion displacement, was altered, causing the corresponding capacitance to undergo further change. DNA detection is demonstrated by comparing the results of the capacitance measurements using bare, immobilized, and hybridized electrodes [11]. As observed in Figure 13, the immobilization gave rise to a maximum of 50% capacitance reduction when 20-mer thiolated oligonucleotides were self-assembled at the gold electrode surface. A further 20% reduction in capacitance is also observed after hybridization, implying that the double layer has changed due to the hybridization event.

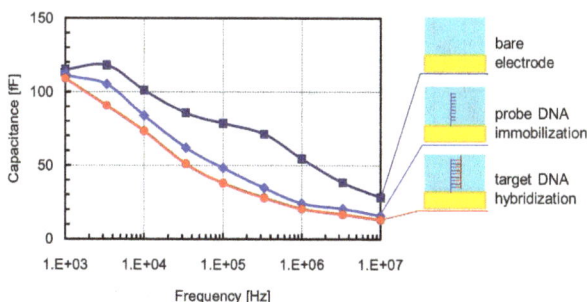

Figure 13. Measured results of capacitance against frequency for bare electrode, after DNA immobilization, and after DNA hybridization.

4.2. Impedance Sensor Array

Electrochemical impedance was measured between two disc electrodes of 20 mm in diameter and with 1 mm of separation, as shown in Figure 14. The specific hybridization is characterized by the change in Curie– von Schweidler exponent a of the constant phase element, which implies the structural change of molecules [12]. This shows that the overall impedance characteristic is more important than the capacitance for detecting randomly distributed molecules.

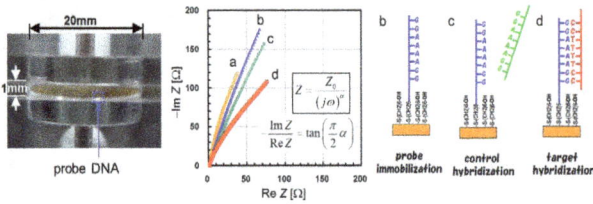

Figure 14. Electrochemical impedance spectroscopy using 20-mm-diameter Au disk electrodes, and Nyquist plot (Cole-Cole plot) of (a) bare electrode, (b) probe/mercaptohexanol immobilization, (c) electrode after noncomplementary binding, and (d) target hybridization. Impedance of constant phase element is proportional to $(j\omega)^{-a}$, where j is the imaginary unit, ω is the angular frequency, and a is the Curie– von Schweidler exponent. The ratio of the imaginary part to the real part becomes a constant $-\tan(\pi a/2)$.

We have designed an on-chip impedimetric sensor unit, which measures the amplitude of impedance at frequencies up to 10 MHz. The sensor unit and peripheral circuitry are shown in Figure 15. To eliminate the effect of turn-on resistance (~20 kΩ) of switch SWA and bit line capacitance (~400 fF), a current amplifier is included in each sensor unit as shown in Figure 15.

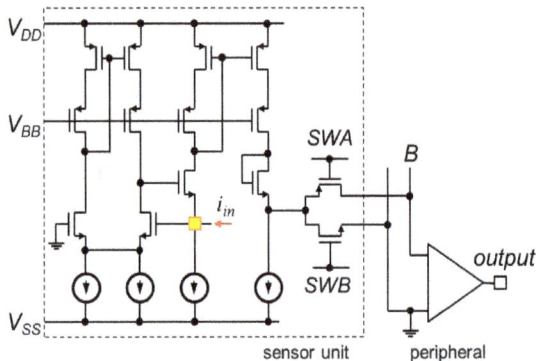

Figure 15. Circuitry of impedimetric sensor unit. AC current is amplified inside a sensor unit. V_{BB} is DC bias voltage. A similar current amplifier was used in [13].

5. Multimodal Sensor Array

In large-scale integration (LSI) circuit fabrication, the initial cost for making a set of photo-masks is quite high. On the other hand, the chip cost is extremely low if a large number of chips are produced. Table 1 shows the typical cost in several technologies. From this table, more than 10,000 chips are necessary to balance the initial cost. This means that standardiza-tion and general-purpose sensor chips are important. Our strategy is to realize a multimodal sensor array for synthetic analysis and standardization. The chip consists of amperometric, potentiometric, and impedimetric smart cells containing an amplifier in the sensor cell, ach-ieving noise reduction and not influencing the measurement system. The chip can be cus-tomized by patterning the insulator layer to cover the unused sensor cells, as shown in Figure 16.

Technology	layers	Cost of a set of photomasks	Cost of 1-cm² chip excluding photomasks
0.6 µm	2P3M	$ 30K	$ 5
0.25 µm	1P4M	$ 100K	$ 7
0.18 µm	1P6M	$ 240K	$ 9
0.13 µm	1P8M	$ 600K	$ 14

Table 1. Typical cost of LSI fabrication

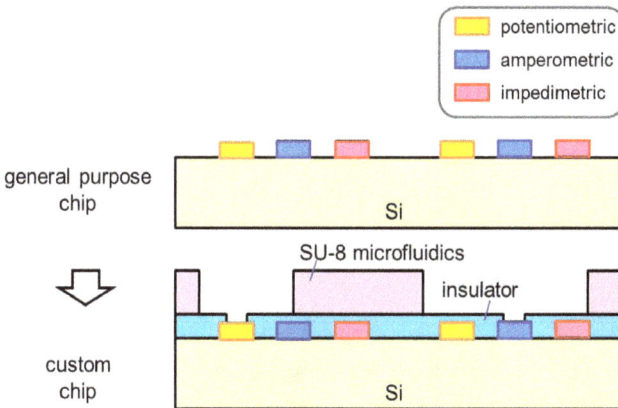

Figure 16. General-purpose sensor chip integrated with potentiometric, amperometric, and impedimetric sensor units. The chip can be customized by the post-CMOS process.

A multimodal sensor unit with potentiometric, amperometric, and impedimetric sensors is shown in Figure 17. The chip was fabricated by using a 1.2-µm 2P2M (2-polysilicon and 2 metal layers) CMOS process. Furthermore, a 16×16 multimodal sensor array with 0.24 mm pitch was designed using a 0.6-μm 2P3M mixed-signal general CMOS process.

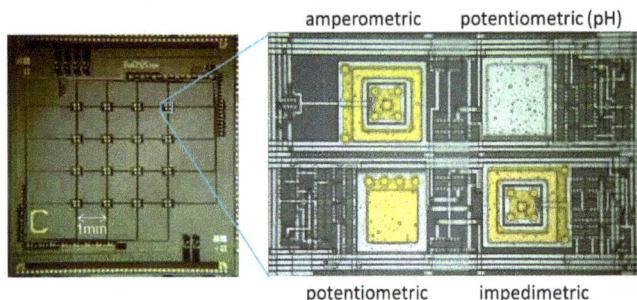

Figure 17. A 4×4 1-mm-pitch multimodal sensor array integrated with potentiometric, amperometric, and impedimetric sensor units.

6. Conclusion

Potentiometric, amperometric, and impedimetric sensor arrays using standard CMOS technology are described. Biomolecular interactions were observed as the time evolution of two-dimensional distribution. The multimodal sensor array with potentiometric, amperometric, and impedimetric sensor units will enable synthetic sensing and standardization of Bio-CMOS LSIs.

Acknowledgements

This study is based on work conducted with Mr. Hiroo Anan, Dr. Yusmeeraz Binti Yusof, and Dr. Shigeyasu Uno of Nagoya in collaboration with Dr. Masao Kamahori and Mr. Yu Ishige at the Central Research Laboratory, Hitachi, Japan. This research was financially supported by a Grant-in-Aid for Scientific Research (No. 20226009) from the Ministry of Education, Culture, Sports, Science and Technology of Japan. The fabrication of CMOS chips is supported by ON Semiconductor Technology Japan Ltd. (1.2 μm process) and TSMC (0.6 μm process), and the VLSI Design and Education Center (VDEC), University of Tokyo in collaboration with Synopsys, Inc. and Cadence Design Systems, Inc.

Author details

Kazuo Nakazato[*]

Address all correspondence to: nakazato@nuee.nagoya-u.ac.jp

Department of Electrical Engineering and Computer Science, Graduate of Engineering, Nagoya University, Nagoya, Japan

References

[1] Bergveld, P. (1970). Development of an ion-sensitive solid-state device for neurophysical measurements. *IEEE Trans. Biomed. Eng.;BME*, 17-70.

[2] Nakazato, K., Ohura, M., & Uno, S. (2008). CMOS cascode source-drain follower for monolithically integrated biosensor array. *IEICE Trans. Electron*, E91C-1505.

[3] Nakazato, K. (2009). Integrated ISFET Sensor Array. *Sensors*, 9-8831.

[4] Rothberg, J. M., et al. (2011). An integrated semiconductor device enabling non-optical genome sequencing. *Nature*, 475-348.

[5] Kagohashi, Y., Ozawa, H., Uno, S., & Nakazato, K. (2010). Complementary Metal-Oxide-Semiconductor Ion-Sensitive Field-Effect Transistor Sensor Array with Silicon Nitride Film Formed by Catalytic Chemical Vapor Deposition as an Ion-Sensitive Membrane. *Jpn. J. Appl. Phys*, 49-01AG06.

[6] Ishige, Y., Shimoda, M., & Kamahori, M. (2009). Extended-gate FET-based enzyme sensor with ferrocenyl-alkanethiol modified gold sensing electrode. *Biosens Bioelectron*, 24-1096.

[7] Kamahori, M., Ishige, Y., & Shimoda, M. (2008). Enzyme Immunoassay Using a Reusable Extended-gate Field-Effect-Transistor Sensor with a Ferrocenialkanethiol-modified Gold Electrode. *Anal. Sci*, 24-1073.

[8] Anan, H., Kamahori, M., Ishige, Y., & Nakazato, K. (2012). Redox-Potential Sensor Array based on Extended-Gate Field-Effect Transistors with ω-ferrocenylalkanethiol-modified Gold Electrodes. *Sens. Actuators: B. Chem.*, in press.

[9] Hasegawa, J., Uno, S., & Nakazato, K. (2011). Amperometric Electrochemical Sensor Array for On-Chip Simultaneous Imaging: Circuit and Microelectrode Design Considerations. *Jpn. J. Appl. Phys*, 50-04DL03.

[10] Tanaka, H., Fiorini, P., Peeters, S., Majeed, B., Sterken, T., de Beeck, M. O., & Yamashita, I. (2011). Sub-micro-liter Electrochemical Single-Nucleotide-Polymorphism Detector for Lab-On-a-Chip System. *Extended Abstract of 2011 ISSDM*, 1109-1110.

[11] Yusof, Y. B., Sugimoto, K., Ozawa, H., Uno, S., & Nakazato, K. (2010). On-chip Micro-electrode Capacitance Measurement for Biosensing Applications. *Jpn. J. Appl. Phys,* 49-01AG05.

[12] Yusof, Y. B., Yanagimoto, Y., Uno, S., & Nakazato, K. (2011). Electrical characteristics of biomodified electrodes using nonfaradaic electrochemical impedance spectrosco-py. *World Academy of Science, Engineering and Technology,* 73-295.

[13] Manickam, A., Chevalier, A., Mc Dermott, M., Ellington, A. D., & Hassibi, A. (2010). A CMOS Electrochemical Impedance Spectroscopy (EIS) Biosensor Array. *IEEE Tran. Biomed. Eng,* 4-376.

Chapter 8

Novel Planar Hall Sensor for Biomedical Diagnosing Lab-on-a-Chip

Tran Quang Hung, Dong Young Kim,
B. Parvatheeswara Rao and CheolGi Kim

Additional information is available at the end of the chapter

1. Introduction

Increased proliferation of infectious diseases stresses the need for immediate development of a state of the art lab-on-a-chip with the capabilities of single biomolecular recognition and parallel processing not only to minimize the death rates but also to enhance the protection from rapid spread of epidemics. Though there are several methods for detection of biomolecules, the magnetic bead sensing technique has been promising and versatile due to its increased ease of fabrication in miniature designs and also its scope for rapid, inexpensive, high sensitive and ultrahigh resolution point of care diagnosis of several human diseases; thus, magnetic biosensors and biochips have become the subject of intense research interest in recent times globally.

In the magnetic bead sensing technique, the detection of biofunctionalized magnetic beads is normally carried out by sensors that are embedded underneath the sensing regions and provide a direct electrical readout proportional to the surface density of immobilized magnetic beads. There are several magnetic sensor principles in operation; namely, anisotropic magnetoresistance (AMR) sensors [1-2], giant magnetoresistance (GMR) and spin valve sensors [3–6], magnetic tunnel junctions (MTJ) [6-7], micro-Hall sensors [8], and planar Hall effect (PHE) sensors [9–11]. Common procedure employed for all these sensor principles is that the magnetic immunoassay of biological sample is introduced to the biofunctionalized sensor array followed by washing steps. In order to establish reproducible conditions under these various incubation and washing steps, it is desirable to integrate the sensor in a microfluidic system, which further facilitates a study of real time response of the sensor as a function of fluid flow, sensor bias current and bead concentrations. Moreover, multi-analyte biosensors integrated with microfluidic systems can be made to perform numerous tasks automatically by way of sensitively and specifically detecting multiple targets from unprocessed sample material, thus creating a compact instrument in the form of "lab-on-a-chip".

With the phenomenal success of GMR-spin valve sensors and MTJ sensors in hard disc drives and magnetic memories, they have become an inspiration for testing their use in other areas including that of magnetic biodetection. Obviously, GMR and MTJ sensors take pride in finding themselves as one of the most widely investigated magnetic sensors for bioapplications [12-13]. They are also successful biosensors commercially as they offer high sensitivities, flexible sensor geometries and large bead-to-sensor ratio with well established integrated circuit fabrication technology. However, relatively low signal to noise ratio of these sensors may often leave scope for erroneous detection. The AMR sensors, in turn, offer greater ease of fabrication but the sensitivity of the AMR signal measured along the longitudinal direction is, however, limited by Johnson noise originating from thermal fluctuations at high frequencies, and by temperature drift at low frequencies [1]. However, the flaws associated with longitudinal AMR measurements can be greatly improved by measuring the voltage change in the transverse direction instead, a phenomenon known as the planar Hall effect [14]. It has been shown that by using the PHE, the temperature drift was reduced by at least 4 orders of magnitude, and nano-Tesla sensitivity has been exhibited [15]. In addition, compared with longitudinal AMR signals, PHE signals are more sensitive to local spin configuration and have much lower background voltage as well.

We propose here a planar Hall sensor array in exchange biased multilayer structure and demonstrate the performance of the sensor with the capability of detection of a single magnetic bead. Also, the sensor is further shown to be capable of single biomolecule detection. Following a brief introduction on the need for exploring magnetic sensors, the book chapter describes the principle of magnetic sensing and highlights the merits of planar Hall sensor in terms of field sensitivity and resolution in the second section.

In the experimental parts, the details of the general procedure for fabrication sequence of the sensor, its characterization and microarray integration were described. Subsequently, an account on the theory and experiments of bead detection using planar Hall resistance (PHR) sensor in different multilayer structures and geometries leading to a complete evolution of novel PHR sensors is elaborately presented in the fourth section. Nevertheless, a hybrid AMR-PHR sensor in ring geometry has been identified for optimum sensor performance towards the end of this section.

In the fifth section, apart from a brief description on the magnetic beads and their functionalization, a description of sensor performance and its capability for detection of magnetic beads including a single magnetic bead is given. This section also presents an account on the integration of microarray sensors with the aid of microfluidics for performing biomolecule experiments while showing the possibility of the planar Hall sensor for a sensitive detection of even single biomolecule. And, finally, it concludes the processes involved with a specific mention on future trends to cater the needs of the society in general.

2. Principle of magnetic sensors

The principle of detection employed by the magnetic sensors for magnetic bioassay involves a magnetic transduction mechanism which uses the magnetic micro- or nanoparticles as labels. The biomolecules are commonly detected by attaching them to highly specific magnetic labels that can, upon their binding, produce an observable quantitative electrical signal, as compared to the cumbersome detection of light signal from fluorescent labels. The specificity is traditionally achieved through a biomolecular recognition mechanism, such as antigen-antibody affinity which can be accomplished by label functionalization, as demonstrated in Fig.1.

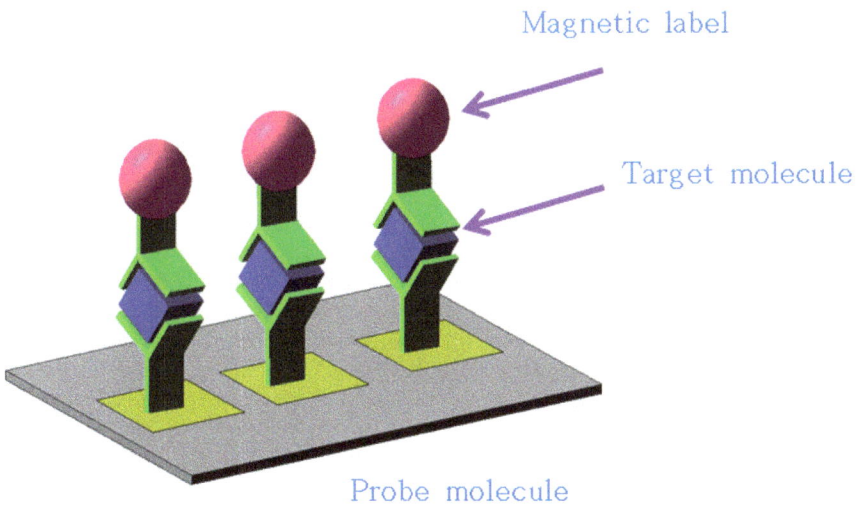

Figure 1. Procedure for the immobilization of probe molecule on the sensor surface and hybridization of the target molecules through Streptavidin coated Dynabeads.

Magnetoresistance (MR) is the property of magnetic materials that results in a change of resistance with applied magnetic field. The MR materials are being developed for the applications such as hard-disk read heads, magnetic random access memories and magnetic field sensors. The materials that display MR property with desired characteristics can replace the inductive coil sensors in a variety of applications including that of biomolecule recognition.

2.1. Magnetoresistive materials for sensing

Thin films and multilayer structures in different geometries show MR characteristics specific to their geometry and all these structures are found to be suitable for one or the other applications. The ferromagnetic single layer exhibits an AMR which is measured in the current direction, while the planner Hall resistance (PHR) effect can be measured in current perpendicular direction in AMR materials [16]. The AMR and PHR effects are due to the anisotropic magnetoresistivity in ferromagnetic layers. The magnetic thin film multilayer structures can give giant magnetoresistance [17] and tunneling magnetoresistance [18-19] effects and the MR property of these structures are superior to that of the AMR structure. The GMR structure consists of two layers of ferromagnetic metal separated by ultra-thin non-magnetic metal spacer layers. The TMR structures are similar to GMR except that they utilize an ultra-thin insulating layer to separate two magnetic layers rather than a conductor. The GMR and TMR effects occur mainly due to the spin-dependent scattering as the current passes from one layer to the other through the spacer layer. The usual figure of merit of the MR ratio is traditionally defined as

$$MR(\%) = \frac{R_{max} - R_{min}}{R_{min}} \times 100 \tag{1}$$

where R_{max} and R_{min} are the maximum and minimum resistance, respectively. The AMR materials typically have MR ratios about 2 - 6 %, and GMR structures exhibit 10 - 50 % while the TMR structures commonly can achieve over 200 % of MR ratio using MgO tunnel barrier instead of the usual Al_2O_3.

A majority of the studies in MR effect in thin films are devoted to the research of multilayered structures showing the largest possible sensitivity of the resistivity for the magnetic field, and consequently a large number of transition metal-based multilayered structures exhibiting large MR ratios have been found. In connection with the technological problems to be solved, this book chapter devotes to a number of MR sensor designs using the planar Hall effect and tested to linearize the transducer signal, to enhance the resolution limited by the MR ratio.

2.2. Relevant sensor characteristics

Though there is a wide choice for sensor designs, optimum sensor performance in each design can be ensured only when specific sensor characteristics are satisfactorily addressed. Among these characteristics, the field sensitivity and the sensor resolution are of utmost concern particularly for a PHR sensor and, thus, they are considered to be described here for elucidating their importance.

2.2.1. Field sensitivity

The field sensitivity of PHR sensor, i.e., the differential of measured PHR voltage versus applied field, can be obtained as

$$\frac{\partial V_{PHR}}{\partial H} = \frac{I(\rho_{//} - \rho_{\perp})}{t} \frac{\cos 2\theta \sin(\gamma - \theta)}{H_K \cos 2\theta + H_{ex} \cos \theta + H \cos(\gamma - \theta)} \tag{2}$$

where I is the current passing through the sensing layer, t is the thickness of the ferromagnetic layer, γ is the angle between applied magnetic field and easy axis, H_K is the uniaxial anisotropy field and $\rho_{//}$ and $\rho_{\perp}$ are the resistivity parallel and perpendicular to the magnetization, respectively. In Eq. (2), the field sensitivity depends not only on the intrinsic parameters $\Delta\rho = \rho_{//} - \rho\, H_K$, and exchange bias field (H_{ex}), but also on the extrinsic parameters such as the magnetization angle at an instant applied field, θ, and the applied magnetic field H.

In a typical curve of PHR signal with applied magnetic field, the maximum field sensitivity of PHR sensor is appeared at low (near zero) magnetic fields. Therefore, the PHR sensor can be used as low magnetic field sensors. Also, the PHR signal does not depend on the sensor size such as the width and length, and therefore, the micro or nano meter order of sensor size is possible by maintaining the same output signal. Thus, the PHR sensor is one of the good candidate bio-sensors for the micro- or nano- bead detector.

2.2.2. Resolution (S/N)

A comparative study of the sensor's characteristics was made systematically and summarized in Table 1 for some of the sensors in practice [12]. It must be mentioned that all the compared sensors have similar active areas and were normally designed for detection of single or small number of micro-size particles.

In Table 1, the part (A) represents the dimensions and properties of different sensor devices compared. The represented thickness is that of the sensing volume used in 1/f noise calculations. Whereas the part (B) shows calculated signals obtained from a single 2 μm bead at the center and on the top of the sensor (the center of the bead is 1.2 μm away from the sensing element), when a 15 Oe rms field is applied. Also represented are the 1/f noise and the thermal noise contributions, and the minimum detectable field as calculated from the expressions in the text, and the signal-to-noise ratio under the conditions described in the text.

Sensor type	W (μm)	h (μm)	t^* (nm)	R_{sq}, Ω/RA, $\Omega\mu m^2$	$\Delta R/R$ (%)	H_k (kA/m)	$\Delta R/\Delta(\mu_0 H)$ (V/T A)
(A)							
SV	3	2	10	18	8	2.4	358
PH-AMR	2.5	2.5	20	6.5	3	2.4	32
AMR ring	$R_{int} = 1$; $R_{ext} = 2$	1	20	6.5	3	2.4	152
GMR	3	2	88	5.3	9	8	71
Hall	2.4	2.4	NA	NA	NA	NA	175
MTJ	2.5	2.5	1	80 (RA)	20 (at 10 mV)	2.4	424

Sensor type	I (mA)	S (μV_{rms})	γ_H, α	N_f (nV$_{rms}$)	N_{therm} (nV$_{rms}$)	$\mu_0 H_{min}$ (nT)	S/N_f
(B)							
SV	10	86	0.1	193	0.61	54	442
PH-AMR	10	15	0.01	10	0.30	32	1453
AMR ring	10	2	0.01	39	0.65	26	50
GMR	5	13	1.2	33	0.33	93	382
Hall	0.3	0.033[a], 4[b]	NA	NA	11	210	3[a], 367[b]
MTJ	1	10	10^{-8} (μm^2)	85	0.42	202	114

[a] Particle–sensor separation of 7 μm.
[b] Particle–sensor separation of 1.2 μm.

Table 1. (from ref.12)

The comparison results have shown that the PH-AMR or PHE sensor has prominent advantages over others such as very high signal-to-noise ratio (S/N$_f$) as well as very high resolution ($\mu_o H_{min}$) in the detection of the magnetic field. Furthermore, the voltage profile of a PHE sensor responds linearly to the magnetic field at the small values. This is a prominent advantage in detection of small stray field induced from magnetic labels. Therefore, we have chosen and mainly focused on the development of the PHE sensors for bio-applications.

3. Sensor fabrication and characterization

3.1. Fabrication procedure of a novel planar Hall sensor

Nowadays, with the advancement of the accurate sputtering and lithography technologies, the sensor with desired composition in micro-size can be easily fabricated by using a lift off method. The general fabrication procedure of a novel exchange biased planar Hall sensor, for example in typical spin valve geometry, using the lift off method is shown in a simplified description in the figure 2 below. However, the same procedure is applied for fabrication of other PHR sensors too in different geometries mentioned in this book chapter.

The SiO$_2$ wafer is first cleaned in the acetone and methanol solutions while placing in the ultrasonic bath, then the SiO$_2$ wafer is covered by a commercial photoresist such as Az (5214E, 9260,…) or SU8-(2000, 3000,…) by using a spin coating system with a defined thickness. The blank cross-junctions are stenciled out on the photoresist coated on SiO$_2$ wafer *i.e.*, the sample is aligned and exposed by a mask aligner system. The short wavelength of ultraviolet source *i.e.*, 456 or 654 nm is used for the exposure, and then the sample is developed by an appropriate developer followed by cleaning the same in DI water.

The sensor materials, *i.e.*, spin-valve structure Ta(5)/NiFe(10)/Cu(1.5)/NiFe(2)/IrMn(10)/Ta(5) (nm), is deposited on the stenciled photoresist layer by using magnetron sputtering system. The base pressure of the system is less than 10^{-7} Torr and the Ar working pressure is 3 mTorr. During the deposition, a uniform magnetic field of 200 Oe was applied in the thin film plane to induce magnetic anisotropy of the ferromagnetic pinned layer and to define the unidirectional field of the thin films. After the thin film deposition, the sample was lifted off in acetone and methanol solutions in order to remove the photoresist as well as the sensor material on this photoresist, so that the sensor material exists on the stenciled junctions only.

After fabricating sensor junctions, the electrodes made by Au are connected with sensor junction to establish the external circuitry and to measure the sensors' response. Further, the sensor junctions and the electrodes are passivated with a SiO$_2$ or a Si$_3$N$_4$ layer coated on top of the sensor junctions and electrodes to protect them from the corrosion and fluid environment during the experiments. Finally, the sensor is activated by a very thin Au layer for biomolecule immobilization. All these steps are carried out at the same way for all the sensors as the steps for the sensor junction fabrication.

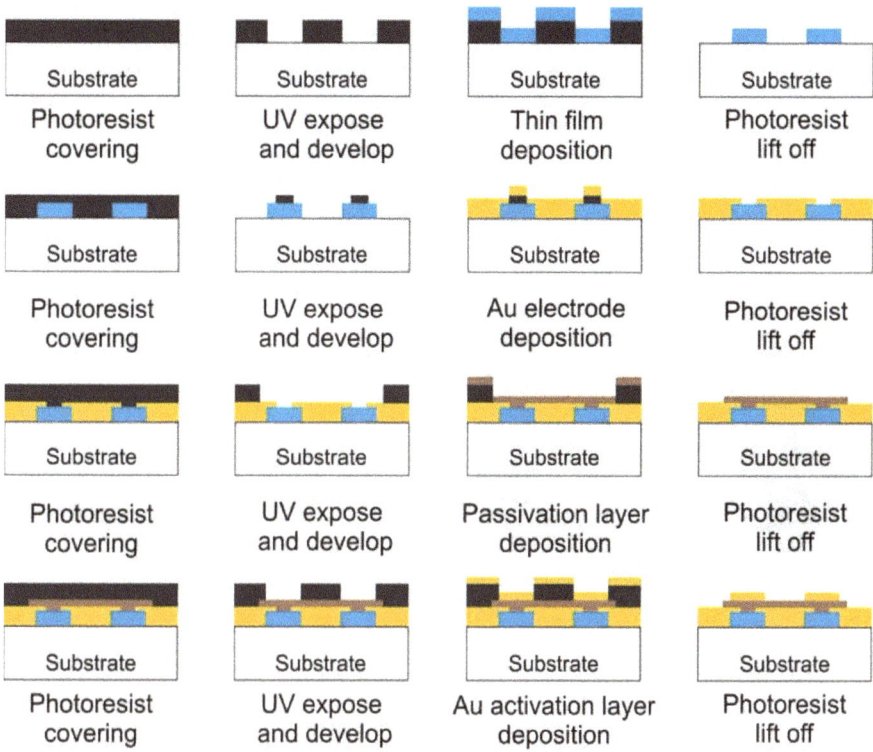

Figure 2. The pattern processes for fabricating the sensor junction, electrodes, passivation and Au activation layers of a planar Hall sensor, this sensor is ready for bio-manipulation.

3.2. Sensor characterization

The quality of the spin-valve structure, Ta(3)/NiFe(10)/Cu(1.5)/NiFe(2)/IrMn(10)/Ta(3) (nm), as observed by a cross sectional transmission electron microscope (TEM) image and also by an energy dispersive X-ray (EDX) spectrum along its thickness, are shown in Fig. 3. The TEM-specimen was prepared by polishing the Si/SiO$_2$ substrate mechanically to a thickness of about 100 μm. After that, the dimpling and hollowing steps were performed at the optimum conditions to ensure that the sample is undamaged by using a GATAN-691 precision ion polishing system (PIPS), *i.e.*, using the Ar-ion beam with an energy of 4.3 keV and under an angle of 6°.

It is evident from Fig. 3(a) that the existence of a multilayer structure is clearly revealed as pointed out by an arrow for each layer. The both seed and top layers of Ta have amorphous behavior and the thickness is about 3 nm. However, there is a difference in the color of the two layers; this can be assumed that the Ta top layer is slightly oxidized [20]. The IrMn layer is well defined, and its thickness is about 10 nm as the nominal thickness when the layer is deposited. In the NiFe/Cu/NiFe region, it is clearly seen that the diffusion of Cu takes place into the adjacent NiFe layers. This kind of diffusion is known to influence the anisotropy significantly [21]. There is also an existence of a rumpling or even a rupture of the layers in some parts. This can be explained when considering the roughness of the NiFe layers; the roughness of NiFe layer is normally about 1.5 – 2 nm [22].

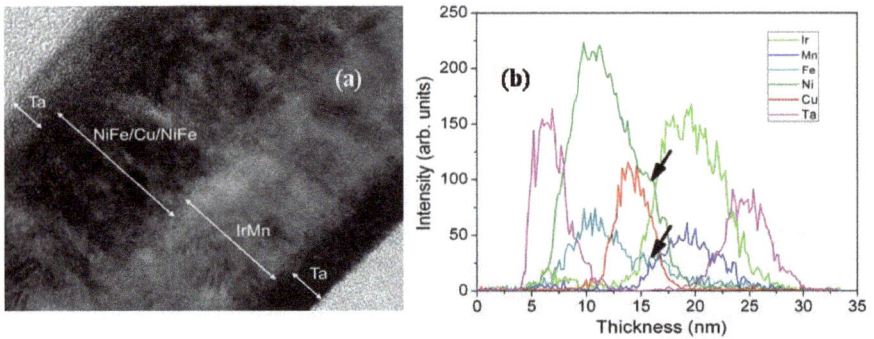

Figure 3. A cross sectional TEM image (a) and an EDX spectra (b) of a spin-valve structure Ta(3)/NiFe(10)/Cu(1.5)/ NiFe(2)/IrMn(10)/Ta(3) (nm).

The result of the cross sectional TEM image is supported by the EDX patterns of the same sample shown in Fig 3(b). It can be seen that the peak of Cu is mixing inside the Ni and Fe peaks indicating the diffusion of Cu in NiFe layers. The overlap between the peaks of Ta and Ni (Fe), of Ta and Ir (Mn), and of Ir (Mn) and Ni (Fe) confirms the roughness at the surface of the NiFe and IrMn layers. Moreover, the shadow in Ni and Fe peaks (black arrows in Fig. 3(b)) indicates the separation of the NiFe pinned and NiFe free layers.

The magnetic property of the fabricated spin-valve structure used for sensor material is characterized by a vibrating sample magnetometer (VSM) of the make Lakeshore 7407 series with a sensitivity of 10^{-6} emu. The external magnetic field is swept in the film plane.

In order to achieve the magnetic anisotropy of the free layer in the fabricated spin-valve structure, we measured the magnetization as a function of external magnetic field in the range of ± 80 Oe in both the easy and hard axis, which is presented in Fig. 4. The shift along the external magnetic field axis of the magnetization profile ($M(H)$) in the easy axis indicates an effective uniaxial anisotropy field of the spin-valve structure (H_{Keff}) by incorporating the free layer shape anisotropy field ($H_{demag.}$) and its uniaxial anisotropy field (H_K) analyzed from the shift of the $M(H)$ profile in the hard axis ($H_{Keff} = H_K + H_{demag.}$) [23]. This indicates that the free NiFe layer (active layer) has very good anisotropy characteristic for further study of the PHE sensor. In addition, the inset in Fig. 4 exhibits a two-step hysteresis loop;

one is from the interlayer coupling and the other is from the exchange bias coupling. The magnetization of the first hysteresis loop (contributing from 10 nm NiFe free layer) is five times larger than the second one (contributing from 2 nm NiFe pinned layer). The interlayer coupling between the ferromagnetic (F)-free and F-pinned layers separated by a non-magnetic layer (Cu) is determined from the first step of the hysteresis loop. Whereas, the exchange bias field due to the interface between the F-pinned and antiferromagnetic (AF) layers is determined from the second step of the hysteresis loop. The obtained interlayer and interfacial coupling fields are 11 Oe and 550 Oe, respectively. This result elucidates that the NiFe pinned and NiFe free layers are separated by a Cu layer.

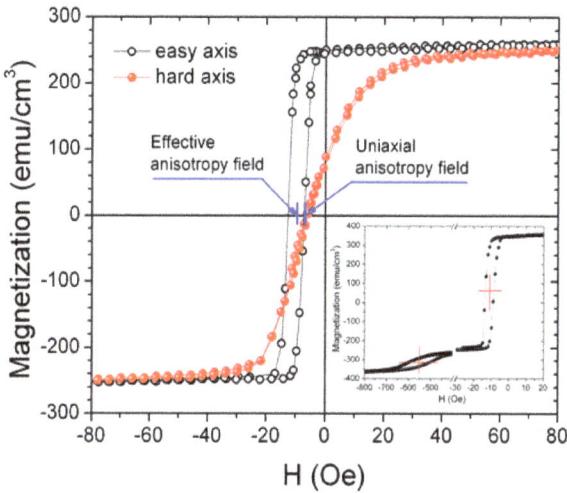

Figure 4. Hysteresis loops of the spin-valve thin film, Ta(3)/NiFe(10)/Cu(1.5)/NiFe(2)/IrMn(10)/Ta(3) (nm), characterized in the easy and hard axis in the field interval from + 80 to -80 Oe. The inset shows the hysteresis loop characterized in the easy direction in the field range of -800 to 20 Oe.

3.2.1. Microarray of the magnetic sensors

Fig. 5 shows a complete micro-array of planar Hall resistance (PHR) sensors. In the figure, it was shown that the unidirectional field, H_{ex}, and/or the uniaxial field of the thin film were aligned parallel to the terminals a–b, and a sensing current of 1mA was applied through these terminals. The output voltages were measured from the terminals c and d at room temperature under a specific range of external magnetic field applied normal to the direction of the current.

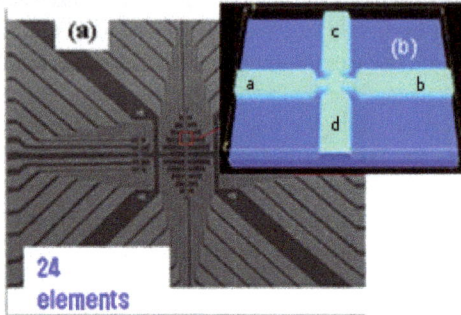

Figure 5. Complete micro-array of a 24 element PHR sensor(a), which can even detect a single micro-paramagnetic Dynabead M®-280. Inset of the figure (b) shows a single micro sized cross-junction.

4. Evolution of novel PHR sensors

Among all the developed magnetoresistive sensors for bioapplications, we mainly focus on the development of PHE sensor because it has prominent advantages compared with others such as signal-to-noise ratio, linearity signal *etc*. Various structures will be used for planar Hall sensor *i.e.*, bilayer, trilayer and spin-valve. Therefore, a short introduction of these multilayer structures will be given in this section. Also, the theoretical approach of planar Hall effect in different sensor geometries, such as cross-junction, tilted cross-junction and ring junction, will be discussed. Finally, the description leads to evolution of hybrid AMR-PHR sensor with optimized sensor characteristics for effective use in bioapplications.

4.1. AMR sensor

The magteoresistive anisotropy in ferromagnetic material depends on the direction of magnetization. The electric field due to the magnetoresistivity is expressed as follows [24];

$$\vec{E} = \rho_{\perp}\vec{j} + (\rho_{\parallel} - \rho_{\perp})\vec{m}(\vec{j} \cdot \vec{m}) \tag{3}$$

where $\vec{m}$ is magnetization vector in single domain, and $\vec{j}$ is current density direction. The ρ and $\rho_{\parallel}$ are the resistivity when the magnetization vector and current density direction are perpendicular and parallel, respectively. The $\Delta\rho = \rho_{\parallel} - \rho$ is defined by the anisotropic resistivity, which is the intrinsic resistivity by the spin-orbit scattering in ferromagnetic materials. In Eq. (3), the electric field can be measured in the current direction as well as perpendicular to current direction due to the anisotropic resistivity, which are called, as mentioned, as the AMR and PHR, respectively.

The AMR properties have been discovered at ferromagnetic material by William Thomson in 1857 [25]. In the AMR response, varying differences between the direction of the magnetizing vector in the ferromagnetic film and the direction of the sensing current passing through the film lead to varying the resistance in the direction of the current. The maximum resistance occurs when the magnetization vector in the film and the current direction are parallel to one another, while the minimum resistance occurs when they are perpendicular to one another. The resistance change by AMR effect in the patterned film with thickness t, width w and length l can be expressed from Eq. (3).

$$V_{AMR} = I(R_\perp + \Delta R \cos^2 \theta) \tag{4}$$

where $\Delta R = (\rho_{\|} - \rho_\perp) l / \omega t$ is the anisotropic magnetoresitivity and the θ is the angle between the magnetization vector and current, I. In AMR effect, the MR ratio is expressed as $\Delta R / R_\perp \times 100$. The AMR effect has an offset resistance of $R_\perp$. This offset resistance must be reduced to improve the performance by using a compensating voltage or a Wheatstone bridge circuit [26].

4.2. Planar Hall resistance sensor

The planar Hall resistance (PHR) in ferromagnetic thin films was considered when the resistivity depends on the angle between the direction of the current density j and the magnetization m. For magnetization reversal of the single domain when m makes an angle θ with j, the electric field is described as follows;

$$E_{PHR} = j(\rho_{\|} - \rho_\perp) \sin \theta \cos \theta \tag{5}$$

The PHR effect also varies when there is a difference between the direction of the magnetizing vector in the ferromagnetic film and the direction of the sensing current passing through the film; however, it leads to varying the resistance in the perpendicular direction of the current only. The longitudinal component of PHR voltage is related to E_{PHR} in Eq. (3) and can be revealed when anisotropy of resistivity exists. On the other hand, in this sensor, the measured PHR voltage was described as follows:

$$V_{PHR} = \frac{I(\rho_{\|} - \rho_\perp)}{t} \sin \theta \cos \theta \tag{6}$$

where t is the thickness of ferromagnetic film. The PHR in Eq. (6) varies with the angle θ. The PHR does not impose the offset resistance. Therefore, it has the advantage of obtaining a large PHR ratio and a linear response characteristic when the angle θ having a small value. The PHR effect depends on the intrinsic magneto-resistivity, $\Delta \rho = \rho_{\|} - \rho_\perp$ and the sample thickness, t. This means that the PHR signal does not depend on the sensor size (width ω and length l). Therefore, the PHR sensor can be used as the micro- or nano sized sensor for the micro- or nano- bead detection maintaining the large output signal voltage.

In order to analyze the PHR signal with magnetic field, we must know the angle θ between the magnetization vector and current direction, which depends on the magnetic field. Fig. 6 shows the general coordinates used to describe the rotational magnetization process under the applied magnetic field in ferromagnetic/antiferromagnetic (F/AF) coupled samples. H_{ex} is the exchange coupling field due to the antiferromagnetic layer, and it shows a biasing field effect. K_u is the effective in-plane anisotropy constant with an angle γ from H_{ex}.

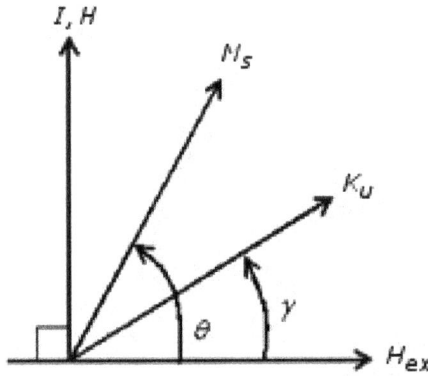

Figure 6. The coordinates for domain rotation process. Here, γ and θ are the angles of the anisotropy constant, and magnetization from the exchange-coupling field, H_{ex}, respectively, and I is the measuring current.

The applied magnetic field H is directed perpendicular to H_{ex}, and force the magnetization to rotate by an angle θ towards H. We introduce the modified Stoner-Wolfforth model with magnetic energy density, E_T for the F layer in the F/AF sample, which can be written in the following simple form [12, 27]

$$E_T = K_u \sin^2(\theta - \gamma) - HM_s \sin\theta - H_{ex}M_s \cos\theta \qquad (7)$$

where M_s is the saturation magnetization. The angle θ determines the orientation of the magnetization in an equilibrium state with minimum total energy, whose values are calculated under the conditions of $\partial E_T/\partial\theta=0$.

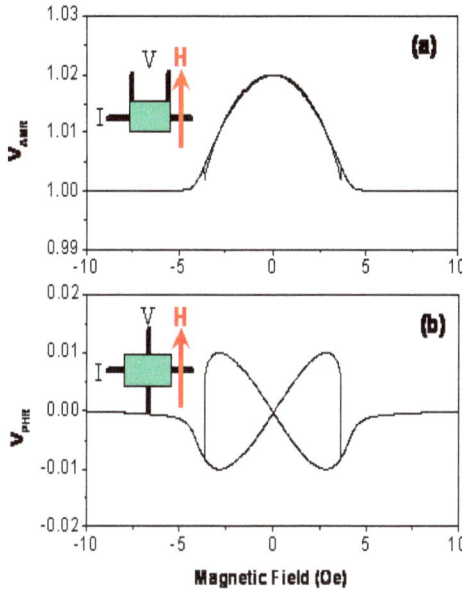

Figure 7. (a) Calculated V_{AMR} and (b) V_{PHR} with applied magnetic field in single ferromagnetic film

Fig. 7 shows the calculated V_{AMR} and V_{PHR} in ferromagnetic single layer without exchange bias field, H_{ex}. The measuring configuration of AMR and PHR voltage is shown in the inset of the figures. The current was applied parallel to the easy axis and the magnetic field was applied parallel to the hard axis of the magnetic thin film. The AMR voltage was measured in the direction of the sensing current passing through the film, while the PHR voltage was measured in the perpendicular direction of the sensing current. In the case of AMR effect, the signal shows the symmetric behavior in the functions of applied magnetic field with offset voltage of $R_{\perp}$. The PHR signal shows the linear behavior in the functions of applied magnetic field with zero offset voltage.

Therefore, the PHR sensor justifies that it can be used as the micro- or nano- sized magnetic field sensor for the detection of the micro- or nano- bead. The PHR signal in ferromagnetic single layer shows large hysteresis behavior. The hysteresis effect is due to the switching of the magnetization in ferromagnetic layer. In order to remove the hysteresis of PHR, the exchange biased F/AF bilayers are considered.

4.3. PHR effect in exchange biased F/AF multilayer structures

The single ferromagnetic layer with high AMR ratio such as NiFe, CoFe and NiCo alloys has the uniaxial anisotropy. The easy axes for stable magnetization direction are 0 and 180 degrees. If one cycle of magnetic field is applied in the perpendicular direction to the easy axis in ferromagnetic films, the magnetization direction changes from 0 to 90 degrees as the magnetic field increases, and 90 to 180 degrees as the magnetic field decreases. And then the direction of the magnetization changes from 180 to 270 and then to 360 degrees as the reversed magnetic field increases and decreases. In that case, the AMR and PHR, which are dependent on the angle between the current and magnetization directions, can show the large hysteresis loop. On the other hand, the exchange biased F/AF bilayers induce the unidirectional anisotropy, which rotates the magnetization direction from 0 to 90 and 90 to 0 degrees as the magnetic field increases and decreases, respectively. It means that the AMR and PHR signal in exchange biased F/AF bilayers show the reversal behavior and the hysteresis can be disappeared.

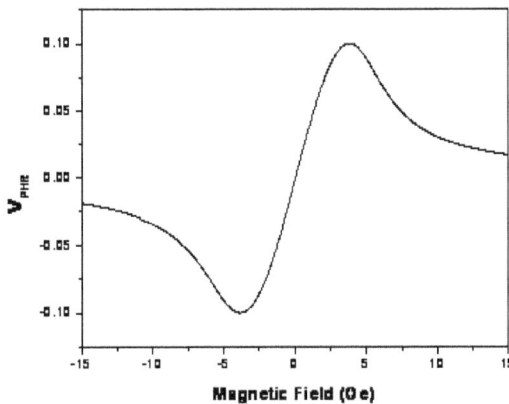

Figure 8. V_{PHR} signal with applied magnetic field in exchange biased F/AF bilayers

Fig. 8 shows the calculated PHR signal for the exchange biased F/AF bilayers. By comparing the PHR signal of the exchange biased F/AF bilayers in Fig. 8 with that of the single ferromagnetic layer in Fig. 7(b), we can clearly confirm that no hysteresis behavior of PHR signal takes place in exchange biased F/AF bilayers. The exchange bias field, H_{ex} plays the role of the reversible rotation of the magnetization as the magnetic field changes, which is due to the unidirectional anisotropy compared with the uniaxial anisotropy in single ferromagnetic layer. Also the reversible rotation of the magnetization in exchange biased F/AF bilayers can reduce the Barkhausen noise, which is usually dominated in the irreversible domain motion. Therefore, the signal to noise ratio (S/N ratio) of PHR sensor can be increased by using the exchange biased F/AF bilayers. Further, the PHR effect in exchange biased bilayers shows good linearity and thus it has the advantage for magnetic field sensor application. In the case of GMR or TMR materials, though they have high MR ratios, however, theirs' linearity is not good compared with the PHR signal. Therefore, PHR effect in the exchange biased F/AF bilayers has advantages in use as a bio-sensor for micro or nano bead detection.

4.3.1. Bilayers

There exists an interfacial coupling in F/AF bilayers. The hysteresis loop of the F layer, instead of being centered at zero magnetic field, is now displaced from $H = 0$ by an amount noted as the exchange field H_{ex}, as if the F layer is under a biased magnetic field. Hence, this phenomenon is also known as exchange bias [28]. In such a structure the anisotropy may behave as unidirectional anisotropy. Technologically, exchange bias is of crucial importance in the field-sensing devices. An example $M(H)$ loop of Ta(3)/NiFe(10)/IrMn(10)/Ta(3) (nm), which is usually the structure being used for fabricating a sensor, is used for this study. The center of the hysteresis loop of this bilayer, as shown in Fig. 9, is shifted from zero applied magnetic field by an amount H_{ex}, the exchange bias field.

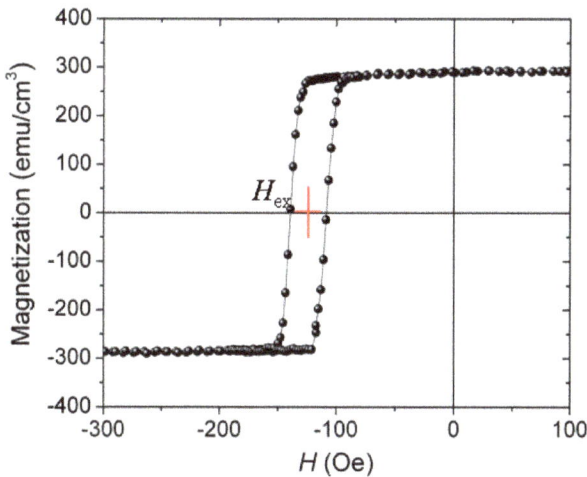

Figure 9. The shifted hysteresis loop in an exchange biased bilayer thin film

In a bilayer structure, the exchange coupling between the F and AF layers can easily induce the unidirectional magnetic anisotropy of the F layer. In addition, the F layer is improved to be constrained to the magnetization in coherent rotation towards the applied fields, so the sensor can prevent Barkhausen noise associated with the magnetization reversal, and improves the thermal stability [29]. Because of these advantages, a bilayer structure is a good candidate for developing sensor materials.

Bilayer has been used as PHE sensor materials by M.F. Hansen *et al*, C.G. Kim *et al*, and F.N.V. Dau *et al*. It is revealed from the literature that the sensitivity of a PHE sensor is increased with the thickness of ferromagnetic layer up to 20 nm [27].

4.3.2. Spin-valves

The spin-valve structure, as shown in Fig. 10(a), which was known as a simple embodiment of the GMR effect, typically consists of two F layers separated by a nonmagnetic conductor whose thickness is smaller than the mean-free path of electrons. The magnetic layers are uncoupled or weakly coupled in contrast to the generally strong AF state interaction in Fe-Cr-like multilayer; thus the magnetization of F layer with uniaxial anisotropy can be rotated freely by a small applied magnetic field in the film plane, while the magnetization of other magnetic layer had unidirectional anisotropy and was pinned by exchange bias coupling from AF layer. If the relative angle between the magnetization of the two layers changes, a giant magnetoresistance change occurs.

In an illustrative demonstration of the operation of the spin-valve, the applied magnetic field is directed parallel to the exchange biased field and cycled in magnitude. The $M(H)$ loop and the corresponding magnetoresistance curves are shown schematically in Fig. 10(b) and (c), respectively.

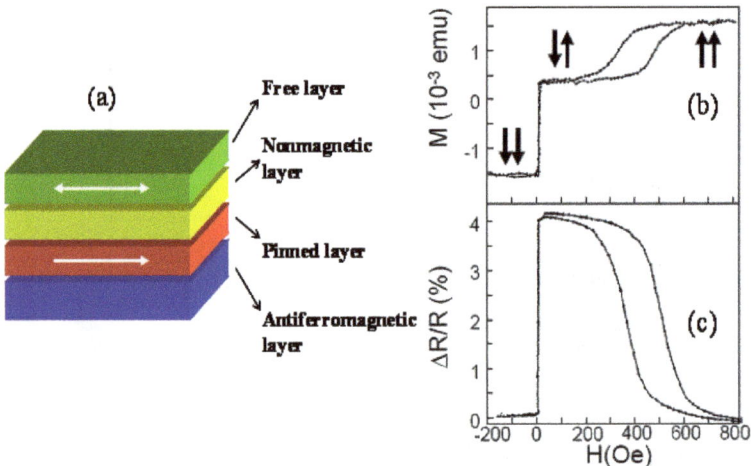

Figure 10. (a) Schematic of a typical spin-valve structure, (b) Hysteresis loop, and (c) magnetoresistance of a spin-valve sample of composition, Ta(5)/NiFe(6)/Cu(2.2)/NiFe(4)/FeMn(7)/Ta(5) (nm), at room temperature [27, 30].

The sharp magnetization reversal near zero magnetic field is due to the switching of the free magnetic layer in the presence of its weak coupling to pinned magnetic layer. The more rounded magnetization reversal at higher magnetic field is due to the switching of the pinned magnetic layer, which overcomes its exchange biased coupling to an AF layer for these fields. Therefore, it was emphasized that a spin-valve here makes use of two different exchange couplings; exchange biased coupling from pinned layer to AF layer and interlayer exchange coupling between two magnetic layers, which in origin, was tentatively assigned to a Ruderman–Kittel–Kasuya–Yosida (RKKY) interaction. The relative orientations of two magnetic layers were indicated by the pairs of arrows in each region of the $M(H)$ curve where the resistance is larger for antiparallel alignment of the two magnetic layers.

In order to optimize the spin-valve structure for high sensitivity PHE sensor, Kim's group has investigated systematically the effect of the thickness of F-pined and F-free layers (t_f and t_p) in the spin-valve structure Ta(5)/NiFe(t_f)/Cu(1.2)/NiFe(t_p)/IrMn(10)/Ta(5) (nm) with t_f = 4, 8, 10, 12, 16, 20 nm, and t_p = 1, 2, 6, 9, 12 nm. The results show that the sensitivity is increased linearly with t_f and is decreased exponentially with t_p in the investigated range. As the result, the optimized spin-valve structure for highest sensitivity is Ta(5)/NiFe(20)/Cu(1.2)/NiFe(1)/IrMn(10)/Ta(5) (nm). The details explanation could be found in Ref 13.

4.3.3. Trilayers

The origin of interlayer coupling in F/spacer/AF trilayer structure is totally different from interlayer coupling induced in F/spacer/F multilayer thin films. The observation of F/AF exchange coupling across a nonmagnetic layer by Gökemeijer et al., [31] demonstrates that the exchange bias is a long-range interaction extending to several tens of Å. This coupling is not oscillatory but decays exponentially as $J \sim \exp(-t/L)$. The range of F/AF exchange coupling is specific to the spacer material, and thus most likely electronic in nature.

In our experiment, we choose Cu as spacer layer in the trilayer structure, Ta(3)/NiFe(10)/Cu(0.12)/IrMn(10) (nm), because it gives a small exchange coupling with a thin Cu layer. In the sensor application, it can reduce the shunt current resulting in enhanced sensitivity. The exchange coupling of the trilayer structure, determined by the shift of the hysteresis loop in the magnetic field direction and is measured in the order of few tens of Oe, is one order smaller compared with the exchange coupling in a typical bilayer structure (in order of hundred Oe) as shown in Fig. 9. A comparison of the PHE voltages generated by the bilayer, spin-valve and trilayer structures and their corresponding sensitivities are shown in Fig. 11. Thus, it can be easily seen from the figure that the trilayer structure can improve the field sensitivity of a sensor better than those of the bilayer and spin-valve structures [32].

4.4. Sensor geometry

The performance of the sensor depends largely on its physical geometry. There were several geometries reported in the literature in the design of planar Hall sensor. Among these geometries, the cross-junction and circular geometries need special mention as they result better performance of the sensor. Thus, it is intended to present the results of the sensor for better understanding of the sensor performance when the geometries are explored in the form of cross-junction, tilted cross-junction and circular ring junction.

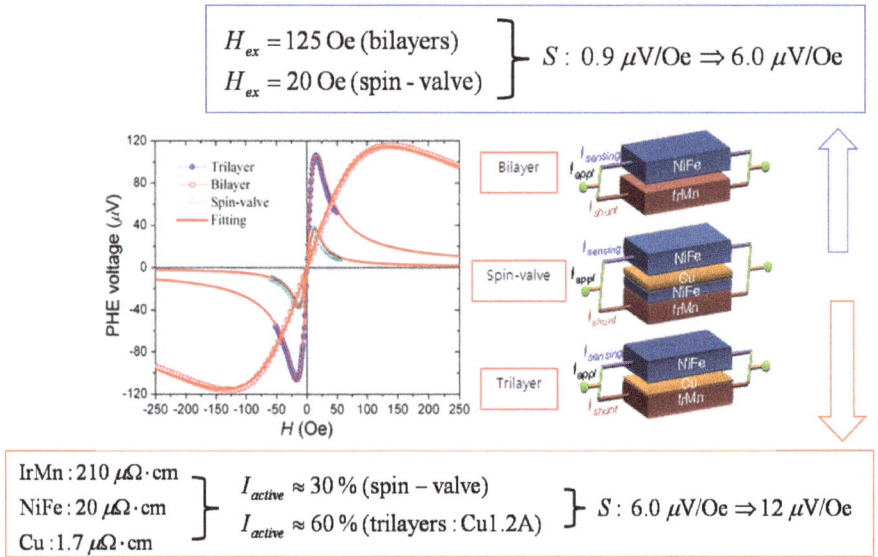

$$H_{ex} = 125 \text{ Oe (bilayers)}$$
$$H_{ex} = 20 \text{ Oe (spin - valve)}$$
$$\Big\} \quad S: 0.9 \, \mu V/Oe \Rightarrow 6.0 \, \mu V/Oe$$

IrMn : $210 \, \mu\Omega \cdot cm$
NiFe : $20 \, \mu\Omega \cdot cm$
Cu : $1.7 \, \mu\Omega \cdot cm$
$\Big\}$ $I_{active} \approx 30\,\% \text{ (spin - valve)}$
$I_{active} \approx 60\,\% \text{ (trilayers : Cu1.2A)}$
$\Big\}$ $S: 6.0 \, \mu V/Oe \Rightarrow 12 \, \mu V/Oe$

Figure 11. Comparison of the PHE performance between the bilayer, spin-valve and trilayer structures

4.4.1. Cross-junction

In this part, we discuss the effect of the sensor size on the output voltage of a cross-junction PHE sensor.

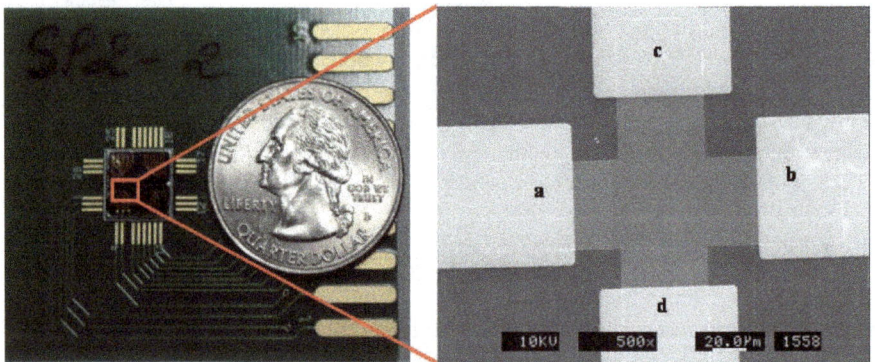

Figure 12. (left) Illustration of a fabricated PHR sensor, (right) top view micrograph of a single 50 μm × 50 μm PHE sensor junction

Fig. 12 (left) shows the illustration of a fabricated PHR sensor and the Fig.12 (right) shows the SEM image of the passivated single sensor cross-junction of the size 50 μm × 50 μm. The terminals a-b represents the current line and c-d represents the voltage line. The unidirectional anisotropy field, H_{ex}, and the uniaxial anisotropy field of the thin film are aligned parallel to the long terminals a-b. Planar Hall effect (PHE) profiles were measured by the electrodes bar c-d with a sensing current of 1 mA applied through the terminals a-b and under the external magnetic fields ranging from – 50 Oe to 50 Oe applied perpendicular to the direction of the current line and in sensor plane. The induced output voltages of cross-junctions were measured by means of a Keithley 2182A Nanovoltmeter with a sensitivity of 10 nV. All these sensor characterizations were carried out at room temperature.

For studying the size effect in planar Hall sensor, cross-junctions with various sizes of x × 50 μm^2 and 50 × x μm^2, (x = 30, 50, 100) using spin-valve structure Ta(3)/NiFe(10)/Cu(1.5)/NiFe(2)/IrMn(10)/Ta(3) (nm) were fabricated. For estimating the free layer magnetic anisotropy of the fabricated spin-valve structure, we measured the magnetization as a function of the external magnetic field in the range of ± 80 Oe in both the easy and hard axis (refer to Fig. 4). As mentioned, the shift along the field axis of the magnetization profile in the easy axis indicates that the free NiFe layer (active layer) has very good anisotropy characteristic for further studying the PHE voltage profiles of the sensor.

The PHE voltage profiles of the fabricated sensors with various junction sizes are given in Fig. 13. Analogous to the other PHE results, the PHE voltage in all the sensor junctions initially changes very fast and appears linear at low fields, reaches a maximal value at H ~ 11 Oe and finally decreases with further increase in the magnetic fields.

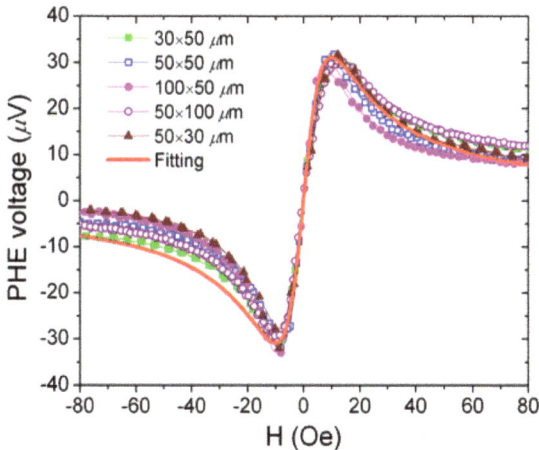

Figure 13. The PHE voltage profiles of the various size sensor junctions based on the spin-valve thin film Ta(3)/NiFe(10)/Cu(1.5)/NiFe(2)/IrMn(10)/Ta(3) (nm)

It is noteworthy that the maximum value of the PHE voltage profile is obtained at the field close to the effective uniaxial anisotropy field, H_{Keff}, of the free layer. This finding was studied systematically in a spin-valve structure and has been reported, previously [31-34]. Moreover, it is observed in the linear response region (at the field range from -11 Oe to 11 Oe) only despite having variation in the junction size, and the slope of the PHE voltage profile remains constant. That means there is no change in the field-sensitivity when the sensor junction is varied either in length or width.

The theoretical voltage profile of the fabricated PHE sensor was also calculated with a set of following parameters: $K_u = 2 \times 10^3$ erg/cm^3, $M_s = 800$ emu/cm^3 for the NiFe, $J = 1.8 \times 10^{-3}$ erg/cm^2 ($J = t M_s H_{int}$), $H_K = 2 K_u / M_s$, $I = 1$ mA and $V_o = \dfrac{I(\rho_{//} - \rho_{\perp})}{t} = 62$ μV and the calculated result is represented as solid line in Fig. 13. The excellent agreement between the theoretical and experimental results confirms the point that the field-sensitivity of the PHE sensor is independent of the size of the cross-junction.

This result is important for the bio-applications because the sensitive detection of low bimolecular concentration is proportional to the junction size.

4.4.2. Tilted cross-junction

The idea behind the study of the tilted cross-junction is to combine some of the magnetoresistive effects, such as GMR, AMR and PHE and to explore how beneficial the sensor could be in its performance [35]. Therefore, the spin-valve structure which has GMR effect causing by spin scattering of electron between two F layers through a spacer layer, AMR and PHE effects causing by the spin-orbit coupling in the F layer are the best candidates for a sensor material.

To study the tilted cross-junction bars, 100 μm × 50 μm, with various tilt angles of $\zeta = 0°, 4°, 8°,$ 10°, 30°, 45° using Ta(5)/NiFe(6)/Cu(3)/NiFe(3)/IrMn(15)/Ta(5) (nm) spin-valve structure are fabricated. The tilted cross-junction bar with a tilt angle ζ is shown in Fig. 14, in which the angle between the electrodes a–b and c–d is deliberately altered from 90° to 45°. The unidirectional field, H_{ex}, and the uniaxial field of the thin film were aligned parallel to the long terminals a-b, and sensing current of 1 mA was applied through these terminals. Output voltages were measured from the short terminals c and d at room temperature under the external magnetic fields ranging from - 45 Oe to 45 Oe applied normal to the direction of the current bar.

In general, the GMR and AMR effects could be obtained from the parallel direction to the current bar or longitudinal part while the PHE can be obtained from the transverse part of the sensor junction. Therefore, in a novel design of the sensor based on the tilted cross-junction the longitudinal and transverse contributions could be combined together in one sensor. In this tilted junction, we observed that there is an enhancement of PHE sensitivity and better linearlity of MR longitudinal component.

In Fig. 15 we demonstrated the output voltage profiles of the sensor junctions with different tilted angles. It clearly shows an increase in amplitude of the output voltage profile with increasing tilted angle ζ and the upward shift of the drift voltage. In particular, a significant en-

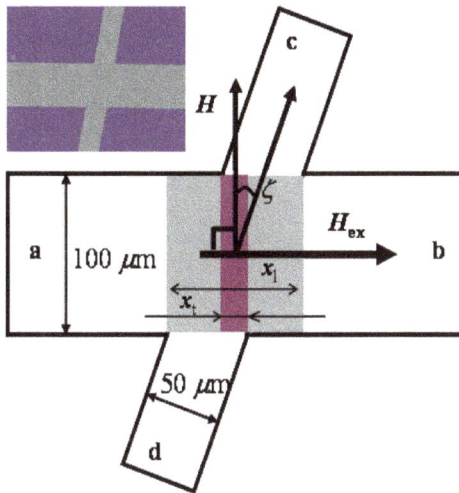

Figure 14. The geometry of a tilted cross-junction. The width of current and voltage bars are 100 μm and 50 μm, respectively. The inset shows the micrograph of the cross-junction with tilt angle $\zeta = 10°$.

hancement of sensor sensitivity by about 30% is observed when the cross-junction is tilted with an angle of 45°, and in this case, the sensitivity about 9.5 $\mu V/Oe$ is reached. It is also noteworthy to observe a gradual change in the shape of the output voltage profile from asymmetric to symmetric which implies a corresponding increase of longitudinal MR voltage due to the increment of tilted angle in the cross-junction, *i.e.*, for the first case when $\zeta = 0°$, the voltage profile corresponds to the PHE only. In the other tilted cross-junctions ($\zeta \neq 0°$), the output voltage profiles consist of the PHE, AMR and GMR components.

In order to understand the voltage contribution from each effect in a tilted cross-junction quantitatively, we have performed systematic investigations on the role of the MR and PHE in the tilted cross-junction. In such case, it was noticed that the active PHE region and active MR region are from the transverse part and longitudinal part of the sensor, respectively. When the tilt angle of cross-junction increases, the length of the transverse part (x_t in Fig. 14) decreases and the length of longitudinal part (x_l in Fig. 14) increases accordingly.

It is observed that the PHE voltage is independent of the junction size irrespective of its change in the length or the width in previous part. Therefore, the PHE voltage component in the tilted cross-junction is always a constant. Then the transverse PHE component (corresponding to $\zeta = 0°$) is decomposed from experimental data for different tilted cross-junctions. The decomposed results are illustrated in Fig. 16 for the sensor junction with $\zeta = 10°$. Clearly, a strong contribution of the longitudinal MR component is evidenced. However, the PHE dominates good linearity and high sensitivity at low magnetic fields.

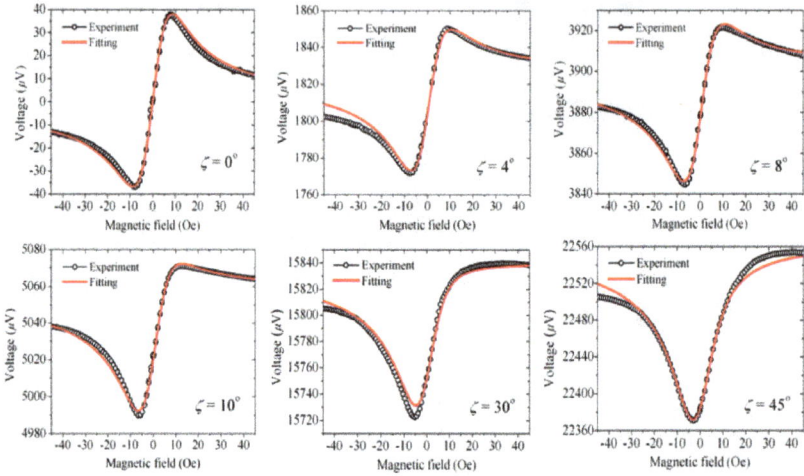

Figure 15. The experimental and theoretical voltage profiles of cross-junctions with different tilt angles of 0°, 4°, 8°, 10°, 30°, 45°.

Applying the above mentioned decomposition procedure for all investigated sensor junctions, one can derive the values of the drift (minimal) voltage (V_{MRmin}), the MR voltage (or the MR voltage change in external magnetic fields) ($\Delta V_{MR} = (V_{MRmax} - V_{MRmin})$) and the percentage of voltage change of the longitudinal MR voltage profile ($\Delta V_{MR}/V_{MRmin}$).

The results are listed in Table 2. Note that, V_{MRmin} and ΔV_{MR} increases as the tilted angle increases and thus the ΔV_{MR} enhances the total output voltage profiles.

ζ (°)	S (μV/Oe)	V_{MRmin} (μV)	ΔV_{MR}(μV)	$\Delta V_{MR}/V_{MRmin}\times100$ (%)
0	7.4	-	-	-
4	7.5	1799	11.0	0.61
8	7.6	3877	24.0	0.62
10	7.7	5021	30.5	0.61
30	9.1	15752	94.5	0.60
45	9.5	22385	136.0	0.60

Table 2. The sensor sensitivity (S) and values of the minimal voltage (V_{MRmin}), MR voltage change in the applied fields (ΔV_{MR}), relative voltage change of the longitudinal MR voltage profile ($\Delta V_{MR}/V_{MRmin}$) of different tilted cross-junctions

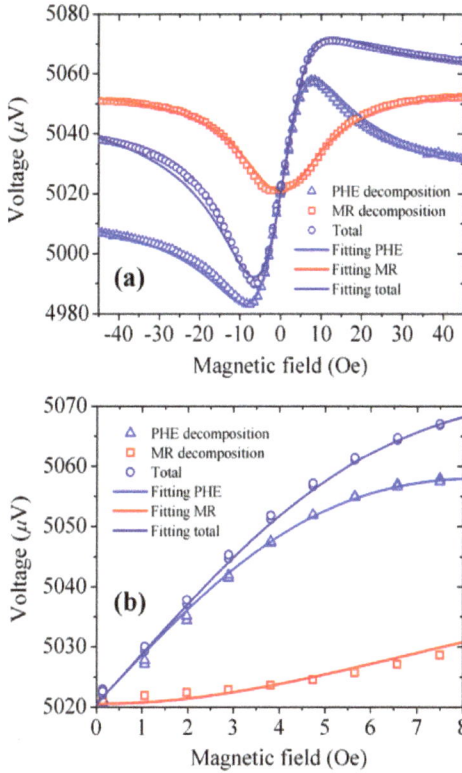

Figure 16. PHE and MR voltage components are decomposed from the experimental voltage profile of the sensor junction with the tilt angle $\zeta = 10°$ (a) at the field range of ±45 Oe and (b) at the field ranging from 0 to 8 Oe to illustrate the linearity of the sensor. In this figure, the origin of the PHE voltage component is adjusted to the minimum voltage of the MR components.

Generally, the longitudinal MR component was contributed from AMR and GMR effects [25,36]. The total output voltage induced from these effects satisfies the following equation [33]:

$$V_{MR} = I \times R_s \times \sin\zeta \times (1 + 0.5 \times GMR \times (1 - \cos(\theta - \theta_p)) + AMR \times \cos^2\theta) \qquad (8)$$

In this equation, θ_p is the angle between the magnetization direction of the F-pinned layer and the easy axis of F-free layer, and the drift voltage term ($I \times R_s \times \sin\zeta$) was modified from Ref. [33] in accordance with the investigated sensor junctions, because it depends on the length of the sensor junction. The increased length of the active region of the MR components depends on the sinusoidal function of tilt angle ζ. From Eq. (8), if the sensor junction has no tilt angle, V_{MR} is zero, in which case the sensor has only the PHE contribution. When the junction starts to tilt, the MR components contribute to the total sensor output voltage. The drift voltage and then the MR voltage depend on sinusoidal function of the tilt angle ($\sim I \times R_s \times \sin\zeta$) [37].

The decomposed MR voltage profiles can be described with values of the sheet resistance R_s = 28.5 Ω, GMR = 1.8 % and AMR = 0.4 %. Other parameters are kept the same as for the PHE voltage profile calculations. The trend of the calculated results of representative sensor junction with $\zeta = 10°$ is presented by the red solid line in Fig. 16.

Finally, the total output voltage profiles of the tilted junctions are calculated by combining both the PHE and MR components represented in Eq. (8). The results are shown by solid lines in Fig. 16, where the calculated drift voltages are adjusted to the experimental drift voltages. It is clearly evident that a rather good consistence between the experimental and the calculated data is obtained. Thus, the tilted cross-junction exhibited not only a better sensitivity in comparison with individual PHR sensor but also a better linearity compared with individual MR sensor.

4.4.3. Ring junction

The idea to develop the sensor based on a ring is to combine both the PHE and AMR components in one ring junction [38]; thus, the output voltage of the sensor can be enhanced. In the following, the role of the output signal as well as the optimization results will be discussed.

Firstly, for studying the role of the signal in the ring junction, we design the ring with different configurations. These rings have the same diameter of 300 μm and the same width of 20 μm. The illustration schemes and the tested results corresponding to each configuration using exchange biased structure Ta(3)/NiFe(50)/IrMn(10)/Ta(3) (nm) are given in Fig. 17

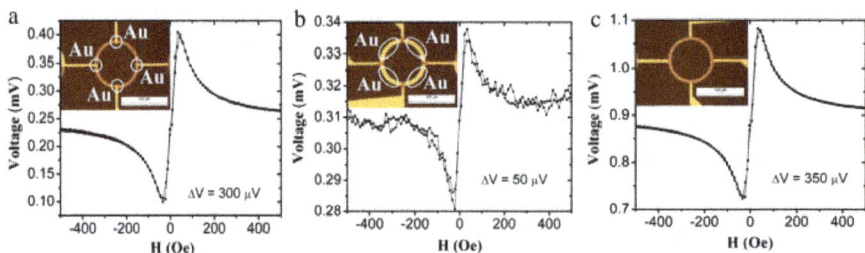

Figure 17. Designed rings with different Au electrode configurations and their corresponding output voltage profiles for the case of AMR arms (a), PHR elements (b) and a full ring (c) in the exchange biased structures shown in the inset.

It is evident from Fig. 17 that the signal change in the case of a full ring (350 μV) is close to the sum of the signals in the cases of a AMR arms (300 μV) and PHR elements (50 μV). Based on these obtained results we assume that, in the full ring junction, there exist two components AMR (Fig. 17(a)) and PHR (Fig. 17(b)).

4.5. Hybrid AMR and PHR ring sensor – Optimized performance

In order to optimize the performance of the sensor using a ring junction, efforts were made to design an hybrid AMR and PHR ring sensor. It is known that the maximum voltage of the AMR and PHR voltages in the ring can be calculated using:

$$V_{AMRo} = \frac{r}{\omega}\frac{I\Delta\rho}{t}$$
$$V_{PHRo} = \frac{I\Delta\rho}{t}$$

(9)

where r and ω are the radius and the width of the ring junction, I is the applied current, t is the thickness of the sensor material.

It is clear from the above that the PHR component is always constant while the AMR component increases linearity with the increase in r/ω ratio. It is noteworthy that when $r/\omega = 1$ the AMR voltage is equal to the PHR voltage, in which case the ring becomes the full disk. By fixing $I\Delta\rho/t = 1$, the output signal of the sensor is calculated, and the result is shown in Fig. 18.

The results in Fig. 18 ensure that the higher the r/ω ratio the larger the output voltage of the ring. To increase the r/ω ratio, basically, we can increase the radius, r, or reduce the width, ω, of the ring. However, for integrating with the other devices using present silicon technology, the ring size must be restrained to a certain limit. We assume that the ring size should be limited to about 300 μm, corresponding to the radius of $r = 150$ μm. The second problem that must be considered for optimizing the sensor performance is the width of the ring; the thinner the width, the higher the resistance, therefore, the higher output voltage can be achieved. But the width can not be made so thin, because the heat generated during the working time will burn the sensor junction. By considering these parameters, the optimized ring will have the radius of 150 μm and the width of 5 μm ($r/\omega = 30$).

The results of the sensitivity versus r/ω of the ring sensor using bilayer and trilayer structures (Ta(5)/Ru(1)/NiCo(10)/IrMn(10)Ru(1)/Ta(5) and Ta(3)/NiFe(10)/Cu(0.12)/IrMn(10)/Ta(3) (nm) are illustrated in Fig. 19. It is abundantly clear from the figure that the ring sensor using trilayer structure has higher sensitivity compared to that of bilayer structure. So the best performance of the ring is obtained using the trilayer structure, in which case the sensitivity is about 340 μV/Oe, and this is a much improved sensitivity compared to the sensitivity of an AMR or a PHR sensor (normally, the sensitivity of PHR sensor < 15 μV/Oe).

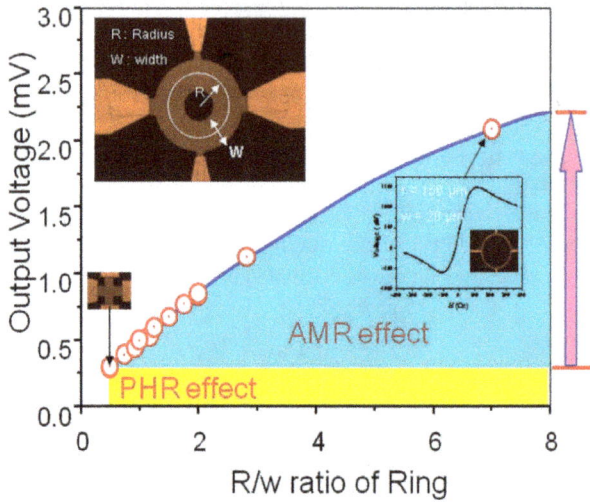

Figure 18. The calculation and experimental results of PHR and AMR output voltage components versus r/ω ratio of the ring. The insets show schematics of a ring junction with defined r and ω, and a representative PHE voltage profile of ring sensor for $r = 150\ \mu m$, $\omega = 20\ \mu m$.

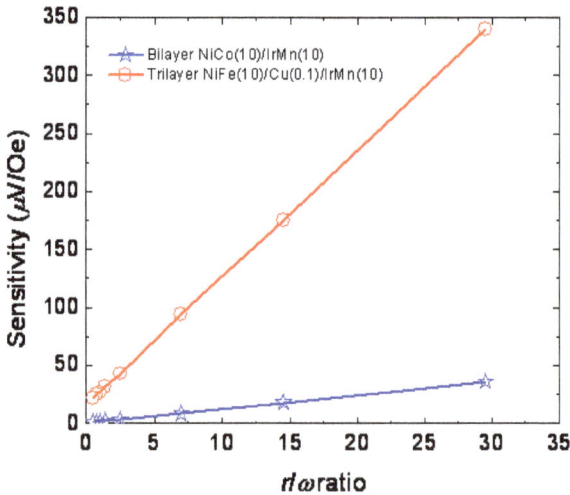

Figure 19. Experimental results of the sensitivity versus r/ω ratio of the rings using a Ta(5)/Ru(1)/NiCo(10)/IrMn(10)Ru(1)/Ta(5) (nm) bilayer thin film and trilayer thin film Ta(3)/NiFe(10)/Cu(0.12)/IrMn(10) /Ta(3) (nm).

It can be summarized from the above that the systematic investigations on the ring junctions revealed that there exist both PHR and AMR voltages contribute to the output voltage profile. The PHR voltage component is always kept constant when varying the size of the ring, while the AMR voltage component linearity increase due to the increasing the r/ω ratio of the ring. For practical and application aspects, the ring must be optimized both in terms of its size and performance. The optimized radius and the width of the junction are 150 μm and 5 μm, respectively. By using the trilayer structure, the best performance of the sensor is obtained. In such case, the highest sensitivity sensor is about 340 μV/Oe. This hybrid sensor is very much improved in the sensitivity compared to an AMR or a PHR sensor.

5. Biofunctionalized magnetic bead detection for state of the art lab-on-a-chip

Ever since the report of Baselt *et al.* on a magnetoresistive-based biochip with magnetic labels instead of fluorescent labels [3], the magnetic biochip has been extensively investigated as an advanced tool for sensitive detection of low bio-target concentration in body fluids for early diagnostics. Obviously, the focus in these investigations lies in development of a high sensitive magnetic field sensor that is optimized for magnetic label detection, and therefore different magnetoresistive sensing approaches, including the one that has just been described above i.e., hybrid AMR and PHR ring sensor, were adopted subsequently for this purpose. All these magnetic biosensors detect the stray field of magnetic particles that are bound to biological molecules. Since the biological environment is normally non-magnetic, the possibility of false signals being detected is negligible. In addition, the properties of magnetic particles are also stable over time and they may also be manipulated via magnetic forces, which can be produced by current lines that are fabricated into the chip itself. The advantages of magnetic labeling techniques have ultimately led the researchers to intensify their efforts in developing modern technologies for on-chip integration of micro- and nano-scale magnetics with molecular biology with a final goal of realizing highly sensitive, fast, reliable, cost-effective, portable and easy-to-use biomolecular sensor, the so called *magnetic lab-on-a-chip*.

5.1. Magnetic beads

Superparamagnetic nanoparticles coated with Streptavidin make ideal labels in bio-applications using magnetic sensors, because they can be readily magnetized to large magnetic moments. Most of our experiments were carried out with Dynabeads®M-280, which are composed of ultra small Fe_2O_3 nanopaticles embedded in a polymer matrix and the Streptavidin was conjugateed with the surface of the beads. The magnetization curve of the magnetic beads is shown in Fig. 20 [27, 39].

M-280 streptavidin

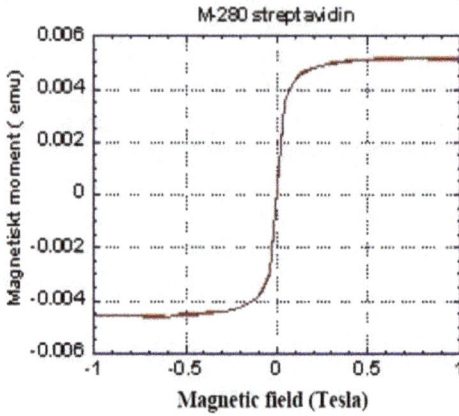

Figure 20. Magnetization curve of Dynabeads M-280 Streptavidin. This is supported by Dynal company.

When the magnetic bead appears on the sensor surface under an external magnetic field the magnetic field strength produced by a single bead can be estimated as [12, 14]

$$H = \frac{MR^3}{3r^3}(3(\hat{M} \cdot \hat{r})\hat{r} - \hat{M})$$ (10)

where M, $\hat{M}$ are the magnitude and unit vector of magnetization. R is the bead radius, and r, $\hat{r}$ are the magnitude and unit vactor of the distance from the center of the bead to observation point as shown in Fig. 21.

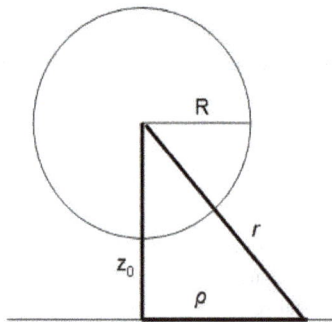

Figure 21. Schematic of a bead with the radius R placed above the sensor, r is the distance from the center of the bead to the observation point, z_0 is the vertical distance from the center of the bead to the sensor, ρ is the distance in the sensor plane from the center of the bead to the observation point.

Assuming that the applied field is in x direction in a polar coordinate system then Eq. (10) can be rewritten as:

$$H_x = H\hat{x} = \frac{MR^3}{3r^3}(3\sin^2\theta\cos^2\varphi - 1) \tag{11}$$

with $\hat{x}\hat{r} = \sin\theta\cos\varphi$ and $\hat{x}M = 1$ when converting from polar coordinate system to spherical coordinate system.

Substituting $\sin\theta = \frac{\rho}{r}$ and $r = \sqrt{\rho^2 + z^2}$ into Eq. (11), H_x can be rewritten as

$$H_x = \frac{MR^3}{3}\frac{3\rho^2\cos^2\varphi - (\rho^2 + z_0^2)}{(\rho^2 + z_0^2)^5} \tag{12}$$

Following the Eq. (12), stray field of magnetic bead reaches maximum when $\rho = 0$, in this case $(H_x)_{max} = -\frac{MR^3}{z_0^3}$. This field reaches maximum at a right angle to the magnetization of the bead ($r \equiv z_0$ in Fig. 21) and decreases at other points on the sensor plane. The effective field of a bead influences the sensor, H_x is integrated over a general sensor area, A.

$$\langle H_x \rangle = \frac{1}{A}\int H_x dA \tag{13}$$

If the sensor geometry is considered as a circle, the effective field of a bead can be calculated from Eq. (13) as

$$\langle H_x \rangle = -\frac{MR^3}{3z_0^3}\frac{1}{(1 + \frac{\rho_s^2}{z_0^2})^{3/2}} \tag{14}$$

Here ρ_s is the radius of the circular ring sensor

And if the sensor geometry is quadrate the effective field is given by

$$\langle H_x \rangle = -\frac{MR^3}{3z_0^3}\frac{1}{(1 + \frac{\omega^2}{4z_0^2})(1 + \frac{\omega^2}{2z_0^2})^{1/2}} \tag{15}$$

here ω is the width of the cross-junction sensor

It is revealed from Eq. (14) and Eq. (15) that the field effect to the sensor is very much depending on the size of the sensor, , it is proportional to the invert cube of radius of circular sensor or of the width of a quadratic sensor ($\frac{1}{\rho^3}$ or $\frac{1}{\omega^3}$).

5.2. Biofunctionalization of the beads

It is known that the biotin-streptavidin is one of the strongest non-covalent biological interaction systems having a dissociation constant, 'K_d', in the order of 4×10^{-14} M leading to the strength and specificity of the interaction to be one of the most widely used affinity pairs in molecular, immunological and cellular assays [40]. Usually in most assays, streptavidin is coupled to a solid phase such as a magnetic bead, or a biosensor chip, while biotin is coupled to the biomarker of interest, often a nucleic acid or antibody. Taking advantages of magnetic labels and specific ligand-receptor interactions of the biomolecules one can manipulate, separate and detect specific biomolecules.

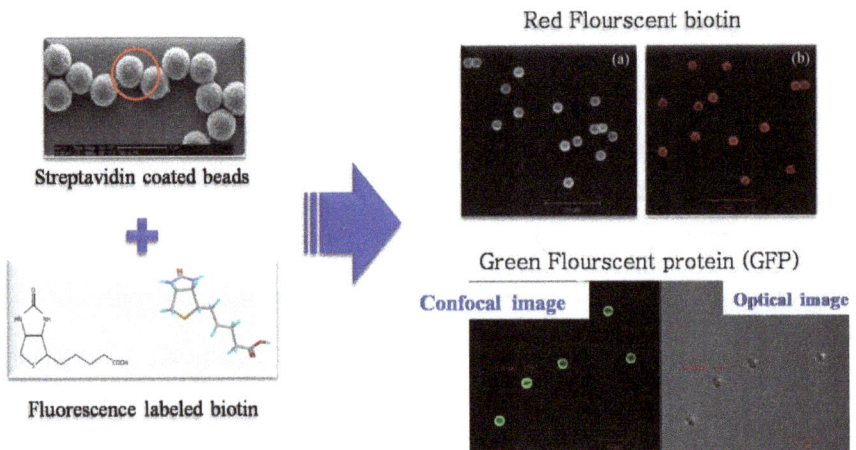

Figure 22. Procedure for the immobilization of fluorescent labeled biotin on the Streptavidin coated dynabeads measured by confocal optical microscope.

To demonstrate the translocation of streptavidin-biotin magnetic labels using the micro system, we have chosen the commercially available streptavidin coated magnetic beads (Dynabead® M-280) of 2.8 μm size to bind with fluorescent labelled biotin. Atto 520 is a new label with high molecular absorption (110.000) and quantum yield (0.90) as well as sufficient stokes shift (excitation maximum 520 nm, emission maximum 524 nm). Due to an insignificant triplet formation rate it is well suited for single molecule detection applications. In this experiment, Atto 520 biotin is attached on the streptavidin coated magnetic beads and observed the fluorescence signal through confocal microscope. In order to attach the Atto 520 biotin on streptavidin coated magnetic labels, we have taken, 5 μl of streptavidin coated magnetic labels (Dynabead® M-280) mixed with 0.1 M PBS buffer solution (90 μl) with pH of 7, and 5 μl of fluorescent label biotin (chemical concentration of fluorescent label was 1 mg/200 μl in EtOH) also added to the previous mixing solution and continuously stirring the solution for 2 hours at room temperature for the reaction completion.

Fig. 22 provides the direct evidence of protein immobilization which was obtained by immobilizing green fluorescent protein (GFP) and observed the fluorescence through confocal laser microscopy.

5.3. Sensor size and bead detection capability

For the micro-bead detection using a PHE sensor, it is noted that the magnetization of the magnetic sphere is purely a dipole at the center of the sphere with a magnetic field at a distance identified by the dipole field from Eq. (15). The stray field of a single bead on the sensor surface could be crudely calculated by [41]

$$H_{bead} \approx -\frac{\chi V}{4\pi r^3} H \qquad (16)$$

where V is the volume of magnetic bead, χ is the volume susceptibility of magnetic beads. This stray field is in the opposite direction to the applied field, thus it reduces the effective field on the sensor surface. Under the experiment conditions, the stray field of N beads on the sensor surface reduced the sensor output signal as follows:

$$V_{bead} = V_{PHR}(H_{eff}) \approx V_{PHR}(1 - NH_{bead}) \approx V_{PHR} + \Delta V_{bead} \qquad (17)$$

where H_{eff} is the effective field on the sensor surface, S is the sensor sensitivity of PHR sensor. The voltage signal, ΔV_{bead} generated by the magnetic bead themselves can be expressed as

$$\Delta V_{bead} = V_{bead} - V_{PHR} \approx V_{PHR}\left(\frac{N\chi V}{4\pi r^3} H\right) \qquad (18)$$

By substituting the value $\chi = 0.13$ [39] and $r = 1.55$ μm (the distance including the radius of Dynabeads® M-280 and the thickness of passivated SiO and Ta layers) into Eq. (17), the stray field of single bead is estimated to be $H_{bead} \sim 0.03$ H under the applied field. The number of bead separately placed on the sensor surface can be calculated using the PHR sensor.

Fig. 23 shows the V_{PHR} and the V_{bead} in the functions of magnetic field with number of bead N=1, 5, and 10, respectively. It is clearly shown that the beads on the sensor surface modify the PHR signal due to the small stray field compared with applied magnetic field.

In the PHR sensor, the V_{PHR} can be used the reference signal. The difference voltage ΔV_{bead} between the V_{PHR} and V_{bead} can be estimated, which is shown in Fig. 24.

The pure bead signal ΔV_{bead} is small compared with V_{PHR}. However, the ΔV_{bead} changes with the applied magnetic field and show maximum and minimum values at special magnetic field, which is due to the PHR sensor performance. Therefore, the bead detection capability can be determined at the maximum and minimum ΔV_{bead}. If we set the applied field at the maximum or minimum value of ΔV_{bead}, we can detect the magnetic bead with high signal voltage.

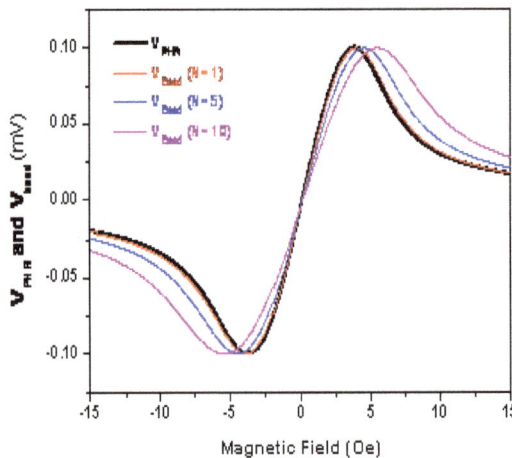

Figure 23. V_{PHR} without bead (black solid line) and V_{bead} with bead (N=1, 5 and 10) by using the F/AF bilayers in the functions of applied magnetic field H.

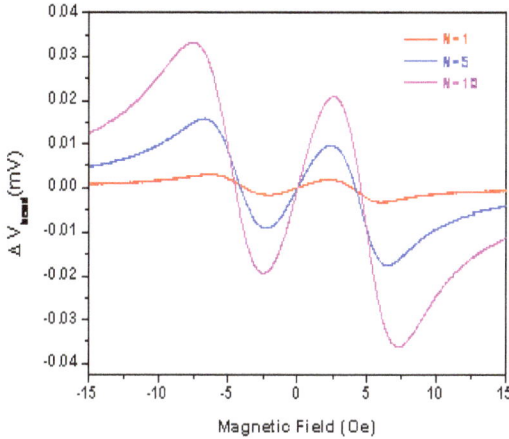

Figure 24. Calculation of ΔV_{bead} of the PHR sensor with N=1, 5 and 10.

5.3.1. Multi-bead detection

We performed the magnetic bead detection using PHR sensor using Ta(3)/NiFe(16)/Cu(1.2)/NiFe(2)/IrMn(10)/Ta(3) (nm) to demonstrate the feasibility of magnetic bead detection for bio applications. The diluted 0.1 % magnetic bead solution streptavidin coated Dynabeads® M-280 is used for bead drop and wash experiments on the sensor surface. The real-time profile measurements of the PHE voltage for magnetic bead detection are carried out in the optimum conditions, that is, in an applied magnetic field of 7 Oe and with a sensing current of 1 mA. The results are illustrated in Fig. 25 for three consecutive cycles. The lower state represents the signal change in sensor output voltage after dropping the magnetic bead solution on the sensor surface whereas the higher state represents the sensor output voltage after washing magnetic beads from the sensor surface. Total output signal annuls in three consecutive cycles were found to be about 7.1 μV, 16 μV and 21.8μV for the first step and 11.3 μV and 16.7 μV in the second step of the second and third cycles, respectively. It is clearly shown from the figure that for the first cycle, the signal changed by one-step and the signal was further changed into two steps in the second and third cycles.

This two step-type profile is due to the aggregation process of the magnetic beads on the sensor surface. The aggregation of the magnetic beads occurs at the drying stage. That is, after dropping the bead solution on the sensor surface, it needs some time to dry. The first step changes of the signals are assumed to be due to the viscous flow motion for stabilization as well as the Brownian motion of the beads. When the solution dries, the beads rearrange. During this time, some beads aggregate and become clusters on the sensor surface.

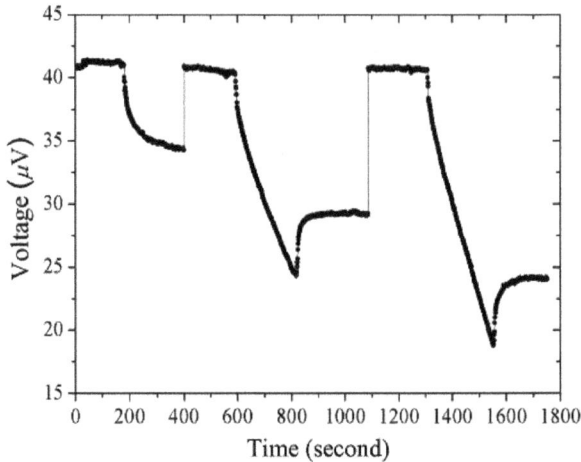

Figure 25. Real-time profile of PHR sensor under an applied magnetic field of 7 Oe with the sensing current of 1 mA

This lessens the total stray field on the sensor surface and hence, the second step in the second and third cycles was observed in the real-time profile.

In the process of analyzing the micro-bead detection using PHE sensor, it is noted that the direction of magnetic field H and the stray field of magnetic bead on the sensor surface H_{bead} (Eq. (16)) are oppositely aligned, and thus the effective field on the sensor surface is reduced.

Thus, a rough estimate of the number of magnetic particles on the sensor surface in this identical experiment based on the reduced stray field and sensor output signal can be expressed from Eq. (18) by rephrasing it again here for better clarity:

$$\Delta V_{bead} = V_{PHR}\left(\frac{N\chi V}{4\pi r^3}H\right)$$

By substituting the value $\chi = 0.13$ and $r = 1.55$ μm (the distance including the radius of Dynabeads® M-280 and the thickness of passivated SiO_2 and Ta layers) into Eq. (16), the stray field of single bead is estimated to be 2.2×10^{-2} Oe under the applied field of 7 Oe. Theoretically, with the sensor sensitivity $S = 7.6$ μV/Oe and the sensing current $I = 1$ mA, the number of beads separately placed on the sensor surface can be calculated in the first step of the three cycles by using Eq. (18), which are estimated to be about 4, 10 and 13 beads, respectively.

These estimated results strengthen our explanation. It is clearly shown in the first cycle, the number of beads on the sensor surface is estimated to be small, and the distance among beads on the sensor junction is far enough to avoid the effect from the rearrangement of beads during the drying stage. In the second and third cycles, the number of magnetic beads on the sensor junction is larger; they easily aggregate to become clusters under applied magnetic field due to short bead-bead distance

5.3.2. Single bead detection

We performed single magnetic bead detection experiments on several kinds of sensor structures such as spin-valve and bilayer exchange biased thin films [27, 42 - 44], and the representative results are being presented here. For the purpose of performing single micro-bead detection, the PHR sensor with the junction size of 3 μm × 3 μm was fabricated using Ta(5)/NiFe(16)/Cu(1.2)/NiFe(2)/IrMn(15)/Ta(5) (nm). This is the optimized spin-valve thin film for the PHR sensor in our lab. A droplet of 0.1 % dilute solution of the Dynabeads® M-280 was introduced on the surface of the sensor. A single micro-bead was isolated and positioned on the center of the sensor junction by using a micro magnetic needle which is known as a tweezer method. The magnetic needle was prepared by using a soft magnetic micro wire, the wire is magnetized by attaching a permanent magnet to one end of the wire, the single magnetic bead is attracted with the other end due to the magnetic field of the wire and it is dragged and positioned to the center of the sensor junction. It is noteworthy that the magnetic bead is attracted by the magnetic force; this force is strong enough to compensate the Brownian motion during the experiment. The experiment was carried out under the observation of an optical microscope. When the solution dried, the bead was fixed on to the sensor surface.

Since the magnetic properties of the MR as well as the PHE response to the magnetic field are described in the previous section, the results of single bead detection using 3 μm × 3 μm PHR sensor will be discussed here.

The SEM image of a single bead on the center of the sensor junction is shown in the Fig. 26(a). The voltage profiles of the PHE sensor in the absence and presence of a single micro-bead are presented in Fig. 26(b) by black circle and red rectangle ones, respectively. It is shown from the figure that in the increasing region of the PHE voltage profile (in the field ranging from 0 Oe to 10.6 Oe), the $V_{PHE}(H)$ is decreased when the magnetic bead exists on the sensor surface and vice versa for the decreasing region of the PHE voltage profile (at the fields exceeding 10.6 Oe).

For understanding the role of a single micro-bead detection using a PHE sensor, we consider the voltage drop by stray field of a single magnetic bead. The calculation method is the same as deduced for Eq. (18). And when considering that the magnetic bead is located on the center of sensor junction, the stray field affects the PHE voltage as follows:

$$V_{stray} = I \times S \times \left(1 - k\frac{\chi V_{bead}}{4\pi z^3}\right) \times H_{app} \tag{19}$$

where V_{stray} denotes the voltage change due to the stray field of magnetic bead, $S = \dfrac{\partial V_{PHE}}{\partial H}$ is the sensitivity of the sensor at instantaneous applied fields.

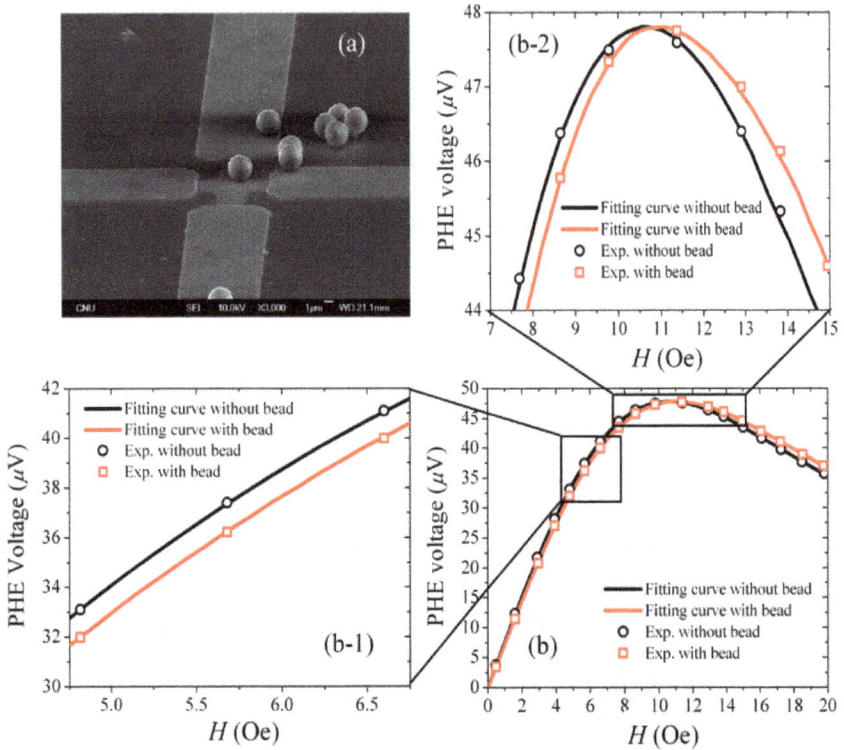

Figure 26. (a) The SEM image of the sensor junction in the presence of a single micro-bead, (b) the theoretical and experimental PHE voltage profiles in the absence and presence of a micro-bead, (b-1) enlarged picture of the increasing PHE voltage region at the field range of 4.75-6.74 Oe and (b-2) enlarged picture of the PHE voltage profiles around the maximum PHE voltages [44].

By substituting $\chi = 0.13$ [39] in Eq. (19) with active fraction of $k = 0.62$ and $z = 1.55$ μm (along with 150 nm thick SiO_2 passivation layer and 1.4 μm of magnetic bead radius), the PHE voltage is calculated at instant applied fields for the presence of a micro-bead. The solid lines in the Fig. 26(b) illustrate the calculated profiles for the case of absence and presence of a micro-bead, respectively. These calculated results are in good agreement with the experimental results.

By comparing the PHE voltage profiles in the absence and presence of a micro-bead, one can find that (*i*) at low magnetic fields, the PHE voltage increases with the field increase, *i.e.* the sensitivity of the sensor is positive. In this case, the presence of magnetic bead lessens its PHE voltage as illustrated in Fig. 26(b-1). (*ii*) In the presence of the magnetic bead, the maximum PHE voltage shifts to a higher field with an amount of H_{bead} as presented in Eq. (16); at

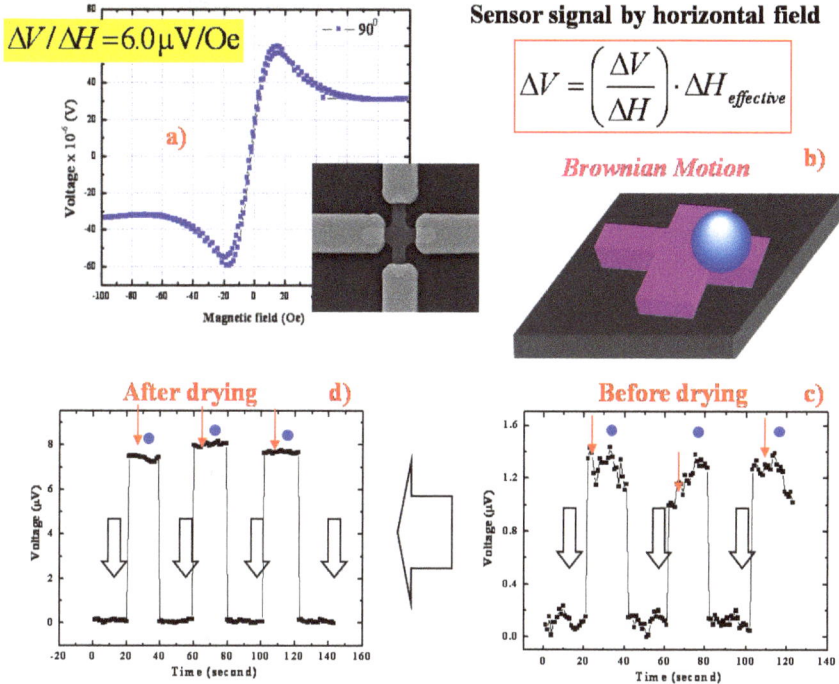

Figure 27. Voltage change of the PHR sensor versus applied field when a single magnetic bead appear on the sensor surface.

about 10 Oe this stray field strength is approximately 0.43 Oe. (*iii*) At higher applied fields (> 10 Oe), the PHE voltage decreases with the field increase, *i.e.* the sensitivity of the sensor is negative. In this case, the presence of magnetic bead increases the PHE voltage with an amount of $k \frac{\chi V_{bead}}{4\pi z^3} I \cdot S \cdot H_{app}$. This is clearly evident in the Fig. 26(b-2) and thus the PHE signal satisfies Eq. (19).

In particular, at low field range, a very good linear and large change of the PHE voltage always occur, so this field range is usually chosen to demonstrate the feasibility of the digital detection of the magnetic beads [10-14]. In our approach for this sensor, the signal change versus the applied field is extracted from PHE voltage curves in the presence and absence of magnetic bead, the result is drawn in Fig. 27, the maximum change of $V_{PHE}(H)$ about 1.14 μV can be obtained at the applied field ~ 5.6 Oe. This calculated result satisfies Eq. (19). Further, Fig 26(b) shows that there is a very good agreement between the single bead measurement data and the theoretical curves. There is only a very small noise scatter of experiment data from the fitting curve, this is the evidence showing that the fabricated PHE sensor has high

SNR. Therefore, the PHE sensor has advantages for more accurate detection of the small stray fields of magnetic beads.

This simple calculation is suitable for the effect of a single bead on the center of small size sensor junction. When the area of the sensor junction is larger than the area of magnetic beads, the calculation must be considered the effect of the magnetic bead from different positions of sensor junction and the contribution of nearby beads or chains of beads on the sensor. In such a case the output signal changes negative for the bead inside of the sensor junction and changes positive for the beads outside of the junction. Moreover, the signal change does not depend on the number of magnetic beads proportionally. This was studied systematically and was reported by P.P. Freitas *et al.*,[23], L. Ejsing et al., [9] and Damsgaard *et al.*, [45].

5.4. Integration of magnetic sensors/microfluidic channels

In this part, we design and optimize the planar Hall ring sensor for detecting the hydrodynamic magnetic labels. Once the magnetic labels appear on one arm of the ring sensor, the resistance of the sensor will be changed, the role of resistance change obey the Wheatstone bridge circuit geometry hence the sensor is very sensitive to detect the magnetic labels.

Planar Hall ring sensor was fabricated by photolithography technique. Sensor material Ta(3)/NiFe(10)/IrMn(10)/Ta(3) (nm) was fabricated by using a DC sputtering system with the based pressure of 7×10^{-8} Torr. The field sensitivity of the ring sensor based on the bilayer thin film was found to be about 0.3 mV.Oe^{-1}. The sensor was integrated with a microfluidic channel, which can produce the laminar flow of the magnetic labels (beads and/or tags) in the specific arms of the ring sensor by hydrodynamic flow focusing technique. This magnetic platform can detect even a single magnetic bead of 2.8 μm motion in real time by the measurement system with a sampling rate of 5 kHz.

The schematic representing the integrated magnetic platform is shown in Fig. 28. In magnetic bead separation experiments initially the magnetic beads with different sizes are injected into the main stream of the microfluidic channel with certain fluidic flow rate. Then the beads are gathered at the weir in the fluid channel and then sorted according to the attractive force exerted on the magnetic bead by the magnetic elements/magnetic pathways. Therefore, the labeled magnetic beads of same kind will attract to one of the magnetic pathways in the sub channel. The weir at the entrance of the sub-channels opposes the beads temporarily for magnetic beads whose magnetization is insufficient to be attracted by the magnetic elements. But, the beads whose magnetization is sufficient to be attracted by the poles of the saturated ellipses due to the external rotating magnetic field can overcome the weir.

After successful separation of the magnetic beads of different sizes we wish to adopt two types of different sensing techniques such as an array of PHR biosensors and multi-segmented nanowires. The planar array of PHR sensor can detect magnetic beads with micron size only. But in case of nanometer size magnetic beads, we wish to use simple read out technique of multi-segmented nanowires. We are also planning to combine magnetic pathway method with the microwire and coil method.

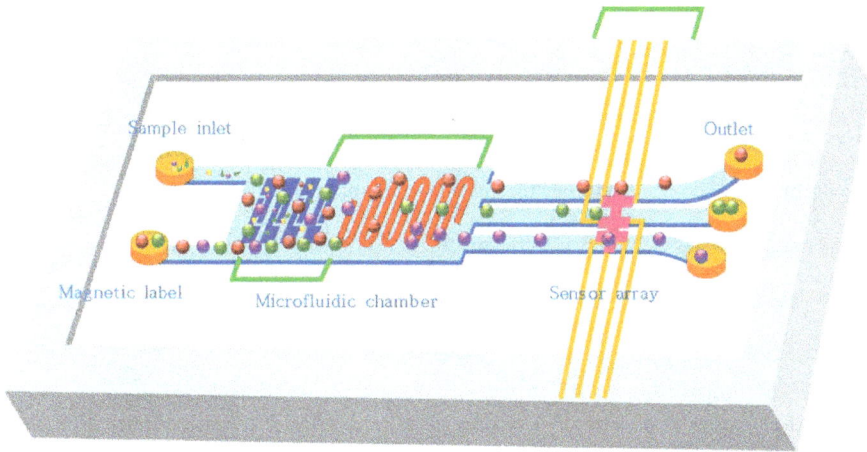

Figure 28. Schematic represents the magnetic platform integrating an array of planar Hall ring sensors and a microfluidic channel.

6. Conclusions

The underlying principle for magnetic biosensing has been elaborately described at first with the examples of different magnetoresistive sensing techniques. Then, the planar Hall resistance sensor has been shown as one of the best sensors for conducting magnetic bead detection experiments. While making an in depth study on the capabilities of a PHR sensor in different configurations and geometries, the sequence of narration ultimately has lead towards describing the evolution of hybrid AMR and PHR ring sensor in spin-valve configuration with optimized performance for precise detection of even single magnetic bead. Biofunctionalization experiments were also conducted to ensure that our PHR sensor is capable of biomolecule recognition. Therefore, our present sensor can be used to promote for the biomolecular recognition and other molecular interaction detection. This novel planar Hall effect based sensor has been further demonstrated that it can be easily integrated into a lab-on-a-chip and is feasible for bead detection in the sensing current generated magnetic field (without the external applied magnetic field) so as to ensure it an efficient tool for high sensitive biomolecules recognition.

Author details

Tran Quang Hung[1,4], Dong Young Kim[2], B. Parvatheeswara Rao[3] and CheolGi Kim[1*]

*Address all correspondence to: cgkim@cnu.ac.kr

1 Department of Materials Science and Engineering, Chungnam National National University, Daejeon, Korea

2 Department of Physics, Andong National University, Andong, Korea

3 Department of Physics, Andhra University, Visakhapatnam, India

4 Laboratoire Charles Coulomb, CNRS-University Montpellier 2, Montpellier, France

References

[1] M.M. Miller, G.A. Prinz, S.F. Cheng, S. Bounnak, Appl. Phys. Lett. 81, 2211–2213 (2002).

[2] . Andreev, P. Dimitrova, Journal of Optoelectronics and Advanced Materials 7, 199-206 (2005).

[3] D.R. Baselt, G.U. Lee, M. Natesan, S.W. Metzger, P.E. Sheehan, R.J. Colton, Biosens. Bioelectron. 13, 731–739 (1998).

[4] D.K. Wood, K.K. Ni, D.R. Schmidt, A.N. Cleland, Sensors and Actuators A 120, 1–6 (2005).

[5] .C. Rife, M.M. Miller, P.E. Sheehan, C.R. Tamanaha, M. Tondra, L.J. Whitmana, Sensors and Actuators A 107, 209–218 (2003).

[6] . Schotter, M. Panhorst, M. Brzeska, P. B. Kamp, A. Becker, A. Pühler, G. Reiss, H. Brueckl, Nanoscale Devices - Fundamentals and Applications, edited by R. Gross et al.(Springer) 35–46 (2006). GMR-TMR

[7] .S. Moodera, L.R. Kinder, T.M. Wong, R. Meservey, Phys. Rev. Lett. 74, 3273 (1995).

[8] P.A. Besse, G. Boero, M. Demierre, V. Pott, R. Popovic, Appl. Phys. Lett. 80, 4199–4201(2002).

[9] L. Ejsing, M. F. Hansen, A. K. Menon, H. A. Ferreira, D. L. Graham and P. P. Freitas, Appl. Phys. Letts. 84, 4729 (2004).

[10] N.T. Thanh, K.W. Kim, C.O. Kim, K.H. Shin, and C.G. Kim, J. Magn. Magn. Mater. 316, e238 (2007).

[11] N. T. Thanh, L. T. Tu, N. D. Ha, C. O. Kim, CheolGi Kim, K.H. Shin, and B. Parvatheeswara Rao, J. Appl. Phys. 101, 053702 (2007).

[12] P. P. Freitas, H. A. Ferreira, D. L. Graham, L. A. Clarke, M. D. Amaral, V. Martins, L. Fonseca, J. S. Cabral, in Magnetoelectronics, edited by M. Johnson (Elsevier, Amsterdam, 2004) and references therein.

[13] (a) B. D. Tu, L. V. Cuong, T. Q. Hung, D. T. H. Giang, T. M. Danh, N. H. Duc, C. G. Kim, IEEE Trans. Magn. 45, 2378 (2009); b) T. Q. Hung, S. J. Oh, B. D. Tu, N. H. Duc, L. V. Phong, J.-R. Jeong, and C. G. Kim, IEEE Trans. Magn. 45, 2374 (2009).

[14] E. Ejsing, the dissertation of doctor of philosophy, Technical University of Denmark, Denmark (2003).

[15] a) F. Nguyen Van Dau, A. Schuhl, J.R. Childress, M. Sussiau, Sensors and Actuators A, 53, 256-260 (1996); b) N. T. Thanh, B. Parvatheeswara Rao, N. H. Duc, and CheolGi Kim, phys. stat. sol. (a) 204, 4053–4057 (2007).

[16] T. R. Mcguire and R. I. Potter, "Anisotropic magnetoresistance in ferromagnetic 3d alloys," IEEE Trans. Magn. 11, 1018-1038 (1975).

[17] B. Dieny, J. Magn. Magn. Mater. 136, 335 (1994).

[18] S. S. Parkin, C. Kaiser, A. Panchula, P. M. Rice, B. Hughes, M. Samant and S. H. Yang, "Giant tunneling magnetoresistance at room temperature with MgO (100) tunnel barriers," Nature Mat., 3, 862-867 (2004).

[19] S. Yuasa, T. Nagahama, A. Fukushima, Y. Suzuki, and K. Ando, "Giant room temperature magnetoresistance in single crystal Fe/MgO/Fe magnetic tunnel barriers," Nature Mat., 3, 868-871, (2004).

[20] S. J. Oh, Le Tuan Tu, G. W. Kim, and CheolGi Kim, Phys. Stat. Sol. (a) 204, 4075 (2007).

[21] G. H. Yu, M. H. Li, F. W. Zhu, H. W. Jiang, W. Y. Lai, and C. L. Chai, Appl. Phys. Lett. 82, 94 (2003).

[22] T. Lucinski, G. Reiss, N. Mattern, and L. van Loyen, J. Magn. Magn. Matter. 189, 39 (1998).

[23] P. P. Freitas, R. Ferreira, S. Cardoso, and F. Cardoso, J. Phys.: Condens. Matter. 19, 165221 (2007).

[24] C. Parados, D. Garcia, F. Lesmes, J. J. Freijo, and A. Hernando, Appl. Phys. Lett. 67, 31 (1995)

[25] D. Y. Kim, C. G. Kim, B. S. Park, C. M. Park, "Thickness dependence of planar Hall resistance and field sensitivity in NiO(30 nm)/NiFe(t) bilayers" J. Magn. Magn. Mater. 215-216, 585–588 (2000).

[26] (a) www.ssec.honeywell.com; b) Candid Reig, María-Dolores, Cubells-Beltran and Diego Ramirez Munoz, sensors. 2009, 9, 7919 – 7942; c) Hoffmann, K., Applying the Wheatstone bridge circuit, HBM S1569-1.1en, HBM, Darmstadt, Germany.

[27] N.T.Thanh, the dissertation of doctor of philosophy, Chungnam National University, Korea (2006)

[28] (a) W. H Meiklejohn; C. P Bean, Physical Review, 105, 904–913 (1957); (b) J. Nogués, Ivan K. Schuller, J. Magn. Magn. Mater., 192, 203 (1999).

[29] (a) William C. Cain and Mark H. Kryder, IEEE Trans. Magn. 25, 2787 (1989); b) Z. Q. Lu, G. Pan, and W. Y. Lai, J. Appl. Phys. 90, 1414 (2001).

[30] B. Dieny, V. S. Speriosu, S. Metin, S. S. P. Parkin, B. A. Gurney, P. Baumgart, and D. R. Wilhoit, J. Appl. Phys. 69, 4774 (1991).

[31] N. J. Gökemeijer, T. Ambrose, and C. L. Chien, Phys. Rev. Lett. 79, 4270 (1997).

[32] T.Q. Hung, Sunjong Oh, Brajalal Sinha, J.R. Jeong, Dong-Young Kim and CheolGi Kim, J. Appl. Phys. 107, 09E715 (2010).

[33] B. Dieny, J. Phys.: Condens. Matter. 4, 8009 (1992).

[34] B. Dieny in: Magnetoelectronics, Ed. M. Johnson, (Elsevier, Amsterdam 2004).

[35] T. Q. Hung, J.-R.Jeong, D.Y. Kim, N. H. Duc and C. G. Kim, J. Phys. D: Appl. Phys., 42, 055007.1-5 (2009).

[36] M. Li, S.-H. Liao, C. Horng, Y. Zheng, R. Y. Tong, K. Ju, and B. Dieny, IEEE Trans. Magn., 37, 1733 (2001).

[37] D. Y. Kim, B. S. Park and C. G. Kim, J. Appl. Phys. 88, 3490 (2000).

[38] Sunjong Oh, P.B. Patil, T.Q. Hung, B. Lim, M. Takahashi, Dong Young Kim and CheolGi Kim, Solid State Commn., 151, 1248-1251 (2011).

[39] See physical characteristics of Dynabeads® M-280 streptavidin at http://tools.invitrogen.com/content/sfs/manuals/112.05D06D%20602.10%20Dynabeads %20M280%20Streptavidin%20(rev%20012).pdf

[40] Anders Holmberg, Anna Blomstergren, Olof Nord, Morten Lukacs, Joakim Lundeberg and Mathias Uhlén, Electrophoresis, 26, 501–510 (2005).

[41] L. Ejing, M. F. Hansen, and A. K. Menon, "Planar Hall effect magnetic sensor for micro-bead detection," in Conf. Proc., Eurosensors 2003, 1095–1098 (2003).

[42] Bui Dinh Tu, Tran Quang Hung, Nguyen Trung Thanh, Tran Mau Danh, Nguyen Huu Duc and CheolGi Kim, J. Appl. Phys. 104, 074701 (2008).

[43] Tran Quang Hung, the dissertation of doctor of philosophy, Chungnam National University, Korea (2010).

[44] (a) Tran Quang Hung, Sunjong Oh, Jong-Ryul Jeong, and CheolGi Kim, Sensors and Actuators A, 157, 42–46 (2010); b) Brajalal Sinha, S. Anandakumar, Sunjong Oh, CheolGi Kim, Actuators A, 182, 34–40 (2012).

[45] C.D. Damsgaard and M.F. Hansen, J Appl. Phys. 103, 064512 (2008).

Chapter 9

Impedimetric Biosensors for Label-Free and Enzymless Detection

Hilmiye Deniz ErtuğruL and Zihni Onur Uygun

Additional information is available at the end of the chapter

-

1. Introduction

Currently, Biosensor technology has provided a number of benefits to detect both biological and chemical molecules. Abiosensor is a promising device, which is combination of sensitivity of electrochemistry and specificity of biological recognition, enables to detect any kind of molecules in a short time with selectively and sensitively. Likewise many analytical methods, it has also limitations, such as high oxidation potentials lead to detection of non-target molecules, furthermore non-electroactive species cannot show electroactive signal for measurement or some biomolecules cannot be transformed by enzymes, even if they can be transformed, they require secondary molecules such as mediators, coenzymes or labels. In order to detect molecules without electrochemical reaction, electrochemical impedance spectroscopy (EIS) can be employed as a measurement technique "to see electrode surface modifications just by looking impedance curves". As it is known, electrochemical impedance spectroscopy is an electrochemical technique that provides the examination of electrical properties of electrode surface and binding kinetics of molecules between electrolyte and electrode surface. Therefore it can be used for biomolecular recognition, biomolecular bindings and biomolecular interactions between molecules such as DNA-DNA, DNA-protein, Receptor-Ligand, Protein-Ligand, Antibody-Antigen, and Ion Channels-Ligands. As a consequence of this affinity provides label-free detection without chemical transformation and this binding property can be monitored by electrochemical impedance spectroscopy expeditiously. In this chapter, the information will answer a number of questions about the development of impedimetric biosensors. In fact that is focused on the usage of impedance for biosensor technology, and to demonstrate impedance curves and the meaning of electrical elements for obtained Nyquist Plots especially faradaic impedance. Employed biorecognition receptors and not yet employed biorecognition receptors, which have different chemical residues, are discussed.

2. Theory of electrochemical impedance spectroscopy for biosensors

Electrochemical impedance spectroscopy and the method of impedance are widely used for corrosion, batteries, bioelectrochemistry and electrochemistry. EIS provide electrochemical examination of electrical properties of electrode surface; on the other hand it can be called as electrochemical surface characterization. Therefore it is possible to realize the differentiation of electrode surface alterations easily. In biosensor technology it is used for monitoring bio-sensor modifications, layer formation on electrode surface and binding kinetics between molecules such as DNAs, receptors, antibodies, antigens, proteins, ions etc. This advantage provides examination binding kinetics of molecules, just by using obtained impedance spec-trums for binding kinetics of molecules leads to label-free detection. As it is known, for en-zyme based biosensors, a molecule needs to be transformed into another molecule by enzyme for obtain electroactive molecules or electrons for gain an electroactive signal elec-troactive signal can be disturbed by other molecules, which oxidation reduction potentials are same as analyte molecule. Electrochemical impedance spectroscopy overcomes this problem and provides non-electroactive detection of molecules. There is only one condition, which is the most important handicap, is to find a most specific biorecognition receptor for analyte. Likewise all electrochemical measurement techniques, by employing EIS for biosen-sor technology has the same fundamentals, which are composed of electrical circuits in or-der to determine electrochemical measurement. This electrical circuit is affected by AC current, which is generally used for impedimetric experiments. Alternative current (AC) is a wave shaped and has a frequency; therefore both potential and current oscillate (Fig. 1). This oscillation causes differentiation in time because AC excitation signal and sinusoidal current response are both based on Ohm law. As it is known the Ohm law includes; a potential, a current and a resistance for ideal DC circuits. However, for AC, some mathematical units must be added because of the frequency of AC.

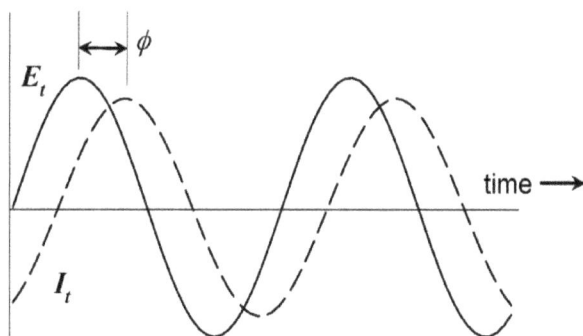

Figure 1. Alternative current; E_t and I_t.

As it can be seen in figure 1, the sinusoidal fluctuation of both current and potential show a difference, this difference, Φ, is determined as impedance which is an alternative current system resistance. Mathematical equationof this system is transformed into this equation 1 (Z; impedance, E_t; potential in a time, I_t; current in a time, E_0; potential at zero point, I_0; current at zero point, ω; frequency, t; time)

$$Z = \frac{E_t}{I_t} = \frac{E_0 Sin(\omega t)}{I_0(Sin(\omega t + \Phi))} = Z_0 \frac{Sin(\omega t)}{Sin(\omega t + \Phi)} \tag{1}$$

In this equation, impedance is represented as Z, and Z is a phase shift of AC, furthermore this phase shift is angle of impedance curve of Nyquist plot.

This theory has been performed for biosensor technology for a long time, its aim is examination of electrical characteristics of electrode surface for every layer formation and every interaction between molecules, and the obtained signal variations. In fact that charged groups of molecules have effect on impedance curves, layer has influence on electrical characteristic of electrode, this causes distribution of electrode surface charge, subsequently capacitive current varies, hence electrical circuit of the system keeps it balance and impedance increases or decreases [1].

EIS has an advantage over the other electrical measurement technique, which is an opportunity to design electrical circuit, according to obtained Nyquist plot curve. For figure 2, there is an electrical circuit model for obtained impedance curve. As you can see there are resistances and capacitance, in figure 2 there are both parallel capacitance and resistance, which they represent electrode surface, and a resistance is serial over this circuit, capacitance represents electrical double layer of electrode, R_2 represents resistance of the electrode, and R1 represents resistance of the solution in cell which is located electrodes inside of it. The Nyquist plot of this electrode starts not zero point which means that the solution in the cell shows a resistance (If it started at zero point, that means the resistance of solution(R_1) does not exist); therefore a resistance element(R_1) is added in circuit model. The rest of the curve shows a characteristic sinusoidal impedance curve, which means only a resistance occurs, and R_2 is added on circuit, and capacitance always occurs as a function of capacitive current, which represents in homogenity on electrode surface, because electrical double layer occurs.

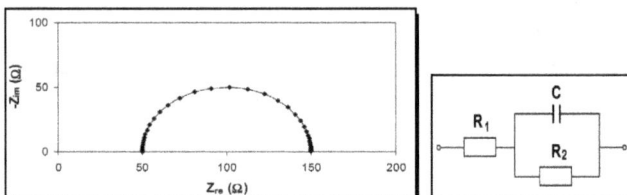

Figure 2. Non-faradaic impedance curve, R_2 electrode surface resistance and capacitance, and R_1 cell surface resistance.

Variation on impedance curve changes the electrical circuit model, an alteration especially on Nyquist curve an circuit element is added after R_2 circuit element as serial [2].

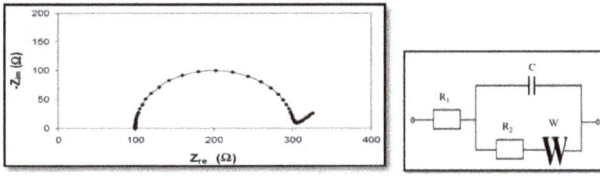

Figure 3. Faradaic impedance curve, R_2; electrode surface resistance, W; Warburg impedance and C; capacitance, and R_1; cell solution resistance.

For figure 3, there is an additional circuit element, Warburg impedance(W) which represents mass transfer to electrode surface. This resistance occurs when an interaction formation, which is formed by electrical interaction, adsorption e.g., between electrode surface and solution a mass transfer occurs towards electrode surface, this transfer cannot be calculated as diffusion because there is an accelerated mass transfer by affection. Therefore Nyquist curve varies and becomes linear; this linearity represents Warburg impedance, which means mass transfer resistance. The interaction between electrode surface and solution, keep the balance between Warburg impedance and electrode surface resistance(R_1), this balance can be unbalanced by mass transfer and after any increase on mass transfer the Warburg impedance shows dominancy on electrode surface resistance (R_1) [3]. This domination shields resistance and sometimes resistance doesn't occur. A study was performed by Uygun and Sezgintürk, in this study gold film modified glassy carbon electrode was modified by SAM of Cysteamine layer and positively charged Cysteamine attracted negatively charged redox probe($Fe(CN)_6^{-3}/Fe(CN)_6^{-4}$), this mass transfer and this attraction shadowed no resistance and Warburg impedance showed dominancy on circuit [4]. On the other hand a redox probe as $[Ru(NH_3)_6]^{3+}$ will be repelled by positively charged modification. As you can see there are a number of conditions for impedance to design a circuit model, which surface of electrode, content of solution, characteristics of redox probe are important on electrical circuit modelling.

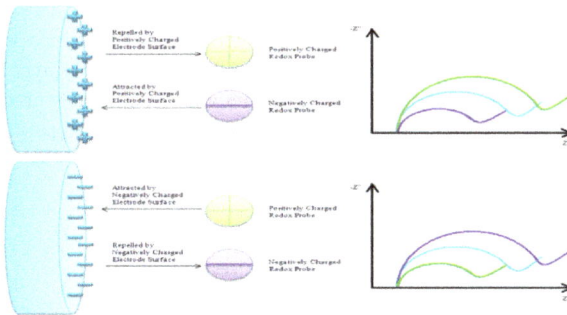

Figure 4. A schematic representation of electrode surface and redox probe interaction and their impedance spectrums.

For electrochemical impedance spectroscopy based biosensor systems, frequency scanning between two frequencies were chosen according to the solution, on the other hand electrical conductivity of solution is very important for choosing frequencies, in higher electrical conduction ability of solution, which means the solution is highly concentrated by ions, lowest frequencies can be chosen especially in the presence of redox probe (lower than 0.1 Hz). A potential must be applied to gain a proper signal, this potential is called as AC excitation signal [2]. Its magnitude depends on the solution of measurement system's cell. When the solution includes redox probe such as ferricyanide, osmium complexes or ferrocene, according to the oxidation or reduction potentials of these materials, the beginning of the electrochemical transformation potential is chosen. An unknown solution or unknown potential can be measured by cyclic voltammetry to find out the beginning oxidation/reduction potential of the electrolyte solution.

3. The importance of label-free detection

Most affinity biosensor systems require secondary molecule for amperometric or voltammetric experiments, which is attached to analyte or biorecognition receptor molecule to obtain an electroactive signal for measurement. It is necessary that labeling process, because some biomolecules or molecules cannot be transformed or cannot give electroactive signal for measurement; therefore another electroactive molecule must be used for electroactive signal. These labels must be easily applicable to gain proper signal such as fluorophores, nanoparticles, enzymes, quantum particles. This labeling process may need an extra molecule and extra preparation process to detect the real analyte. Moreover, this process can change properties of biorecognition receptor or analyte molecule and their affinity to each other and the most important thing these processes increase the expenses. Because of all these problems, electrochemical impedance spectroscopy becomes a phenomenon technique for label-free detection. In order to obtain electrochemical impedance signal, only interaction between two molecules suffices for measurement. By using enzymes some molecules can be detected easily, however instead of enzymes, transporters such as glucose transporters, ion channels, receptors, antibodies or aptamers can be used for impedimetric measurement because when these biomaterials attach their target molecules, their structural composition changes, this changing leads to rearrangement of the capacitive double layer of the electrode and the electron transfer resistance will alter. Thus they can be used as biorecognition receptors. Traditional affinity biosensors, antibody-antigen and receptor-ligand couples can be used for label-free detection.

4. Construction of impedance based biosensors

Our chapter mainly focuses on development of non-faradaic and mostly faradaic electrochemical impedance based biosensors, in other words how to construct them, which biorecognition receptor can be employed. This heading gives development of impedimetric biosensors based on different biorecognition receptor to detect different molecules.

4.1. Impedimetric receptor based biosensors

As it is known that, receptors are located on cell surface area for sensing molecules to receive and transfer signal into cell. They are usually found on the outer surface of cells, extending through the plasma membrane.Their sensing mechanism is based on weak interaction between their ligand and receptor. Therefore they can show regeneration potential to sense and induced their roles for biological form, when they are used as biosensor recognition element. This would be an advantage for reusable for biosensor. For extracellular membrane, receptors take the signals such as hormones, growth factors, ions, neurotransmitters etc. Hormones and other factors are very useful, their detection reveal a number dysfunctions, cancer and sicknesses. Excessive and deficient biomolecule concentration in the organism can be a signal of any diseases. Hormones and growth factors are peptides or proteins and their measurement is hard, because they cannot give any electroactive signals by transformation. In this problem, impedimetric biosensor systems are prominence, to immobilize a receptor on electrode surface provides to detect its ligand by biosensor technology. Before we mentioned that the most receptors are located on cell surfaces, their structure integral proteins, this proteins have two domains, which are hydrophobic and hydrophilic. For hydrophobic domains that may be a problem when immobilization process because forming a proper hydrophobic environment is very hard and may harm hydrophobic domain. This complications lead to find a proper immobilization technique, which is low cost for material and less complicated.

Uygun and Sezgintürk have developed an ultrasensitive impedimetric vascular endothelial growth factor receptor 1(VEGFR-1) based biosensor system for vascular endothelial growth factor, which is a protein produced by cells to stimulate cellular growth. It's important that is a biomarker for metastasis and lower and higher levels have disaster meanings for organism. In this study, a protein, VEGF was analyzed by electrochemical impedance spectroscopy, VEGFR-1 used as a biorecognition receptor and this affinity biosensor proved its ability that provide measurement lower concentrations, which 100 femtogram in per milliliter. Gold electrode was coated layer-by-layer; a gold film layer, SAM of Cysteamine layer, Avidin, Biotin and VEGFR-1, respectively. This long immobilization layer provides electrode ultra sensitivity. Because of using a redox probe, $[Fe(CN)_6]^{3-/4-}$, faradic impedance was in progress and Randles circuit model was applied to this impedimetric biosensor. Calibration curve was constructed 100-600 fg/ml. By calculating this calibration curve alteration of electron transfer resistance (ΔR_{et}) was used for calibration curve. In addition, this biosensor was acknowledged by Kramer-Kronig transforms to correlate of biosensors repeatability, stability and linearity [4]. This study was combination of EIS and biological recognition receptor, VEGFR-1, which shows high affinity to its ligand, therefore the study provides lower concentrations, and good selectivity without any electrochemical transformation of analyte.

Kim et al. were constructed an estrogen biosensors, which is based on impedance spectroscopy as well. For this biosensor estrogen receptors were used as biorecognition receptors of

17β-estradiol. As it is known that estrogen shows sexual characteristics of females [5], however estrogen is a carcinogen for tumor initiation and promotion [6,7]. For this reason its measurement, in a good sensitivity, plays a crucial role for female health. In this study, Gold electrode was coated a SAM layer, which is 3-mercaptopropionic acid to bind estrogen receptors via –COOH groups. LOD was $1.0x10^{-13}$M. calibration curve was constructed in a linear range between $1.0x10$-$1.0x10^{-11}$M [8]. This study is another receptor based study, it can be seen that its immobilization method is very easy, and estrogen hormone is a steroid, which is insoluble in water, is detected easily by EIS.

4.2. Impedimetric antibody-antigen based biosensors

Antibodies (Ab), or immunoglobulin, are elements of immune system to take an action in case of contamination of body. Antibodies are most widely used affinity element of biosensor technology, because of their sensitive and selective properties against antigens. Antibody based biosensor technology has a great advantage of affinity biosensor technology, because antibodies have a wide range affinity from low molecular weight molecules to high molecular weight proteins. High affinity is the major factor for an antibody to use as biorecognition receptors for biosensors. As a consequence of Ab characteristic, when an immobilization is on process then it must not be forgotten that stability, matrix effect and susceptibility [9]. In addition antibody production is very useful or any antigen, because it is widely used for antibody production in vitro. Ab and its antigen play a key role on development of an Ab based biosensor, therefore it must be known that the interaction between Ab and antigen, which they are hydrogen bonding, hydrophobic, van der Waals and coulomb interactions [10]. These forces are called as "weak interactions", therefore it is easy to break apart Ab-Ag binds, and moreover this provides regeneration for biosensor technology. Another advantages of Ab usage for biosensor technology,Ab production is very effective for obtaining proper Ab, experimental animals can produce antibody when they counter any xenobiotic, thus antibody production is stimulated for unknown molecule and new antibodies are purified. For further production recombinant technology can be used to obtain antibody. Antibodies can be divided into two main groups, which are monoclonal and polyclonal. In short, monoclonal antibodies show higher affinity to their antigen in comparison with polyclonal antibodies and it is very useful for selective measurements. Using antibody for biosensor technology has another advantage, which is capable of signal amplifying. It is known a common name as "sandwich method"; this method is combination of antigen and a secondary antibody(Figure 5). This secondary antibody can be labeled such as an enzyme, nanoparticles, quantum dots, etc. However, for impedimetric biosensors it is not necessary to label secondary antibody, because binding of antigen and antibody will give impedimetric signal, additional secondary antibody label-free increases the measurement signal that is impedance.

Pichetsurnthorn et al. have proposed a study, which is about impedimetric biosensor for trace atrazine detection from water samples. Atrazine is a small xenobiotic, a pesticide, and it is very harmful any organism when it is in taken. Therefore pesticide analysis plays a crucial role for food and environmental industry. As it is known that a number of biosensor for pesticide detection have been published and continuing, but enzymatic biosensors show less selectivity, because they depend on enzymatic inhibition, which an enzyme can be affected by ionic strength, pH, heavy metals etc. and this enzymatic inhibition is not reliable for pesticide detection. In this manner, electrochemical impedance spectroscopy has become prominence for label-free and enzymless detection. In this study, nanoporous alumina membranes were used just as ELISA method without using enzyme to achieve non-faradaic measurement, in spiked samples, the biosensor gives as low as femtogram/ml. On a gold metal electrode alumina membrane, for biomolecular confinement, was used because of nanomolecular space. For antibody immobilization *thio bis succinimidyl propionate* (DSP), a linker, was used. This provides thiol links with gold surface, and its amine groups were used immobilization of anti-atrazine. Calibration curve was constructed 10fg/ml to 1ng/ml in wide range. This sensor was designed in three ways to increase selectivity, which one of them is used planar sensor, second is 200nm alumina pores and three is 20nm alumina pores. Decreasing in alumina pore increased the selectivity of atrazine molecule. This study provided label-free, selective and sensitive atrazine measurement in ultra low concentrations [11]. Based on antibody efficacy, a cross-reactivity was observed, which was the result of antibody specification against other molecule, malathion.

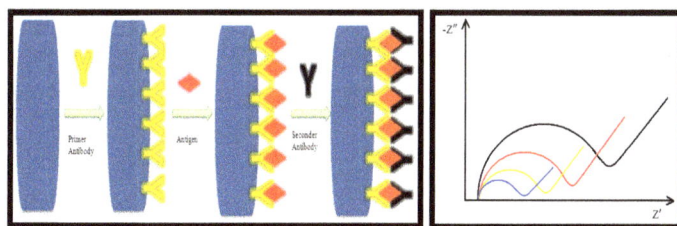

Figure 5. Schematic representation of Sandwich Method and its Impedance Spectrums

Huang et al. have developed allergen biosensors, which is based on mite allergen Der f2 and its antibody by using electrochemical impedance spectroscopy. Type I allergic reaction is a reaction of immune dysfunction and represents a problem of health of organism. Development of industry, especially on home industry, increases the allergic reactions of bodies because the natural environment of body has been changing and adaptation is hard. The allergic reactions occurs when body comes across an allergenic material and this plays a crucial role that the detection of allergenic reaction. Huang et al. find out a solution for point of care allergenic detection, which is based on impedance spectroscopy to determine Der f2. Firstly a glassy carbon electrode was modified by gold electrodeposition, subsequently Der f2 solution dropped on electrode and waited for a night to ensure that the active sides of nanogold were all occupied by allergen molecules. Then it was exposed to different concentration of murine monoclonal antibody solution. Electrochemical impedance spectrums were employed for calibration curve from 2µg/ml to 300µg/ml [12].

4.3. Impedimetric DNA-aptamer based biosensors

DNA is known as double-stranded (ds) source of genetic information. This genetically occurred biomolecule includes base pairs, which are Adenine-Thymine and Guanine-Cytosine,
phosphate backbone, which provides negative charge properties. These pairs occur on the
dsDNA strand is only if specific its complementary DNA strand. This specificity is very important for label free molecule detection or DNA sequence detection. This specificity leads
to utilize DNAs for biosensor, which DNA is used as biorecognition receptor. By using
DNAs specificity, a new method has become revealed that is Aptamer usage as a biorecognition receptor. Aptamers are artificial nucleic acids with specific binding affinity to molecules, proteins, amino acids, drugs, pesticides, etc. Their advantages proposed them as
alternatives of antibodies [13-16]. On the other hand their specific affinity to wide range of
molecules, an aptasensor is only specific for one target. Thus, this limitation must be improved for detection of other molecules. Detection of more than one molecule by using aptamer is a challenging technique, because of its limitations. To detect one molecule a strand
is enough, but to detect more than one molecule, DNA strand must be prolonged to bind
two or more molecules. This prolongation, which contains more than one aptamer units, on
DNA increases the flexibility of DNA strand and affects formation of aptamer-analyte molecule complex [17], or hybridization on same strand occurs. Aptasensors are widely used,
however their ultrasensitive detections are limited because of their low association constants, for the signal amplification many methods have been proposed for aptasensor usage,
such a rolling circle amplification [18,19], strand displacement amplification [20,21], enzyme
label [22-24]. These methods are very advantageous for signal amplification, besides they
are complicated and expensive.

Deng et al. have proposed a bifunctional aptasensor to detect lysozyme and adenosine. In
this method, two DNA strands, which were used adenosine contained DNA and lysozyme
aptamer. $[Ru(NH_3)_6]^{3+}$ was used as signaling transducer, moreover gold nanoparticles
(AuNPs) were used to increase signal. DNA strands, which were used as biorecognition receptors, were modified thiol-terminated. According to self-assembly method (SAM), DNA
(SAM) layers formed on gold electrode layer. Main idea of this study was a DNA was used
as capture; a DNA was used as linker, and a DNA-AuNPs as used increase the signal. Negatively charged DNA strands were treated with $[Ru(NH_3)_6]^{3+}$ complexes. In any consideration, which adenosine was added on electrode surface, aptamer lysozyme complex released
from surface, lysozyme was added on electrode surface, aptamer-adenosine complex pre-
leased from surface. DNA-AuNPs complexes increase the selectivity by increasing the lysozyme aptamer formation signal. Analytes, lysozyme and adenosine, bind to electrode
surface $[Ru(NH_3)_6]^{3+}$ complexes were released and amplify the electrical signal. All in all this
bifunctionalaptamer biosensor is based on structure-switching properties. This signal
changes were monitored by impedance spectroscopy to measure layer changes. Switching
on the layer for any molecule will change the capacitive and resistive properties of the electrode and impedance becomes a prominence measurement technique and it provides label
free and enzymless detection. Lysozyme and adenosine concentrations were detected 0.01µg
mL^{-1} and 0.02nM, respectively [25].

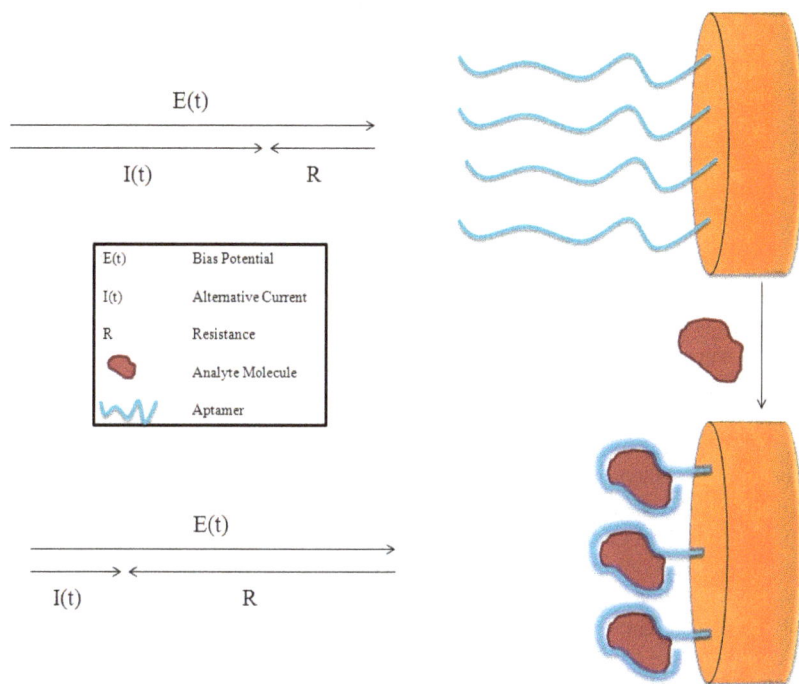

E(t)	Bias Potential
I(t)	Alternative Current
R	Resistance
	Analyte Molecule
	Aptamer

Figure 6. Schematic representation of impedimetricaptamer biosensor.

Ensafi et al. developed a DNA biosensor, which is based on DNA-DNA hybridization. This DNA hybridization was used for detection a cancer type that is chronic lymphocytic leukemia (CLL), which provokes the production of white blood cells called B lymphocytes in bone marrow. This biosensors system is based on porphobilinogendeaminase (PBGD) gene, which is associated with CLL cancer. A probe 25-mer DNA modified with thiol-terminated and binds to AuNPs as SAM. The target, complementary DNA was detected as through hybridization. The PBGD gene hybridizations were monitored impedance spectrums, by using redox probe, which is $[Fe(CN)_6]^{3-}/^{4-}$. Deposited AuNPs on gold electrode, SAM of DNA layer formed and complementary DNA hybridized. This biosensors LOD result was calculated as 1 femtoM. By using impedance spectroscopy mismatched DNA strands was able to observed, even for one base mismatched. This showed that label-free detection is provided in higher sensitivity levels [26].

4.4. Other impedimetric biosensors

After most widely used biomolecules (DNA, Receptors and Antibodies) for biosensors, other biomolecules for impedimetric biosensor usage is presented in this part of the chapter. In this chapter two molecules, which show specific affinity themselves other than receptors,

DNAs and antibodies, will be revealed, such as specific protein-molecule, cell-molecule and protein-cell. When it is mentioned before that to develop an impedimetric biosensor, it is only necessary to find two molecules, which show affinity. Because of this affinity the binding will be come true, and impedimetric signal will be obtained. In order to illustrate some ides, this part will be written by giving examples about other impedimetric affinity biosensors.

Hu et al. has proposed a study, which is Multi-wall carbon nanotube-polaniline biosensor based on lectin-carbohydrate affinity for ultrasensitive detection of Con A. Concavalin A is a kind of allosteric protein, which has four affinity residues on it for binding specific atoms and molecules, which they are calcium and manganese cations to activate binding sides for carbohydrates, one is for hydrophobic recognition, one is for R-mannose or D-glucose [27]. Concavalin A is called in lectin family, which is specific protein family to carbohydrates, moreover it has an ability to activate and proliferate for mature T cells [28]. Lectin binding carbohydrate ligands, in other words carbohydrate specific proteins, show high affinity likewise antibody-antigen interactions. Lectins have been used for biosensor technology for detection of carbohydrate. In this study carbohydrate, D-glucose was used as recognition receptor for lectin detection. Using nonmaterials(PANI and CNT) increased the surface are of electrode. MWNT-PANI nanocomposite material was dropped on GCE and glucose added in 50 C for 6h through Schiff-base reaction. This modified electrode was exposed different concentration on Con A solutions and EIS spectrums were obtained. R.S.D was found as 2.1% and a calibration curve was constructed from 3.3pM to 9.3nM and LOD was 1.0pM [29]. MWNT-PANI nanocomposite layer was very promising modification step, in order to reach lowest limits.

Oliveira et al. published an article, which is an impedimetric biosensor based on self-assembled hybrid cystein-gold nanoparticles and CramoLLlectin for bacterial lipopolysaccharide recognition. As it is known lipopolysaccharides are endotoxines, structural components of gram-negative bacteria, which is common for humans, animals and plants [30]. In this study electrode surface was modified poly(vinyl chloride-co-vinyl acetate-co-maleic acid)(PVM), gold nanoparticles-cystein composite (AuNpCys), CramoLL, respectively. This electrode was specific for lipopolysaccharides. PVM layer was charged positively, this provides that attract the CramoLLlectin electrostatically. LPS detection depended on carbohydrate composition of LPS, and this limited the selectivity and binding abilities of biosensor. In this study, different cells, which had constructed LPS layer, were used to detection of any difference when LPS layer differs. Because of linear LPS layer composition of *S. marcescens* highest impedance value was obtained. The CramoLL-LPS binding process can be significantly inhibited by saccharide, glucose, providing further proof of the sensing mechanism on lectin interface. AuNpCys-PVM complex positively charged surfaces were suitable for CramoLLlectin immobilization. As it is known that cell surfaces are negatively charged [31].

Tlili et al. developed an impedimetric biosensor system, based on cell, which was Fibroblast Cells: a sensing bioelement for glucose detection by impedance spectroscopy. Fibroblast cells were used as biorecognition receptor for glucose detection. Diabetes is one of the most widely studied disaster in the world, in order to find diabetes, its markers must be determined, which is most important glucose. Several studies were carried out glucose biosensing. Using living cells as biorecognition receptors provides an opportunity for high sensitivity in a broad range of biologically active substances that affect the response of the cells. In this study, 3T3-L1 fibroblast cells are able to metabolise glucose through the activation of specific glucose transporters (Glut 1 and Glut 4). Biological cells are very poor conductors at low frequencies, for this reason, force electrical currents to bypass them. Impedance spectrums were proportional to fibroblast cell growth on electrode; especially gaps between cells will affect impedance spectrums. An ITO (Indium Tin Oxide) electrode was used for this biosensor, fibroblast cells were attached on electrode, after 2-4 days a thin layer was formed by fibroblast cells. The effect of additional glucose concentrations, Nyquist diagram was chosen, which allows more sensitive measurement. Glucose and cell interaction depended on the glucose intake ability of cells through Glut 1 and Glut 4. To examine the specificity of the biosensor, D-mannitol was chosen instead of glucose and no variation was observed. No signal variation was obtained by inhibition of glucose transporters, therefore glucose intake provides metabolic process in order to detect glucose molecules, not incorporated by increased osmotic pressure. Addition of carbohydrate not metabolized by cells did not give any signal variation in impedance curves. The calibration curve was constructed from 0 to 14mM glucose concentration [32]. On the other hand for this study, as a biorecognition receptor, cell surface transporters were used for biosensor indirectly. This provided selective measurement.

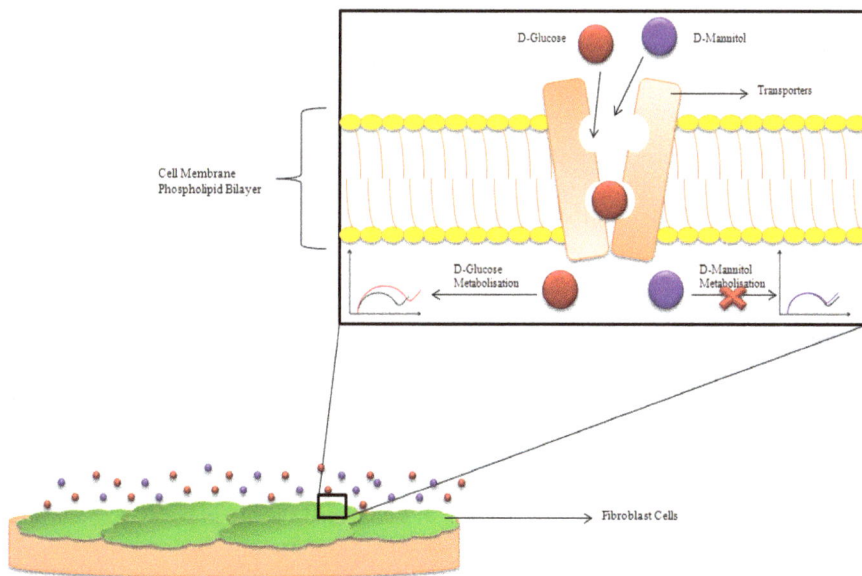

Figure 7. Schematic representation of fibroblast cell based impedimetric biosensor.

Tong et al. have declared an annexin V-based biosensor for quantitatively detecting early apoptotic cells. Apoptosis is programmed cell death in organisms and it is necessary for organism homeostasis. Measurement of apoptosis can give clues about any metabolic process, for example anti-tumor therapies, drugs. Annexin V is a calcium dependent phospholipid-binding protein with high affinity for phosphatidylserine, which is altered in translocation in plasma membrane for early apoptotic cells [33]. Therefore this protein can be used as PS exposure marker upon cell membranes. In this study, the authors described first time annexin V-modified biosensor was developed for quantitatively detecting early apoptotic cells. Gold electrode was modified; 1,6-hexanedithiol, gold nanoparticles, annexin-V, respectively. Then PS exerted cells were detected by using this modified electrode. As you can a slight difference on electrode can be detected by using impedance spectroscopy and this detection provides apoptotic cell existence. This biosensor's signal was dependent upon calcium ion concentrations. Therefore, before PS detection, the modified electrode treated with $5\mu M$ Ca^{2+} for 1 h [34].

5. Some footnotes about impedimetric biosensors

One of the most important points of impedimetric biosensors is to find proper molecule that shows affinity to your analyte. Secondly your measurement process will be considered; especially modification steps and oxidation/reduction ability of redox probe will define experimental parameters such as frequency, bias potential, electrical circuit model. In order to obtain sensitive detection, faradaic impedance is proposed to reach lower frequencies, because of reduction/oxidation properties of redox probe, the electrons, which produced by this oxidation/reduction, can move easier. This transportation can be measured as electron transfer resistance. On the other hand non-faradaic impedance is only measure resistance and surface capacitance, that technique is for whole surface by electrical circuit model.

Electrical circuit model is constructed by characteristic of impedance curve, which depends on the electrical conductibility of electrochemical measurement solution, electrode surface and interaction between electrode-electrolyte. Impedance spectroscopy provides sensitive and label free detection by its sensitive electrode surface characterization ability. Especially for sensitive measurements faradaic experiments are considered that is better, by using redox probes, it is possible to reach lower frequencies, thus lower detection limits can be reached by examination of the modified electrode surface in lower frequencies. Electrode surface modification plays a crucial role for impedimetric measurements that inhomogenicity on electrode surface, pin holes or a direct interaction with bare electrode surface after a failed modification and electrolyte, the electrons, which through electrode surface without confront any electrical resistance, will give a false impedance spectrum. In this case of inhomogenity, capacitance element of circuit model is redefined as constant phase element to solve inhomogenity problem. In case of lower or less altered signals, impedance spectroscopy provides signal amplification by modifying electrode surface with a molecule or electrode surface can be charged same as redox probe's charge. Another challenging factor for impedimetric biosensor is the affinity, low affinity or affinity for wide range will reduce the importance or usefulness of the impedimetric biosensor.

Analyte	Bioreceptors	LOD	Linear Range	Reference
17β-estradiol	Estrogen Receptor-α	1.0×10^{-13}M	1.0×10-1.0×10^{-11}M	[8]
AIV H5N2	Monoclonal Anti-H5 Antibody	$1 \times 10^{1.2}$ELD$_{50}$/ml	$1 \times 10^{1.2}$ELD$_{50}$/ml- $1 \times 10^{5.2}$ELD$_{50}$/ml	[36]
Atrazine	Anti-atrazine Antibody	10 fg/ml	10 fg/ml-1ng/ml	[11]
Chronic Lymphocytic Leukemia Gene Sequence Mutation	25-Mer PorphobilinogenDeaminase (PBGD) Gene	1.0×10^{-12}M	7.0×10^{-12}–2.0×10^{-7}M	[26]
Concavalin A	Glucose	1.0 pM	3.3pM-9.3nM	[29]
Der f2	Murine Monoclonal Antibody	2.0 µg/ml	2-300 µg/ml	[12]
Glucose	3T3-L1 Fibroblast Cells		0-14 mM	[32]
Lipopolysaccaride Layer of S. marcescens	CramoLLLectin	25 µg/ml	25-200 µg/ml	[31]
Lysozyme and Adenosine	Lysozyme and Adenosine Specific Two Aptamers	0.01µg mL^{-1} and 0.02nM, respectively	0.2-40 nM Adenosine	[25]
PS exerted cells	Annexin V	5 Fu		[34]
Thrombin	Thrombin Aptamer	0.013 nM	0.1-30 nM	[35]
VEGF	VEGFR-1	100 fg/ml	100-600 fg/ml	[4]

Table 1. Comparison of Some Impedimetric Biosensor Systems

6. Conclusion

Consequently, biosensor systems have been developing, the most important factors are selectivity and low cost. These specialties are very important of Point-Of-Care usage. Electrochemical impedance spectroscopy is very effective technique for label-free molecule detection. It can provide sensitivity, low cost and selective biosensor systems. As you can read from above, by using specific molecules such as DNA, Aptamer, Receptor, Antibody, Specific Proteins etc. it is possible to construct selective and sensitive impedimetric biosensor systems.

Author details

Hilmiye Deniz ErtuğruL[1] and Zihni Onur Uygun[2]

1 Ondokuz Mayıs University, Turkey

2 Çanakkale Onsekiz Mart University, Turkey

References

[1] Orazem M E, Tribollet B. Electrochemical Impedance Spectroscopy. Wiley; 2008.

[2] Deniels S J, Pourmand N. Label-free impedance biosensors: opportunities and chal-lenges.Electroanalysis 2007; 19 1239-1257 http://www.ncbi.nlm.nih.gov/pmc/articles/PMC2174792/ (accessed 3 July 2012)

[3] Katz E, Willner I. Probing biomolecular interactions at conductive and semiconduc-tive surfaces by impedance spectroscopy: routes to impedimetricimmunosensors, DNA-sensors, and enzyme biosensors. Electroanalysis 2003; 15(11) 913–947. http://onlinelibrary.wiley.com/doi/10.1002/elan.200390114/abstract (accessed 3 July 2012)

[4] Uygun Z O, Sezginturk M K. A novel, ultra sensible biosensor built layer-by-layer co-valent attachment of a receptor for diagnosis of tumor growth. AnaliticaChimicaActa 2011; 706 343-8. http://www.sciencedirect.com/science/article/pii/S0003267011011743 (accessed 3 July 2012)

[5] Jordan V C, Phelps E, Lindgren J U. Effects of antiestrogens on bone in castrated and intact female rats. Breast Cancer Research and Treatment 1987; 10 31-35. http://www.ncbi.nlm.nih.gov/pubmed/3689979 (accessed 3 July 2012)

[6] Ronald K R, Annlia P H, Peggy C W, Malcolm C P. Effect of Hormone Replacement Therapy on Breast Cancer Risk. Journal of National Cancer Institute 2000; 92 328-332. http://jnci.oxfordjournals.org/content/92/4/328.full (accessed 3 July 2012)

[7] Liehr J G. Is Estradiol a genotoxic mutagenic carcinogen. Endocrine Reviews 2000; 21 40-54. http://edrv.endojournals.org/content/21/1/40.full?sid=a39c7ee2-0744-430f-a43f-7111602db0c3(accessed 3 July 2012)

[8] Kim B K, Li J, Im J-E, Ahn K-S, Park T S, Cho S I, Kim Y-R, Lee W-Y. Impedometric estrogen biosensor based on estrogen receptor alpha-immobilized gold electrode. Journal of Electroanalytical Chemistry 2012; 671 106-111. http://www.sciencedir-ect.com/science/article/pii/S1572665712000781 (accessed 3 July 2012)

[9] Velasco-Garcia M N, Mottram T. Biosensor technology addressing agricultural prob-lems. Biosystems Engineering 2003; 84, 1–12. http://www.sciencedirect.com/science/article/pii/S1537511002002362 (accessed 3 July 2012)

[10] Van Oss C J. Antibody-Antigen Intermolecular Forces. Academic Press, London, 1992; 1 97–100.

[11] Pichetsurnthorn P, Vattipalli K, Prasad S. Nanoporousimpedimetric biosensor for de-tection of trace atrazine from water samples. Biosensors and Bioelectronics 2012; 32 155-162. http://www.sciencedirect.com/science/article/pii/S0956566311008189 (ac-cessed 3 July 2012)

[12] Huang H, Ran P, Liu Z. Impedance sensing of allergen-antibody interaction on glassy carbon electrode medified by gold electrodeposition. Bioelectrochemistry

2007; 70 257-262. http://www.sciencedirect.com/science/article/pii/S1567539406001447 (accessed 3 July 2012)

[13]   Ellington A D, Szostak J W. In vitro selection of RNA molecules that bind specific ligands. Nature 1990; 346 818-822. http://www.nature.com/nature/journal/v346/n6287/pdf/346818a0.pdf (accessed 3 July 2012)

[14]   Tuerk C, Gold L. Systematic evolution of ligands by exponential enrichment: RNA ligands to bacteriophage T4 DNA polymerase. Science 1990; 249 505-510. http://www.sciencemag.org/content/249/4968/505.abstract (accessed 3 July 2012)

[15]   Hamula C L A, Guthrie J W, Zhang H, Li X F, Le X C. Selection and analytical applications of aptamers. Trends in Analytical Chemistry 2006; 25 681-691. http://www.sciencedirect.com/science/article/pii/S0165993606001178 (accessed 3 July 2012)

[16]   Huang Y F, Chang H T, Tan W. Cancer cell targeting using multiple aptamers conjugated on nanorods. Analytical Chemistry 2008; 80 567-572. http://pubs.acs.org/doi/full/10.1021/ac702322j (accessed 3 July 2012)

[17]   Xiao Y, piorek B D, Plaxco K W, Hegger A J. A Reagentless Signal-On Architecture for Electronic, Aptamer-Based Sensors via Target-Induced Strand Displacement. Journal of American Chemical Society 2005; 127 17990-1. http://pubs.acs.org/doi/full/10.1021/ja056555h (accessed 3 July 2012)

[18]   Zhou L, Ou L J, Chu X, Shen G L, Yu R Q. Aptamer based rolling circle amplification: a platform for electrochemical detection of protein. Analytical Chemistry 2007; 79 7492-7500. http://pubs.acs.org/doi/full/10.1021/ac071059s (accessed 3 July 2012)

[19]   Yang L, Fung C W, Cho E J, Ellington A D. Real-time rolling circle amplification for protein detection. Analytical Chemistry 2007; 79 3320-9. http://pubs.acs.org/doi/abs/10.1021/ac062186b (accessed 3 July 2012)

[20]   Shlyahovsky B, Li D, Weizmann Y, Nowarski R, Kotler M, Willner I. Spotlighting of cocaine by an autonomous aptamer-based machine. Journal of American Chemical Society 2007; 129 3814-5. http://pubs.acs.org/doi/full/10.1021/ja069291n (accessed 3 July 2012)

[21]   Weizmann Y, Beissenhirtz M K, Cheglakov Z, Nowarski R, Kotler M, Willner I. A virus spotlighted by an autonomous DNA machine. AngewandteChemie, International Edition 2006; 45 7384-7388. http://onlinelibrary.wiley.com/doi/10.1002/anie.200602754/pdf (accessed 3 July 2012)

[22]   Patolsky F, Katz E, Willner I. Amplified DNA Detection by ElectrogeneratedBiochemiluminescence and by the Catalyzed Precipitation of an Insoluble Product on Electrodes in the Presence of the Doxorubicin Intercalator. AngewandteChemie 2002; 114 3548-3552.   http://onlinelibrary.wiley.com/doi/10.1002/1521-3757%2820020916%29114:18%3C3548::AID-ANGE3548%3E3.0.CO;2-Q/pdf (accessed 3 July 2012)

[23] Patolsky F, Katz E, Willner I. Amplified DNA Detection by ElectrogeneratedBioche-miluminescence and by the Catalyzed Precipitation of an Insoluble Product on Elec-trodes in the Presence of the Doxorubicin Intercalator. AngewandteChemie, International Edition 2002; 41 3398-3402. http://onlinelibrary.wiley.com/doi/10.1002/1521-3757%2820020916%29114:18%3C3548::AID-ANGE3548%3E3.0.CO;2-Q/pdf (accessed 3 July 2012)

[24] Alfonta L, Bardea A, Khersonsky O, Katz E, Willner I. Chronopotentiometry and Far-adaic impedance spectroscopy as signal transduction methods for the biocatalytic precipitation of an insoluble product on electrode supports: routes for enzyme sen-sors, immunosensors and DNA sensors. Biosensors and Bioelectronics 2001; 16 675-687. http://www.sciencedirect.com/science/article/pii/S0956566301002317 (ac-cessed 3 July 2012)

[25] Deng C, Chen J, Nie L, Nie Z, Yao S. Sensitive bifunctionalaptamer-based electro-chemical biosensor for small molecules and protein. Analytical Chemistry 2009; 81 9972-8. http://pubs.acs.org/doi/full/10.1021/ac901727z (accessed 3 July 2012)

[26] EnsafiAA, Taei M, Rahmani H R, Khayamian T. Sensitive DNA impedance biosensor for detection of cancer, chronic lymphocytic leukemia, based on gold nanoparticles/gold modified electrode. ElectrochimicaActa 2011; 56 8176-8183. http://www.science-direct.com/science/article/pii/S0013468611008747 (accessed 3 July 2012)

[27] Yonzon C R, Jeoung E, Zou S L, Schatz G C, Mrksich M, Duyne R P V. A comparative analysis of localized and propagating surface plasmon resonance sensors: the bind-ing of concanavalina to a monosaccharide functionalized self-assembled monolayer. Journal of American Chemical Society 2004; 126 12669-12676. http://pubs.acs.org/doi/full/10.1021/ja047118q (accessed 3 July 2012)

[28] Pongracz J, Parnell S, Anderson G, Jaffrézou J P, Jenkinson E. 10.Con A activates an Akt/PKB dependent survival mechanism to modulate TCR induced cell death in double positive thymocytes. Molecular Immunology 2003; 39 1013-1023. http://www.sciencedirect.com/science/article/pii/S0161589003000440 (accessed 3 July 2012)

[29] Hu F, Chen S, Wang C, Yuan R, Xiang Y, Wang C. Multi-wall carbon nanotube-poly-aniline biosensor based on lectin-carbohydrate affinity for ultrasensitive detection of Con A. Biosensors and Bioelectronics 2012; 34 202-7. http://www.sciencedirect.com/science/article/pii/S0956566312000632 (accessed 3 July 2012)

[30] Preston A, Mandrell R E, Gibson B W, Apicella M A. The lipooligosaccharides of pathogenic gram-negative bacteria. Critical Reviews in Microbiology 1996; 22 139. http://informahealthcare.com/doi/abs/10.3109/10408419609106458 (accessed 3 July 2012)

[31] Oliveira M D L, Andrade C A S, Correia M T S, Coelho L C B B, Singh P R, Zeng X. Impedimetric biosensor based on self assembled hybrid cystein-gold nanoparticles and CramoLLlectin for bacterial lipopolysaccharide recognition. Journal of colloid

and Interface Science 2011; 362 194-201. http://www.sciencedirect.com/science/article/pii/S0021979711007594 (accessed 3 July 2012)

[32] Tlili C, Reybier K, Geloen A, Ponsonnet L, Martelet C, Ouada H B, Lagarde M, Jaffrezic-Renault N. Fibroblast cells: a sensing bioelement for glucose detection by impedance spectroscopy. Analytical Chemistry 2003; 75 3340-3344. http://pubs.acs.org/doi/full/10.1021/ac0340861 (accessed 3 July 2012)

[33] Meers P, Mealy T. Calcium-dependent annexin V binding to phospholipids: stoichiometry, specificity, and the role of negative charge. Biochemistry 1993; 32 11711-11721. http://www.ncbi.nlm.nih.gov/pubmed/8218240 (accessed 3 July 2012)

[34] Tong C, Shi B, Xiao X, Liao H, Zheng Y, Shen G, Tang Di, Liu X. An annexin V-based biosensor for quantitatively detecting early apoptotic cells. Biosensors and Bioelectronics 2009; 24 1777-1782. http://www.sciencedirect.com/science/article/pii/S0956566308004090 (accessed 3 July 2012)

[35] Li L-D, Zhao H-T, Chen Z-B, Mu X-J, Guo L. Aptamer biosensor for label-free impedance spectroscopy detection of thrombin based on gold nanoparticles. Sensors and Actuators: B Chemical 2011; 157 189-194. http://www.sciencedirect.com/science/article/pii/S0925400511002449 (accessed 3 July 2012)

[36] Wang R, Lin J, Lassiter K, Srinivasan B, Lin L, Lu H, Tung S, Hargis B, Bottje W, Berghman L, Li Y. Evaluation study of a portable impedance biosensor for detection of avian influenza virus. Journal of Virological Methods 2011; 178 52-58 http://www.sciencedirect.com/science/article/pii/S0166093411003430 (accessed 3 July 2012)

Chapter 10

Bioelectronics for Amperometric Biosensors

Jaime Punter Villagrasa,
Jordi Colomer-Farrarons and Pere Ll. Miribel

Additional information is available at the end of the chapter

1. Introduction

The Discrete-to-Integrated Electronics group (D2In), at the University of Barcelona, in partnership with the Bioelectronics and Nanobioengineering Group (SICBIO), is researching Smart Self-Powered Bio-Electronic Systems. Our interest is focused on the development of custom built electronic solutions for bio-electronics applications, from discrete devices to Application-specific integrated circuit (ASIC) solutions.

The integration of medical and electronic technologies allows the development of biomedical devices able to diagnose and/or treat pathologies by detecting and/or monitoring pathogens, multiple ions, PH changes, and so on. Currently this integration enables advances in various areas such as microelectronics, microfluidics, microsensors and bio-compatible materials which open the door to developing human body Lab-on-a-Chip implantable devices, Point-of-Care in vitro devices, etc.

In this chapter the main attention is focused on the design of instrumentation related to amperometrics biosensor: biopotentiostat amplifiers and lock-in amplifiers. A potentiostat is a useful tool in many fields of investigation and industry performing electrochemical trials [1], so the quantity and variety of them is very extensive. Since they can be used in studies and targets as different as the study of chemical metal conversions [2] or carcinogenic cells detection, neuronal activity detection or Deoxyribonucleic acid (DNA) recognition, their characteristics are very varied.

For chemical measurement systems in biological applications, potentiostat amplifiers [3] are the electronic interface to a large category of amperometric chemical sensors which are capable of managing many biologically and environmentally important analyses. This characteristic gives us the possibility of building an extremely versatile tool with other interesting specs: portability, accuracy and reliability.

Demand for increased functionality, smaller systems, with smaller electrodes, for ultra-low current detection and versatility will force potentiostat amplifiers to be designed on a system-on-chip (SoC), combining single to multi-sensor measurements, to be implemented in advanced Complementary metal–oxide–semiconductor (CMOS) processes. The scaled supply voltages in these processes [4-8], however, seriously limit the chemical analysis range.

Interesting approaches have been designed in terms of the instrumentation [9-13], attempting to work with low-voltages and low-currents, these electronics are integrated with autonomous powering. In [10] the authors propose an interesting low-power concept of a two-electrodes approach with the capability to measure from 1pA to 200nA with a simple Analog-to-Digital (ADC) approach, but operating at 5V. [11] presents a nice approach based on a Sigma-Delta modulator with the capability to check currents with a sensitivity of 50fA with a power dissipation of 12 μW, operating at 3.3V in this case, but with all the processing electronics being external to the prototype. In [12] a Complementary metal–oxide–semiconductor (CMOS) approach is presented for electrochemical arrays, operating with a bias of 5V, where the array (5X5) is placed in the same substrate with the amperometric detector. The design works in an unipolar fashion detecting pA, with a maximum operating frequency @ 2kHz. Good approach is also presented in [13] where a current-mirror circuit is implemented for three-electrode amperometric electrochemical sensors. There the system has an accuracy of 1%, with a range of currents of 1nA to1μA, operating at 1.8V with a power consumption of70μW. [14] present a 5V, 0.6mA amperometric electrochemical microsystem array (4X4), with a range between 6pA to10μA, where full electronics and array are placed on-chip, also working with an unipolar approach, with a chip size of 2.3X2.2 mm, with the electrode array implemented with electrodes of 2mm of diameter. Also in [15] a similar approach is derived, with an array of 8X8, with circular electrodes of 6μm, and a second implementation with the array implemented in-chip with the electronics. First Cyclic voltammetry (CV) approaches are presented with a typical range between 0.1V to 0.8V, detection currents from-1.5μA to 1.5μA.

Latest advances have been reported in [16] and in [17]. In [16], the novelty resides in the technological approach for the electronics design. Generally, all the implementations are based on silicon electronics, looking for a cheapest massive production and good properties. In this case Poly-Si Thin-Film transistors are used, ideal for their flexibility. In this particular work the system it is just conceived for a single three-based electrode for a cyclic voltammetry to detect diabetes. Electronic performances are not significantly inferior to those based on a CMOS solution. The system operates at 1.2V, working between 0.111V to 0.679V, but not reference is addressed to the power consumption and current range. The last example [17] is quite interesting and a great approach to our concept, but also is an excellent example of the challenge of the present proposal. This system has the capability to define different stimulus for the amperometric and electrochemical measurement. It provides linear sweep, constant potential, cyclic and pulse voltammetry, with enough smart facilities, but the system presents a fixed clock and a maximum power dissipation of 22.5mW, for an array of just 2X2 electrodes. The budget of 22.5mW is a lot of power for the present concept. As a readout circuit, it has good performances, between±47μA, with a linear resolution of 0.5pA.

In terms of the Lock-In Amplifiers, the measurement must be processed in order to extract the impedance variation of the cell. In an electrochemical cell, electrode kinetics, redox reactions, diffusion phenomena and molecular interactions at the electrode surface can be considered analogous to the above components that impede the flow of electrons in an ac circuit. The simplest electrical modelization is based on an equivalent RC circuit, also called a Randles circuit.

In order to proceed with the signal processing there are two main approaches: a) the Fast Fourier Transform (FFT) [18], and b) the Frequency Response Analyzer (FRA) [19]. In the case of the FFT, a pulse, or a step, -the approach to be followed is the ideal Dirac function-, is applied to the sample because it contains a wide frequency content. Then, the response of the sample is digitized and processed in a digital processor, for instance a DSP (Digital Signal Processor), and using the FFT algorithm, the different frequency components are obtained for their analysis. Another possibility is the logarithmic sampling in the DFFT calculus, reducing the data required in the process.

A simpler solution is based on the FRA approach. In this case a sine and cosine signals are adopted and using two multipliers and a filter stage the real and imaginary components of the response are obtained. This measurement must be done for each frequency. Working with just one sensor and in terms of the size of the final product, the FFT option could be adopted, because the response for several frequencies is obtained. The FRA solution is a solution more oriented to multi-sensor approaches but also in the case of single sensors it is also a good option, in terms of the trade-off between complexity and speed, if not too low frequencies are to be measured. This lock-in approach is more feasible.

The perturbation signal provided by the instrumentation system – following the FRA approach – generally uses a sinus wave as the input voltage.

The chapter would present the state-of-the art in the conception of the involved electronics for the potentiostat amplifier instrumentation and EIS approaches, analogue and digital.

2. Amperometric biosensors

Nowadays advances in different multidisciplinary areas like microelectronics, microfluidics, microsensors and bio-compatible materials open the door to developing advanced biomedical solutions for health care, such as Point-of-care devices, combining Lab-on-a-Chip solutions with suitable electronics for measurement, processing and remote telemetry communication [20-22].

In particular, integration of medical and electronic technologies allows biomedical devices solutions capable of diagnosing and/or treating pathologies by detecting and/or monitoring pathogens. There are different approaches for amperometric biosensors instrumentation, based on a potentiostat amplifier with several different amperometric readout configurations [13,23,24].

The interest in electrochemical sensors, and in particular amperometric biosensors, started when changes in the dielectric properties of an electrode surface were detected by potentiostatic methods [1,25,26]. These biosensors are capable of detecting antigens, antibodies, proteins, DNA fragments, PH changes, heavy metal ions, and so on [27-31].

There are different solutions for the electrochemical-based instrumentation that vary from discrete to mixed and full custom-ASICs approaches. The sensor size and the system complexity define the final adopted solution. Electronic designs based on low-cost on the shelf surface mount components can be adopted for electrodes areas greater than 1 mm². Full custom ASIC solutions are valid approaches [8,30,31] for smaller and multiplexed sensors. Moreover, these ASIC solutions are indicated if very low current levels are derived from the sensors. Other key indicators for an ASIC solution near the electrodes are not only the degree of miniaturization, but also the fact that EMI (Electromagnetic Interference), can be reduced, and external disturbances such as vibrations, moisture, sources of electrical noise, etc. are avoided.

The development of multisensory arrays of biosensors, working with low concentration levels and the capability of a multi-bio-analysis in blood (searching for a multi-pathogen detection), is a topic of great importance in the contemporary situation [32,33].

2.1. Electrochemistry and reactions on the electrode

Electrochemical biosensors are the largest group of chemical sensors. These biosensors are normally based on enzymatic catalysis of a reaction that produces or consumes electrons (such enzymes are correctly called redox enzymes). Electrochemical sensors allow three main configurations: voltammetric, potentiometric and conductometric measurements [3]. In the present chapter special interest is focused on the voltammetric sensors. Voltammetric biosensors are those based on the measurement of the current-voltage variations. Amperometric biosensors are a particular case, where determined electrical currents are associated with a redox process where a fixed voltage in the sensor is applied. Some associated current, called faradaic current, is exclusively generated by the reduction or oxidation of some chemical substance at an electrode [34].

Under equilibrium, and in the absence of an externally applied voltage, a polarizable electrode resting in solution will develop a potential based on the ratio of the solution's chemical species [1]. When voltage is applied to the electrode the system is forced out of equilibrium and results in a reduction/oxidation (redox) reaction:

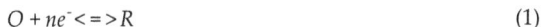

$$O + ne^- <=> R \tag{1}$$

where O is the oxidized form of the species, n is the number of electrons per molecule oxidized or reduced, e- is an electron, and R is the reduced form of the species. This results in a faradic current at the electrode surface for reversible systems also called Nernstian systems [1].

2.2. Electrochemical cell

Typical amperometric sensors configuration is based on two or three electrodes cell topology. A typical two electrodes topology is defined by the working electrode, where the electrochemical reaction takes place and the reference electrode, in addition to the auxiliar electrode which tracks the solution potential and supplies the current required for the reaction. This topology brings some kind of problematic behavior by the auxiliar electrode polarization effects. While the auxiliar electrode is assumed to have a fixed well-known potential some charges are accumulated on the electrode, due to the current supply, making this assumption erroneous. In order to avoid this effect, we need to supply the current using an extra electrode. The three electrodes configuration, which is defined as follows: a) the working electrode (WE), surface where the electrochemical reaction takes place; b) the reference electrode (RE), which tracks the potential solution and c) the counter or auxiliar electrode (CE), which supplies the current required for the electrochemical reaction at WE, is then of special interest.

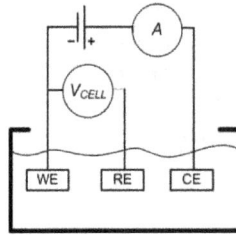

Figure 1. Three electrodes sensor topology.

$$V_{WE} - V_{RE} = V_{CELL} \qquad (2)$$

We must consider the potential solution V_{CELL} voltage to be constant, so in order to keep this condition current through the RE electrode should ideally be zero, avoiding any electrode polarization effect. The current supply is provided by CE, avoiding this undesirable effect.

So, considering the three electrodes topology, once any polarization effect has been avoided, one of the key points to be studied is the theoretical model in terms of the electronic behavior of the electrochemical cell.

An electrochemical sensor must be considered, in an electrical model, as an impedance [28,35], taking as a basic sensor element the presence of a capacitor, used to describe the interface between the electrode and their surrounding electrolyte. This capacitance is based on the electrical double-layer theory [36], so electrodes immersed in an electrolyte solution, can be described as a capacitor storing charge (ions from a solution absorbed on the electrode surfaces).

We must consider this interface model as an electronic representation in terms of electronic passive elements, and it can be very complex. The simplest case uses an equivalent circuit, also called a Randles circuit [1], depicted in Figure 2, which is formed by the double-layer capacitor (Cref), in parallel with a polarization resistor (Rref), which is also described as a charge transfer resistor, and the solution resistor (Raux). This is the electrical model that can be adopted for a mathematical description. Such description is very useful to create an electrical model for the electrochemical cell theoretical impedance analysis, being easy to reproduce in different electronic software emulators or prototype testing, capable of representing the electrochemical cell by means of passive elements like resistors and capacitors. This model can evolve into a more complex one, bringing us the possibility of an easier sensor functionalization, instrumentation design and prototype developing and testing [37].

The impedance measured in the cell is defined by (3), being Z_{CELL} the relation $V_{CELL}(t)/I_{CELL}(t)$.

$$Z_{ref\text{-}work}(j\omega) = Z_{CELL}(j\omega) = \frac{R_{ref}}{1 + j\omega R_{ref} C_{ref}} \tag{3}$$

One of the most interesting features that can be easily achieved with a three electrodes topology is the possibility of developing a multi-bio-analysis system by means of different electrode arrays. These arrays can be used as a multi-purpose system becoming an extremely versatile tool making it feasible to perform at the same time different electrochemical experiences with different biochemical species, average measurements through time or area, etc. [8,38,39].

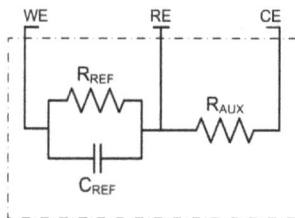

Figure 2. Randles Model for electrochemical cells.

3. Electronics for amperometric biosensors

3.1. The potentiostat amplifier

A potentiostat amplifier is a useful tool in many fields of investigation and industry performing electrochemical measurements [1], so the quantity and variety of them are very extensive, having different characteristics. In this section, we will focus our attention on the development of an analogue potentiostat amplifier approach.

Two different approaches can be followed in the design and implementation of a potentiostat amplifier: a discrete or integrated solution. In order to design a portable system for standard electrochemical assays, a discrete implementation is an extremely good solution in terms of portability, accuracy and economy being a standard on electrochemical experiments. But demand for increased functionality, reduced system size, reduced size of the electrodes, defining complex arrays of sensors, ultra-low current detection and versatility, are introducing a major interest in system-on-chip (SoC) solutions, to be implemented in advanced CMOS processes. The scaled supply voltages in these processes [4-8], however, seriously limit the chemical analysis range.

Driving voltages of amperometric chemical sensors do not scale with electrode size, but are instead defined by the reduction/oxidation (redox) potentials of the analyses being investigated, as stated in [40] many analysis are undetectable using standard potentiostats in a 0.18μm CMOS process due to its maximum supply voltage of 1.8V.

The main tasks of these kind of structures (Figure 3.) are the measurement and recognition of some kind of particles in a media (or the media itself), through the application of an electric signal and the readout and conditioning of an output signal.

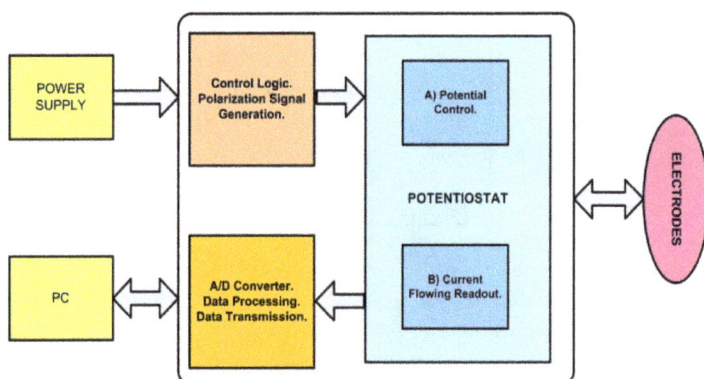

Figure 3. Potentiostat general scheme.

The main functionalities envisaged for a potentiostat amplifier are: a) driving the sensor electrodes with the desired signal V_{CELL}, ensuring that this voltage remains invariable and supplies the current necessary for the electrochemical reaction, potential control module on Figure 3. (section 3.1.1), and b) be able to extract an output signal which is the measuring of the current flowing through the electrochemical cell, current measurement module on Figure 3. (section 3.1.2). Driving the electrochemical cell depends on whether the configuration is based on two or three-electrodes, as stated above. Different approaches are conceived to fulfill this last objective.

3.1.1. Potential control configurations

As has been stated before, one of the main tasks of a potentiostat is the control of voltage difference between working and reference electrodes of the electrochemical cell and supplying the required current from or into the electrochemical cell through the counter electrode.

This task can be realized with two different circuit configurations, grounded working electrode and grounded counter electrode, the first being the most popular configuration depicted in Figure 4 which illustrates the basic implementation of this configuration. As shown, the working electrode is kept at the ground potential and an operational amplifier, called the driving amplifier, controls the cell current I_{CELL}, so the cell potential V_{CELL} is at the desired potential V_{IN}.

The system operation is very simple, but, like other electronic instrumentation circuits, we get potentiostat functionality limitations due to its own driving amplifier limitations. Since current flow in the reference electrode changes the potential of the reference electrode due to polarization effects (see section 2.2), driving amplifier input bias current should be small and input resistance should be very large. Depending on the target, you must consider several limitations on the driving amplifier parameters. Voltage gain and input offset voltage of the driving amplifier define the accuracy and linearity of the potential control. Other important parameters to be considered are the output voltage swing, input referred noise, bandwidth and slew rate, considering the stability as a sensitive issue due to the fact that the electrochemical cell is the load and feedback network of the amplifier.

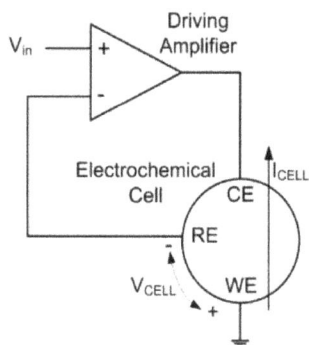

Figure 4. Potential control. General scheme.

In section 2.2. a typical electrochemical cell model has been depicted, where frequency dependent impedances like capacitors are present, the frequency and transient simulations of these impedances being quite complex to study. For this reason, it is necessary that the potentiostat provides stability over wide operation ranges, being able to carry out diverse electrochemical experiences for different biochemical species.

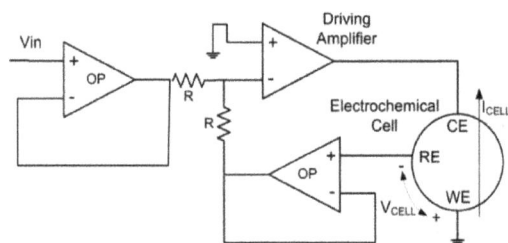

Figure 5. Using voltage buffers to isolate driving amplifier and reference electrode.

3.1.2. Current readout configurations

Some circuits used to fix the electrochemical cell's driving signal have been reported. The next stage concerns circuits related to the flowing current readout. Different approaches can be adopted and are presented in this section.

3.1.2.1. Transimpedance amplifier stage

In the present configuration the measurement is based on the direct conversion of the current generated in the electrochemical cell into a voltage signal using a transimpedance configuration, depicted in Figure 6 dotted rectangle.

In order to read the faradic current generated by the reaction I_{CELL}, a transimpedance ance stage amplifier (TIA), converts it to a voltage signal by means of a single resistor, as is indicated in (4), so the output signal $V_{out,TR}$ is equivalent to the faradic current through working electrode.

$$V_{out,TR} = -I_{CELL} \cdot R_{TIA} \tag{4}$$

The system operation it is simple, but, as is stated in the previous section, we get functionality limitations due to transimpedance amplifier limitations. V_{IN} voltage will be tracked to the electrodes if WE electrode is ground referenced, assuming operational amplifier virtual ground. In that case, input offset voltage and input referred noise must be considered in order to provide a steady virtual ground. Since generated current must flow through trans-impedance amplifier resistor R_{TIA}, trans-impedance amplifier TIA input bias current should be small and input impedance should be very large in order to minimize any current losses through this stage.

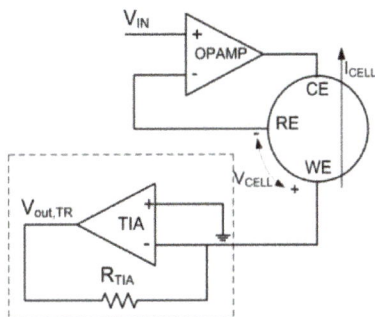

Figure 6. Basic grounded working electrode driving control configuration with a transimpedance amplifier readout stage.

As is shown on the operational amplifier equivalent circuit in Figure 7, the input impedance is the equivalent impedance between the positive and negative inputs. This impedance is linked to some leakage currents (I_{offset}), which could cause problems both in I_{CELL} current readout and V_{CELL} tracking, especially when extremely low faradic currents are generated on the reaction in which case a very high transimpedance resistor is required. This error in both cases can be minimized by reducing amplifier offset and bias current by means of very high input impedance.

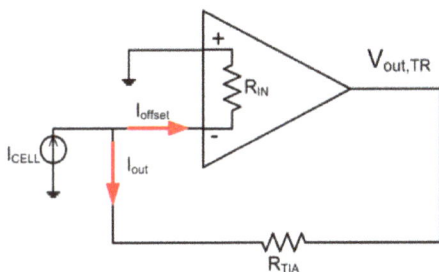

Figure 7. Operational amplifier in transimpedance configuration. General Scheme.

Other parameters to be considered are the amplifier's flicker and thermal noise or inherent current and voltage offset. Flicker and thermal noise are inherent to electronics and are characterized as an output voltage (Vnoise) (5) or as an input current (6) defining the transimpedance amplifier stage resolution, establishing a minimum signal to noise ratio, SNR (7).

$$V_{OUT} = V_{out,TR} + V_{noise} \tag{5}$$

$$I_{noise} = \frac{V_{noise}}{R_{TIA}} \tag{6}$$

$$SNR = \frac{I_{CELL}}{I_{noise}} \tag{7}$$

There are different solutions to maximize resolution of the measurement, increasing the SNR (Signal-to-noise ratio), and sensitivity, reducing thermal and flicker noise. For instance, one of the best solutions is based on chopper modulation, which implies more complexity on the design. A simplest solution is to just place a bandwidth filter. The capacitor on transimpedance amplifier feedback loop should remove the inherent 50 Hz network or any other frequency noise and harmonics but does not permit to avoid flicker noise problems. Figure 8. Depending on the application, if the system is required to work in a limited range of frequencies, like amperometric and voltametric experiences where you apply DC signals, or potential sweeps with an scan-rate of less than 1000 mV/s [1], it is useful to filter low frequencies to remove network powering noise, typically 50Hz and harmonics, and reduce thermal and flicker spectra to improve the SNR.

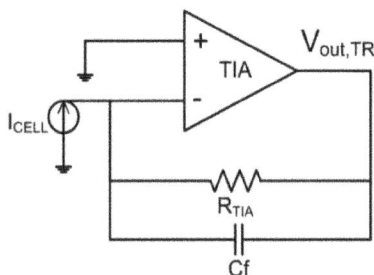

Figure 8. Transimpedance amplifier stage. Low-pass noise reduction capacitor configuration.

3.1.2.2. Instrumentation amplifier stage

This kind of current measurement topology consists in the direct conversion of the current in to a voltage signal by means of a resistor on the counter electrode, and an instrumentation amplifier that measures the voltage difference in the resistance, Figure 9.

We assume that the current through resistor R, is equal to the faradic current developed by the electrochemical reaction (I_{CELL}), and it is considered that voltage between reference and working electrode (V_{CELL}) is more steady than in the transimpedance amplifier stage due to the direct connection of working electrode to ground.

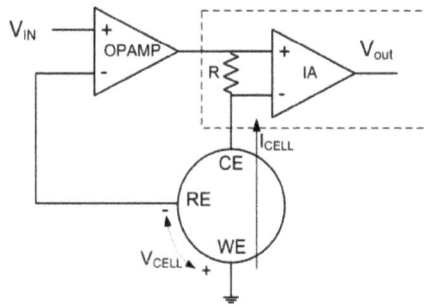

Figure 9. Basic grounded working electrode driving control configuration with a instrumentation amplifier readout stage.

An instrumentation amplifier transfer function is theoretically described by this equation:

$$V_{OUT} = A \cdot (V_+ - V_-) \qquad (8)$$

Where A is the amplifier's gain and (V+ - V-) is the voltage difference on the amplifier's positive and negative, so we can determine that if the voltage (V+ - V-) is (9) the output signal is equivalent to the faradaic current through working electrode (10).

$$(V_+ - V_-) = I_{CELL} \cdot R \qquad (9)$$

$$V_{OUT} = A \cdot I_{CELL} \cdot R \qquad (10)$$

There are some parameters to be considered as a source of noise errors:

Input impedance.

Offset current, bias current.

Amplifiers are a source of noise and non-idealities that are critical, mainly if very low current resolution, such as nanoamperes or picoamperes is desirable. It will be assumed that all the amplifiers have an input bias current that interferes with the current readout system. In order to minimize these effects we need to use an amplifier with very high input impedance, as is depicted in the previous section. There are other parameters to be considered, such as resistor tolerance and resistor thermal noise. This stage is very dependent on resistors, the conversion of the flowing current to a measurable output voltage it depends on the stability of three different resistors. There is a very high probability of getting an error source, making impossible to get a very precise measurement, if the system depends on the tolerance and thermal noise of three different resistors, but it's possible to minimize output voltage dependence of this large number of resistors (Figure 10.).

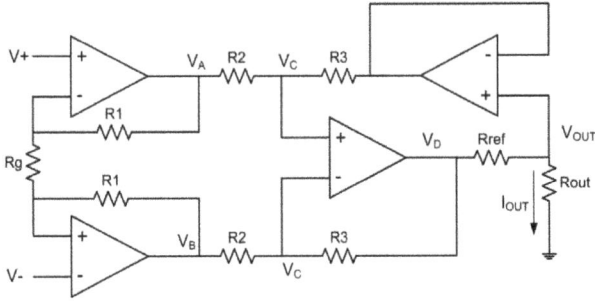

Figure 10. Modified instrumentation amplifier.

The modified instrumentation amplifier transfer function is theoretically described by the following equations:

$$\frac{(V_A - V_-)}{R1} = \frac{(V_- - V_+)}{Rg} = \frac{(V_+ - V_B)}{R1} \tag{11}$$

$$\frac{(V_B - V_C)}{R2} = \frac{(V_C - V_{OUT})}{R3} \tag{12}$$

$$\frac{(V_A - V_C)}{R2} = \frac{(V_C - V_D)}{R3} \tag{13}$$

Combining these three equations and considering that R3 = R2

$$V_{OUT} = I_{OUT} \cdot R_{OUT} \tag{14}$$

$$(V_{OUT} - V_D) = (V_A - V_B) = (V_+ - V_-)\left(1 + \frac{2R1}{Rg}\right) = I_{OUT} \cdot Rref \tag{15}$$

And if we consider

$$Rref = R_{OUT}\left(1 + \frac{2R1}{Rg}\right) \tag{16}$$

$$I_{OUT} = \frac{(V_+ - V_-)}{R_{OUT}} \tag{17}$$

We found that the output current signal is the same current as through electrodes and the evaluation only depends on the value of one resistor.

Regarding other error sources, the fact that there are no active components in the flowing current path, and being both flowing current and measured voltage referenced directly to ground, gives the system better stability than in the transimpedance amplifier stage.

3.1.2.3. Switching capacitors solution

Another feasible solution to current readout is the switching capacitors transimpedance stage avoiding the use of resistors. This topology needs a clock control signal, but, assuming the possibility of using a microcontroller for a later signal processing or data transmission this should not be a problem. The basic circuit approach is depicted in Figure 11.

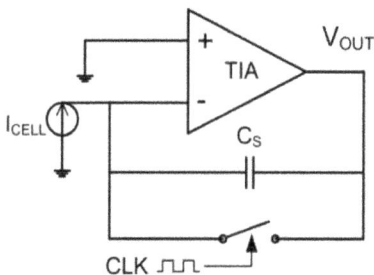

Figure 11. Switching capacitors transimpedance stage.

This stage has a very simple operation system, on the first clock semi cycle the switch is closed and electrochemical cell current, I_{CELL}, charges capacitor C_S to a concrete output voltage V_{OUT} as depicted in equation 18, where T_{CLK} is the clock period.

$$V_{OUT} = \frac{I_{CELL}\ T_{CLK}}{2C_S} \qquad (18)$$

On the second clock semi cycle the switch is opened, and V_{OUT} is directly connected to ground and the capacitor is discharged. Depending on the measurement range, the capacitor or the clock cycle can be modified to a larger or smaller value, giving the possibility of a more versatile stage.

The system operation it is simple, but, as stated in the previous section, we get functionality limitations due to transimpedance amplifier limitations. First of all the WE electrode is ground referenced through amplifier virtual ground. In that case, input offset voltage and input referred noise must be considered in order to provide a steady virtual ground. Since generated current must flow through transimpedance stage, TIA amplifier input bias current should be small and input impedance should be very large in order to minimize any current losses through the stage.

Another important consideration is that the capacitance of the flowing current source, that is the electrochemical cell, must be of a few orders of magnitude higher than the capacitor C_S. If not, the errors due to charge injection will be larger than desirable for any designed application.

This kind of topology is widely used in CMOS processes and microelectronics development due to the difficulty of realizing large resistors at small scales. The fact of using small capacitance values provides the possibility of developing multichannel sensor arrays [11,38,41] on ASIC structures, due to the high degree of integration of small size capacitors.

A more complex development of the switching capacitors technique makes it possible to perform direct A/D conversion by converting the current to variables such as frequency [9,42] or direct current input sigma-delta converter [11,43,44].

3.2. Electrochemical impedance spectroscopy

In section 2.2 the electrochemical cell and its theoretical electrical approximation by circuit modeling was introduced. But this model or other more complex one [45] are just approximations to the reality. A direct measurement of the impedance in a range of frequencies, usually from 1mHz to 1MHz [46], is fitted afterwards to an electrical model, with different elements, that are used to fit the data. In this section the electronics are presented along with the different approaches to extract such measurement, which is defined as Electrochemical Impedance Spectroscopy (EIS).

Current research in biosensor technology has been developed toward better transducers that demonstrate superior sensitivity, portability, accuracy and throughput; where the most promising solution to check the sensor's measurements is based on the use of the EIS technique - the response of an electrochemical cell to small amplitude sinusoidal voltage signal as a function of frequency.

EIS is a more effective method to probe the interfacial properties of the modified electrode, through measuring the change of electron transfer resistance at the electrode surface, caused by the adsorption and desorption of bio-chemical molecules and the antibody-antigen (Ab-Ag) interactions. The measured signal, in this case the signal generated (voltage or current signal) in the experiment, differs in time (phase shift) with respect to the perturbing (voltage or current) wave, and the ratio $V_{CELL}(t)/I_{CELL}(t)$ is defined as the impedance (Z_{CELL}), and accounts for the combined opposition of all the components within the electrochemical cell to the flow of electrons. In this section we will talk about the lock-in amplifier necessary to detect the impedance value of an experience based on an EIS technique.

Measurement of biochemical concentrations is essential for disease diagnose and biological systems characterization. The key electronic component for these measurements, as it is stated before, is the potentiostat amplifier to bias the sensor and read the current produced by the experiment. It is the interface between the biological elements and the lock-in amplifier, which generates the real and imaginary components for the EIS solution rejecting undesirable harmonics and noise interferences [47,48,49,50] even in the presence of high noise level. The block diagram of the whole system is depicted in Figure 12, where a general schematic view of a lock-in amplifier is shown.

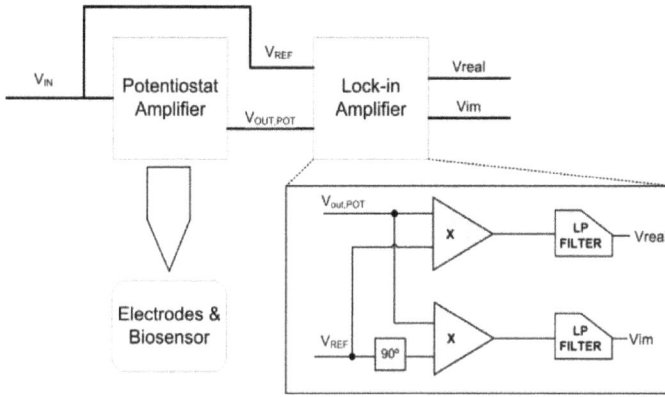

Figure 12. Block diagram view of a complete system of Potentiostat, Biosensor and Lock-in. Schematic view af a Lock-In amplifier.

The basic operation of the Lock-in Amplifier is very simple. $V_{out,POT}$ represents the potentiostat amplifier output signal, so

$$V_{out,POT} = I_{CELL} \cdot R \tag{19}$$

where R is the relation in ohms between the current on the cell and the output potentiostat signal (like R_{TIA} on transimpedance amplifier based potentiostats). Considering the $V_{out,POT}$ function as a frequency dependant, we get:

$$V_{out,POT}(t) = V_{OUT} \sin\left(2\pi ft + \phi_{OUT}\right) =$$
$$= V_{OUT}[\sin\left(2\pi ft\right)\cos(\phi_{OUT}) + \cos\left(2\pi ft\right)\sin(\phi_{OUT})] \tag{20}$$

So, our functions Vreal and Vim are represented by the following equations.

$$Vreal = V_{out,POT} \cdot V_{IN} \sin\left(2\pi ft\right) =$$
$$= V_{OUT} \cdot V_{IN} \cdot \left(\sin^2\left(2\pi ft\right)\cos(\phi_{OUT}) + \frac{1}{2}\sin\left(2\pi ft\right)\cos\left(2\pi ft\right)\sin\left(\phi_{OUT}\right)\right) \tag{21}$$

$$Vim = V_{out,POT} \cdot V_{IN}\cos\left(2\pi ft\right) =$$
$$= V_{OUT} \cdot V_{IN} \cdot \left(\cos^2\left(2\pi ft\right)\sin(\phi_{OUT}) + \frac{1}{2}\sin\left(2\pi ft\right)\cos\left(2\pi ft\right)\cos\left(\phi_{OUT}\right)\right) \tag{22}$$

$$Vreal = \frac{1}{2}V_{OUT} \cdot V_{IN} \cdot \left[\cos(\phi_{OUT}) - \cos\left(4\pi ft\right)\cos(\phi_{OUT}) + \sin\left(4\pi ft\right)\sin(\phi_{OUT})\right] \tag{23}$$

$$\text{Vim}= \tfrac{1}{2}\text{V}_{\text{OUT}}\cdot\text{V}_{\text{IN}}\cdot\left[\cos(\phi_{\text{OUT}}) + \cos(4\pi ft)\sin(\phi_{\text{OUT}}) + \sin\,(4\pi ft)\cos(\phi_{\text{OUT}})\right] \qquad (24)$$

Taking into account only the DC component,

$$\text{Vreal}= \tfrac{1}{2}\text{V}_{\text{OUT}}\cdot\text{V}_{\text{IN}}\cdot\cos(\phi_{\text{OUT}}) \qquad (25)$$

$$\text{Vim}= \tfrac{1}{2}\text{V}_{\text{OUT}}\cdot\text{V}_{\text{IN}}\cdot\sin(\phi_{\text{OUT}}) \qquad (26)$$

The magnitude and phase of $V_{\text{out,POT}}$ are

$$\left|V_{\text{out,POT}}\right| = \tfrac{2}{V_{\text{IN}}}\sqrt{V_{\text{real}}^{2} + V_{\text{im}}^{2}};\ \ \Phi_{\text{Vout,POT}}=\text{arctg}\!\left(\tfrac{V_{\text{im}}}{V_{\text{real}}}\right) \qquad (27)$$

Being the magnitude and phase of impedance Z_{CELL}:

$$\left|Z_{\text{CELL}}\right| = \frac{V_{\text{IN}}\cdot V_{\text{CELL}}\cdot R}{2\sqrt{V_{\text{real}}^{2} + V_{\text{im}}^{2}}};\ \Phi_{\text{ZCELL}}=\text{arctg}\!\left(\frac{V_{\text{im}}}{V_{\text{real}}}\right); \qquad (28)$$

3.2.1. The analogue approach

In the previous section have the idea of a whole system based on a potentiostat and a lock-in amplifier as a complete solution for an electrochemical impedance spectroscopy experiment was introduced. In this section, it is presented the configuration of a lock-in amplifier that generates the real and imaginary components of the impedance (Z_{CELL}) based on an analog instrumentation implementation is presented.

The lock-in amplifier architecture based on an analogue approach [48,49,50], consists in different modules, which are two Synchronous Demodulated Channels which generates DC voltage signals which are proportional to the real (V_{real}) and imaginary (V_{im}) components of the input signal (potentiostat measurement). The circuit schematic, with the demodulator and the low pass filter, is depicted in Figure 13, where $V_{\text{out,POT}}$ represents the potentiostat amplifier output signal (equation 19).

The lock-in amplifier provides real and imaginary components through the DC values Vreal and Vim, respectively, after filtering the rectified signals from the demodulator stage, getting a complete characterization of potentiostat output signal and an accurate estimation of Z_{CELL} (electrochemical reaction characteristics).

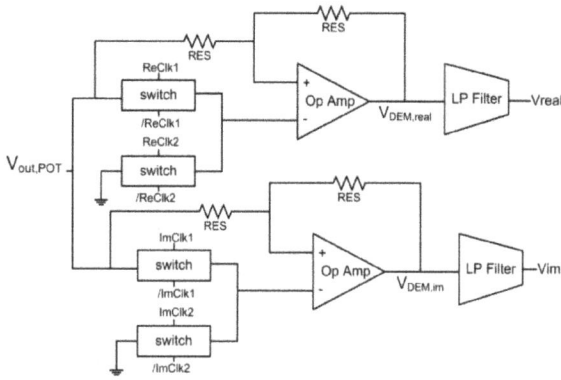

Figure 13. Full schematic view of the implemented lock-in module [58].

Special attention is given to the reference signal used by the demodulator channels which is multiplied by the signal to be measured. The reference (V_{REF}) signal is an ac voltage, of the same frequency of the input signal, which can be either generated by an oscillator, locked to the input signal by a phase locked loop or mainly using the same polarization signal of the previous stages (V_{IN}). A phase shifter allows the reference signal to be trimmed at the following phases: phase ReClk1 = 0º, phase ReClk2 = 180º, phase ImClk1 = 90º, phase ImClk1 = 270º. The clock signals generator is depicted on Figure 14.

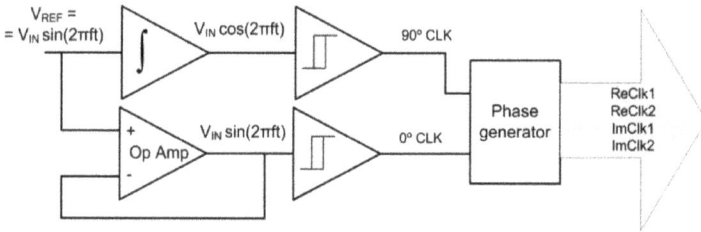

Figure 14. Clock generation module [58].

Clock signals are generated by two hysteretic comparators, one for the 0º phase clock and the other one for 90º phase clock signals, previously generated by an integrator. It's desirable that the different four clock signals be generated with a dead time, DT in Figure 15., between them for each channel, ReClk1 and ReClk2 for V_{REAL} channel and ImClk1 and ImClk2 for V_{IM} channel. The dead time must be implemented in order to avoid undesired spikes at the generated clocks and harmonic distortion coupling on the demodulator channels. Dead time values must be several orders of magnitude less than the clock period to not interfere with the clocks phase shift.

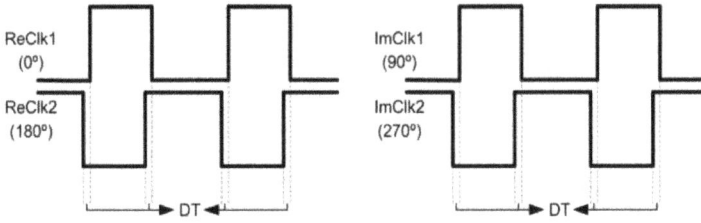

Figure 15. Dead time conceptionClock generation module

Finally, the demodulation stage consists of two simple wave rectifiers. On each rectifier, when the input signal to be measured and the reference signal are of the same frequency, the demodulator output has a dc component proportional to the input signal amplitude [48,49,50]. By adjusting the phase of the reference signal using the phase-shifters present in the reference channel, the phase difference between the input signal and the reference can be brought to zero (null shift procedure). If we get all the four phases; 0°, 90°, 180° and 270°; considering the two different channels on demodulation stage, we have a complete data spectrum to evaluate the whole input signal.

A low pass filter characterized by a low cut-off frequency is necessary to reject the noise and harmonics superimposed to the output demodulation stage and acquire the dc component proportional to the signal. A very interesting architecture is based on a trans-conductance amplifier (OTA), due to the very small trans-conductance values, in the order of nano-siemmens, that can be defined [51]. The basic structure is based on a source degenerated trans-conductance amplifier (OTA) to define the filter. The source degeneration increases the input range of the amplifier and also decreases the equivalent trans-conductance of the OTA amplifier. The ratio between the current mirrors decreases the current level at output, which results in an even minor value [52]. These current mirrors are based on composite transistors, used to reach greater copy factors. The typical transfer equation and cut-off frequency are the following

$$|H_{FILTER}| = \frac{\frac{gm1 \cdot gm2}{C1 \cdot C2}}{s^2 + s \cdot \frac{gm2}{C2} + \frac{gm1 \cdot gm2}{C1 \cdot C2}} \tag{29}$$

$$\omega_0 = \sqrt{\frac{gm1 \cdot gm2}{C1 \cdot C2}} \tag{30}$$

Then, integratable capacitors can be implemented, defining cut-off frequencies in the range of 0.1Hz to 30Hz. The LP Filter on Figure 13. is depicted as a Gm-C second-order low pass-filter in Figure16.

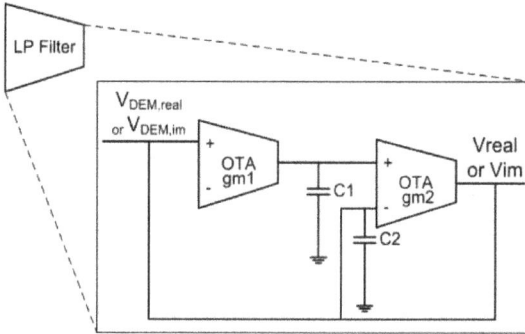

Figure 16. Gm-C Second order filter configuration.

Once we get both real and imaginary components of the measured signal from the analogue lock-in amplifier, we need to process these raw dc values to obtain reliable information about our system to perform an EIS experiment with them. In order to develop a complete system for EIS experiments, the dc raw data on lock-in output must be digitalized in order to carry out the mathematical post-processing, equation 31 to 34, on a microcontroller, DSP or computer. The digitalization of the output data is easiest than in other devices due to the acquisition of only DC signals. The theoretical expression of the module and phase of $V_{out,POT}$, in Figure 13, using the Randles model, are found in following equations:

$$|V_{out,POT}| = \frac{\pi}{2}\sqrt{V_{real}^2 + V_{im}^2} \tag{31}$$

$$\Phi_{Vout,POT} = arctg\left(\frac{V_{im}}{V_{real}}\right) \tag{32}$$

$$|Z_{CELL}| = \frac{2}{\pi} \cdot V_{CELL} \cdot R \cdot \frac{1}{\sqrt{V_{real}^2 + V_{im}^2}} \tag{33}$$

$$\Phi_{ZCELL} = arctg\left(\frac{V_{im}}{V_{real}}\right) \tag{34}$$

Where 2/pi is the mean absolute value of the sine function [48].

Obtaining with equation 36 and 37 a direct measurement of both Z_{CELL} module and phase and Z_{CELL} real and imaginary components and obtaining reliable data for Electrochemical Impedance Spectroscopy experience. In Figure 17 the behavior of the demodulation stage for both channels is shown.

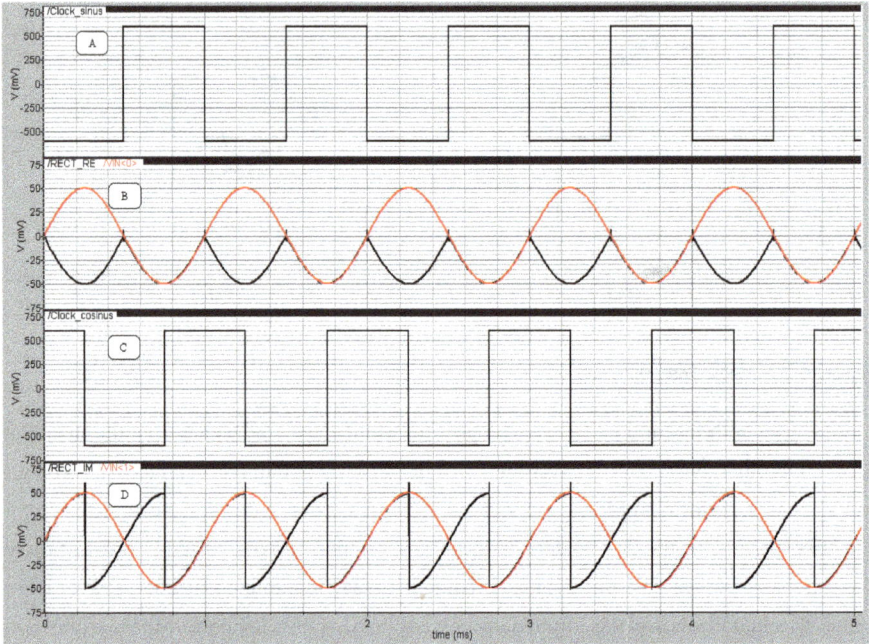

Figure 17. Caption of the rectified signals for the real and imaginary channels before the active filters. Upper trace (A) represents the reference clock signal for the synchronous demodulated channel for the real component, before the filter (Vreal); next figure (B) is the rectified signal. The third signal (C) is the reference clock signal for the synchronoys demodulated signal for the imaginary component (Vim), and the last trace (D) at the bottom, is the trace of the rectified signal at the imaginary channel. These traces are obtained for a 180° condition.

3.2.2. The digital approach

As has been stated in section 3.1.2.3, some potentiostat solutions employ an output digital signal in order to facilitate the data processing and transmission. Since the use of analogue instrumentation processing usually leads to a final data digitalization, the possibility of a direct embedded processing is an interesting approach to developing a lock-in amplifier. The digital lock-in approach is based on an embedded mathematical processing on a microprocessor or DSP device [19,53]. The block diagram of the lock-in software is depicted in Figure 18.

In order to proceed with the signal processing there are two main approaches: a) the Fast Fourier Transform (FFT) [18], and b) the Frequency Response Analyzer (FRA) [19]. In the case of the FFT, a pulse, or a step, -the approach to be followed is the ideal Dirac function-, is applied to the sample because it contains a wide frequency content. Then, the response of the sample is digitalized and processed in a digital processor, for instance a DSP, and using the FFT algorithm, the different frequency components are obtained for their analysis. Another possibility is the logarithmic sampling in the DFFT calculus, reducing the data required in the process [18]. This appears to be simple, but there are several problems in the implementation. First of all, it is very difficult to generate a fast step function and a very fast potentiostat capable of driving this step on the electrodes and extracting the resulting current signal. If the potentiostat rising time is too slow, the resulting frequency components will be distorted. Since the important information is contained in a short period of time after the step is applied, in addition to a very fast potentiostat, very fast ADC with a high precision bit resolution is also required

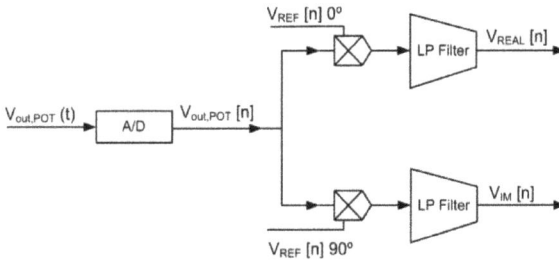

Figure 18. Digital lock-in block diagram.

A simpler solution is based on the FRA approach. In this case a sine and cosine signals are adopted and by means of two multipliers and a filter stage the real and imaginary components of the response are obtained. This measurement must be done for each frequency. Working with just one sensor and in terms of the size of the final product, the FFT option could be adopted, although high speed hardware and heavy algorithm implementation is required, because the response for several frequencies is obtained. The FRA solution is more oriented to multi-sensor approaches but is also a good option in the case of single sensors, in terms of the trade-off between complexity and speed, if not too low frequencies are to be measured.

This lock-in approach is more feasible. The digital lock-in FRA approach [19] is based on the principle that there is no correlation between noise and measured signal. In contrast to the analogue approach, an orthogonal arithmetic multiplying between the incoming potentiostat signal and reference signal are used to get the real and imaginary components, coming close to the theoretical behavior of a lock-in amplifier depicted in section 3.2. A digital lock-in has no low frequency limitations, being capable of working properly at the sub-hertz region. The upper frequency limitation is mainly limited by the ADC conversion time, being able to develop a wide frequency range EIS system. On the other hand, the digital lock-in is limited by area and power consumption. The area and power consumption levels depend on the electronics involved. If a microprocessor is needed, we get typical power consumption, for commercial solutions, of several hundreds of mW, which is far from the desired power waste. But in the recent years a step forward in microprocessors field has been presented in [54] and [55]. [54] Present a microprocessor, in a 180 nm technology, with a power consumption of 226 nW, and area of 915x915 mm^2. It evolved from [55], where the sub-threshold operating region is explored. In the same way there has been an evolution in microprocessor development, in terms of area and power consumption. [19] Present an evolution of the digital lock-in algorithm based on an oversampling solution, simplifying the orthogonal vector arithmetic cutting off all the multiplying operations.

In that way, evolution of both microprocessor hardware and lock-in algorithm software, leads to a whole post-processing embedded system with great throughput, functionality and versatility without involving a high area or power consumption.

4. An example of a CMOS low power potentiostat amplifier

In this section the design of a CMOS low power potentiostat amplifier using a 0.13µm technology is described. Potentiostat architecture [2,6,56,57], using the described electrochemical model, is depicted in Figure 19.

This structure is based on four amplifiers (Opamp) and two resistors. OP4 is the transimpedance amplifier, which defines the virtual ground voltage of the WE electrode, and provides current-to-voltage conversion such that,

$$V_{out,POT} = - I_{CELL} \cdot R_{TIA} \tag{35}$$

where I_{CELL} is the current through the cell and R_{TIA} the gain defined on the transimpedance amplifier.

OP3 is used to ensure minimal current flow through the RE electrode. It senses the voltage difference between the RE and WE electrodes (virtual ground). This difference is compared by OP2 with the desired V_{IN} voltage, changing the voltage at the AUX electrode and defining a current through the cell in such a way that the voltage difference between the RE and WE electrodes follows the defined V_{IN} DC+AC signal that polarizes the sensitive cell.

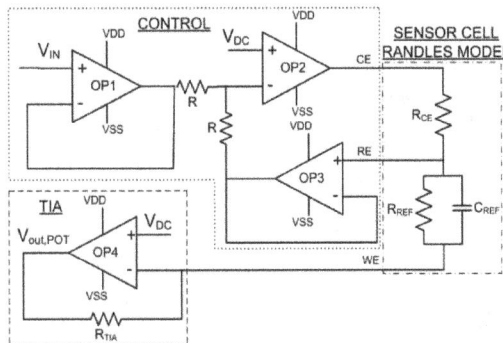

Figure 19. Full schematic view of the potentiostat amplifier with the adopted electrochemical electrical model [58].

If WE electrode is attached at a virtual ground by the transimpedance amplifier, it can be demonstrated that the Zcell impedance, and variations, could be detected continuously by:

$$V_{out,POT} = \frac{R_{TIA}}{Z_{CELL}}(V_{ref} - V_{work}) = \frac{R_{TIA}}{Z_{CELL}}V_{IN} \tag{36}$$

$$Z_{CELL}(j\omega) = \frac{V_{IN}}{V_{out,POT}}R_{TIA} \tag{37}$$

The amplifier adopted to design the potentiostat amplifier is based on a wide-swing, cascode output stage with feedforward class-AB control, Figure 20 and Figure 21, [58-60] for OP1, OP2 and OP3 amplifiers and a full-custom, multi-stage, high input impedance, cascode output stage for transimpedance amplifier OP4. The AB amplifiers (OP1, OP2 and OP3) output transistors have been sized in such a way they can supply the right current for the worst load conditions, defined by the electrochemical model, which has a total value of several MΩ. The input stage of the transimpedance amplifier (OP4) has been designed to increase by three orders of magnitude, the input impedance, minimizing offset and current losses at WE electrode. The power supply is 1.2 V for all the electronics.

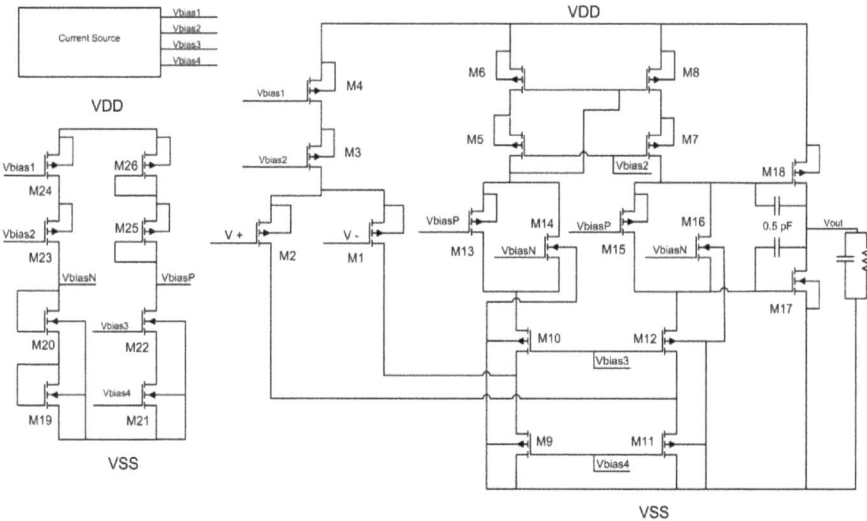

Figure 20. Full schematic view of the AB amplifier [58].

Individual amplifier AB (OP1, OP2 and OP3) is 440μm in height and 500μm in width, with 84 μW of power consumption in nominal conditions (10MΩ@10pF), 108 dB open-loop gain at low frequencies, 300 kHz bandwidth with a PM = 59º and 12 μV of input systematic offset. In Table

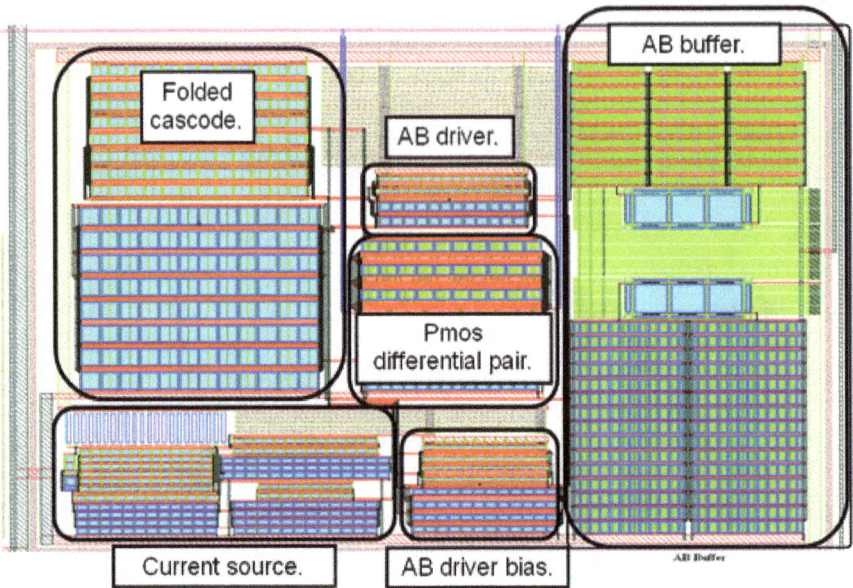

Figure 21. Full layout view of the AB amplifier [58].

1 different results are summarized, based on the typical (TYP), fast (FFA) and slow (SSA)

mobility values of the electrical carriers, and for different simulation conditions, are reported.

TYP							
Parameters	V_{offset} (µV)	F_{0dB} (kHz)	Phase Margin (°)	Gain (dB)	V_{noise} (µV)	THD (%)	P_{supply} (µW)
Schematic	-17,56	402,3	60,19	108,9	118,73	5,56m	78,4
Extracted	-12,04	303,7	58,78	107,9	397,15	3,67m	84,9
FFA							
Parameters	V_{offset} (µV)	F_{0dB} (kHz)	Phase Margin (°)	Gain (dB)	V_{noise} (µV)	THD (%)	P_{supply} (µW)
Schematic	-26,07	335,5	65,79	106,1	62,95	1,74m	90,7
Extracted	-17,27	401,7	54,94	108,9	104,44	2,46m	101,9
SSA							
Parameters	V_{offset} (µV)	F_{0dB} (kHz)	Phase Margin (°)	Gain (dB)	V_{noise} (µV)	THD(%)	P_{supply} (µW)
Schematic	-13,83	390,51k	62,12	110,1	62,12	14,49m	95,4
Extracted	-10,42	158,8 k	71,23	102,9	104,49	4,71m	32,1

Table 1. Folded Cascode AB amplifier characterization @ ± 0.6Vsupply

.

Transimpedance amplifier (OP4) is 500μm in height and 1000μm in width, with 61 μW power consumption in nominal conditions(10MΩ@10pF), 85 dB open-loop gain at low frequencies, 400 kHz bandwidth with a PM = 67º and 26 μV input systematic offset.

Taking account of an extremely low offset requirement for bio-implantable devices, this low offset performance is due to an accurate channel length modulation of transistors size at the differential pairs, applying careful techniques for the analogue layout [61]. The linear range of the potentiostat amplifier has been analysed by simulations and it is expected to be 80% of the supply range. As an example; considering a ±100mV full scale voltage, in the ±75mV range the linearity response is quite good, with a deviation error of less than 0.01% [6], as is depicted in Figure 22. The potentiostat amplifier is 1400μm in height and 1000μm in width, with a power consumption of $400μW@|V_{DD}-V_{SS}|=1.2V$.

The full system has been analyzed based on the extracted views of the design. The ranges of the electrochemical parameters are good enough for the targets of the electrochemical cells. Initial simulations of this approach are presented, with positive results in terms of the potentiostat amplifier and lock-in amplifier response. The potentiostat assures a good linearity, and also ensures that the electrochemical cell follows the input voltage V_{IN} as expected. Current losses on the TIA amplifier stage are totally negligible and enclosed.

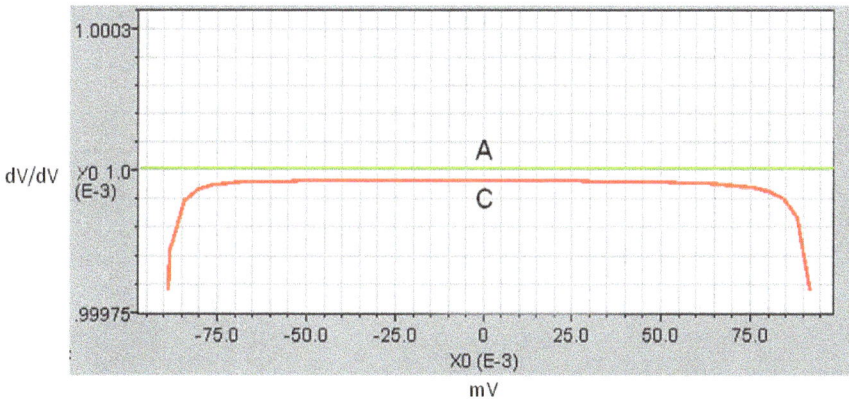

Figure 22. These signals are the derived functions, where A is the input signal (VIN) and C the polarization signal (VCELL).

5. Summary and conclusions

In this chapter, we have introduced the basic principles of biosensors and bioelectronics interfaces specially focused on the design of instrumentation related to amperometrics biosensors, potentiostat amplifiers and lock-in amplifiers. These elements have been introduced with regard to the state-of-the-art and the trends involved in such systems, and the development of custom built electronic solutions for bio-electronics applications, from discrete devices to ASICS solutions.

Discrete systems are useful for implementation in portable point-of-care applications. As we get less area and power restrictions than an ASIC solution, there are several architectures for data acquisition, processing and transmission. Potentiostat amplifiers can be designed with discrete devices or a monolithic ASIC solution while the lock-in amplifier can be designed with discrete devices or monolithic ASIC for an analogue approach, or an external commercial microcontroller or DSP for a digital approach. We must choose our fit depending on portability, accuracy and reliability requirements.

However, for body sensor networks body or implantable devices development, the increased functionality, reduced systems, with smaller multiplexed electrodes, for ultra-low current detection and versatility will require potentiostat amplifiers to be designed on a system-on-chip (SoC), which will force us to implement the system in CMOS technology.

As stated by [62] a trade-off between versatility and power consumption, aggressive digital processing and analogue data processing or general purpose and custom design, must be considered when developing such systems that integrates medical and electronic technologies. These trade-offs have been presented on this chapter split in two different sections, related to the potentiostat amplifier and to the lock-in amplifier. Finally, an example of CMOS low power potentiostat amplifier has been reported.

Structure Topology	Advantages	Disadvantages
Transimpedance amplifier stage. Section 3.1.2.1.	- Simplicity.	- Virtual ground on working electrode. - Active devices through flowing current path. - Current measurement losses due to amplifier input bias current.
Instrumentation amplifier stage. Section 3.1.2.2.	- No virtual ground on working electrode. No active devices through flowing current path.	- Higher flicker and thermal noise due to extensive use of resistors. - Impedance increase on potentiostat feedback loop. - Low integration degree.
Transimpedance switching capacitor stage. Section 3.1.2.3.	- High integration degree. Most versatile analogue output signal to A/D conversion.	- Virtual ground on working electrode. - Active devices through flowing current path. - Highly dependent on electrode morphology and parasitic capacitors.

Table 2. Summary table section 3.1.

In section 3.1, different potentiostat amplifier topologies have been reported. Different approaches to develop a potentiostat amplifier were introduced, taking account of on every particular situation the pros and contras. Table 2. Choosing the best fit analogue instrumentation has repercussions on several benefits in terms of area and power consumption, and is the first step to a solid efficient design. Exploiting the analogue processing before digitization is the optimal way to develop these systems, regarding all the benefits described.

In section 3.2 the lock-in amplifier is shown with two different approaches, the analogue approach and the digital approach. Table 3.

The analogue lock-in amplifier provides several advantages in terms of post-processing requirement as the digitization of the output data, being a DC signal, is easier than in other kinds of devices, decreasing considerably the complexity of the post-processing and data transmission electronics. On the other hand, this analogue approach has some limitations in terms of versatility and bandwidth, which is limited by the whole lock-in electronics. Assuming the implementation of a CMOS monolithic solution for a whole implantable device, versatility and bandwidth limitations can be acceptable in terms of an efficient custom system.

The digital approach of a lock-in amplifier allows us to develop a very versatile and powerful device. The bandwidth of this system is only limited by the analogue to digital converter and data transmission electronics if needed. The digital lock-in approach being a whole mathematical embedded system can be implemented in different processing topologies, such as real time processing by means of a FPGA (Field Programmable Gate Array), or standard processing by means of a DSP or microprocessor. A digital lock-in has no low frequency limitations, being able to work effectively at the sub-hertz region. The upper frequency limitation is mainly limited by the ADC conversion time, being capable of developing a wide frequency range EIS system.

Structure Topology	Advantages	Disadvantages
Analog lock-in. Section 3.2.1.	- Simplest analog output signal A/D conversion. - Less power and area consumption.	- Low and high frequency limitation. - Pure AC signals needed.
Digital lock-in. Section 3.2.2.	- No frequency limitation. Wide frequency range operation. - More versatile system.	- Higher power and area consumption due to ADC or microcontroller.

Table 3. Summary table section 3.2.

We must keep in mind the large area and power increase represented by a microprocessor, DSP or FPGA, which can make this lock-in approach not suitable for low power consumption electronics or implantable devices. However, as it has been reported in section 3.2.2, advances in DSP area and power requirements and advances in digital lock-in algorithms have made the possibility of a digital lock-in implementation on a low-power system-on-a-chip CMOS implantable device feasible.

Author details

Jaime Punter Villagrasa[*], Jordi Colomer-Farrarons and Pere Ll. Miribel

*Address all correspondence to: jpunter@el.ub.edu

Department of Electronics, Bioelectronics and Nanobioengineering Research Group (SIC-BIO), University of Barcelona, Spain

References

[1] Bard A, Faulkner L. Electrochemical Methods, Second Edition. John Wiley & Sons; 2001.

[2] Martin SM, Gebara FH, Larivee BJ, Brown RB. A CMOS-Integrated Microinstrument for Trace Detection of Heavy Metals. IEEE Journal of Solid-State Circuits 2005; 40(12) 2777-2786.

[3] Colomer-Farrarons J, Miribel-Catala P, Rodriguez-Villareal I, Samitier J. Portable Bio-Devices: Design of Electrochemical Instruments from Miniaturized to Implantable Devices. In: Serra PA. (ed.) New Perpspectives in Biosensors Technology and Applications. Intech; 2011. p373-400.

[4] Kakerow R, Kappert H, Spiegel E, Manoli Y. Low-power Single-Chip CMOS Potentiostat. The 8th International Conference on Solid-State Sensors and Actuators, 25-29 June 1995, Stockholm, Sweden.

[5] Kraver KL, Guthaus MR, Strong TD, Bird PL, Cha GS, Hold W, Brown RB. A Mixed-signal Sensor Interface Microinstrument. Journal of Sensors and Actuators A 2001; 91 266-277.

[6] Reay R, Kounaves S, Kovacs G. An Integrated CMOS Potentiostat for Miniaturized Electroanalytical Instrumentation. Proceedings of the 41st IEEE International Solid-State Circuits Conference, ISSCC1994, 16-18 February 1994.

[7] Martin SM, Gebara FH, Larivee BJ, Strong TD, Brown RB. A Low-Voltage, Chemical Sensor Interface for Systems-On-Chip: The Fully-Differential Potentiostat. Proceedings of the International Symposium on Circuits and Systems, ISCAS2004, 23-26 May 2004, Vancouver, Canada.

[8] Stanacevic M, Murari K, Cauwenberghs G, Thakor N. 16-Channel Wide-range VLSI Potentiostat Array. IEEE International Workshop on Biomedical Circuits and Systems, BioCAS2004, 1-3 December 2004, Singapore.

[9] Qisong W, Haigang Y, Tao Y, Chong Z. A High Precission CMOS Weak Current Readout Circuit. Journal of Semiconductors 2009; 30(7) 075011. http://dx.doi.org/10.1088/1674-4926/30/7/075011 (accessed 12 March 2009)

[10] Narula HS, Harris JG. A Time-Based VLSI Potentiostat for Ion Current Measurements. IEEE Sensors Journal 2006: 6(2) 239-248.

[11] Gore A, Chakrabartty S, Pal S, Alocilja EC. A Multichannel Femtoampere-Sensitivity Potentiostat Array for Biosensing Applications. IEEE Transactions on Circuits and Systems 2006; 53(11) 2357-2363.

[12] Ayers S, Gillis KD, Lindau M, Minch B A. Design of a CMOS Potentiostat Circuit for Electrochemical Detector Arrays. IEEE Transactions on Circuits and Systems 2007; 54(4) 736-744.

[13] Ahmadi MM, Jullien G A. Current-Mirror-Based Potentiostats for Three-Electrode Amperometric Electrochemical Sensors. IEEE Transactions on Circuits and Systems 2009; 56(7) 1339-1348.

[14] Yang C, Huang Y, Hassler BL, Worden RM, Mason AJ. Amperometric Electrochemical Microsystem for a Miniaturized Protein Biosensor Array. IEEE Transactions on Biomedical Circuits and Systems 2009; 3(3) 160-168.

[15] Hwang S, Lafratta CN, Agarwal V, Yu X, Walt DR, Sonkusale S. CMOS Microelectrode Array for Electrochemical Lab-on-a-Chip Applications. IEEE Sensors Journal 2009; 9(6) 609-615.

[16] Kimura M, Bundo K, Imuro Y, Sagawa Y, Setsu K. Chronoamperometry Using Integrated Potentiostat Consisting of Poly-Si Thin-Film Transistors. IEEE Electron Device Letters 2011; 32(2) 212-214.

[17] Li L, Liu X, Qureshi WA, Mason AJ. CMOS Amperometric Instrumentation and Packaging for Biosensor Array Applications. IEEE Transactions on Biomedical Circuits and Systems 2011; 5(5) 439-448.

[18] Park S, Yoo J, Chang B, Ahn E. Novel Instrumentation in Electrochemical Impedance Spectroscopy and a Full Description of an Electrochemical System. Journal of Pure and Applied Chemistry 2006; 78(5) 1069–1080.

[19] Li G, Zhou M, He F, Lin L. A Novel Algorithm Combining Oversampling and Digital Lock-In Amplifier of High Speed and Precision. Review of Scientific Instruments 2011; 82(9) 095106. http://link.aip.org/link/doi/10.1063/1.3633943.html (accessed 12 September 2011).

[20] Rios A, Zougagh M, Avila M. Miniaturization Through Lab-On-A-Chip: Utopia or Reality for Routine Laboratories?. Analytica Chimica Acta 2012. http://www.sciencedirect.com/science/article/pii/S0003267012008926 (accessed 12 June 2012).

[21] Schumacher S, Nestler J, Otto T, Wegener M, Ehrentreich-Förster E, Michel D, Wunderlich K, Palzer S, Sohn K, Weber A, Burgard M, Grzesiak A, Teichert A, Branden-

burg A, Koger B, Albers J, Nebling E, Bier F. Highly-Integrated Lab-On-A-Chip system for Point-Of-Care Multiparameter Analysis. Lab on a Chip 2012; 12(3) 464-473.

[22] Yi-Chi W, Shin-Yu S, Lung-Min F, Che-Hsin L. Electrophoresis Separation and Electrochemical Detection on a Novel Line-Based Microfluidic Device. Proceedings of IEEE 25th International Conference on Micro Electro Mechanical Systems, MEMS2012, 29 January – 2 February 2012, Paris, France.

[23] Colomer-Farrarons J, Miribel-Català P, Rodriguez-Villareal I, Samitier J. CMOS Front-end Architecture for In-Vivo Biomedical Implantable Devices. Proceedings of 35th Annual Conference of IEEE Industrial Electronics Society, IECON2009, 3-5 November 2009, Porto, Portugal.

[24] Martin SM, Gebara FH, Strong TD, Brown RB. A Fully Differential Potentiostat. IEEE Sensors Journal 2009; 9(2) 135-142.

[25] Newman AL, Hunter KW, Stanbro WD. Proceedings of the Second International Meeting on Chemical Sensors, 7-10 July 1986, Bordeaux, France.

[26] Bataillard P, Gardies F, Jaffrezic-Renault N, Martelet C. Direct Detection of Immunospecies by Capacitance Measurements. Journal of Analytical Chemistry 1988; 60(21) 2374-2379.

[27] Gardies F, Martelet C, Colin B, Mandrand B. Feasibility of an Immunosensor Based Upon Capacitive Measurements. Journal of Sensors and Actuators 1989; 17(3-4) 461-464.

[28] Bontidean I, Ahlqvistet J, Mulchandani A, Chen W, Bae W, Mehra RK, Mortari A, Csöregi E. Novel Synthetic Phytochelin-Based Capacitive Biosensor for Heavy Metal Ion Detection. Journal of Biosensors and Bioelectronics 2003; 18(5-6) 547-553.

[29] Heer F, Franks W, Blau A, Taschini S, Ziegler C, Hierlemann A, Baltes H. CMOS Microelectrode Array for the Monitoring of Electrogenic Cells. Journal of Biosensors and Bioelecronics 2004; 20(2) 358-366.

[30] Berggren C, Johansson G. Capacitance Measurements of Antibody-Antigen Interactions in a Flow System. Journal of Analytical Chemistry 1997; 69(18) 3651-3657.

[31] Colomer-Farrarons J, Miribel-Catala P, Saiz-Vela A, Rodriguez-Villareal I, Samitier J. A Low Power CMOS Biopotentiostat in a Low-Voltage 0.13 μm Digital Technology. Proceedings of IEEE International Midwest Symposium on Circuits and Systems, MWSCAS2009, 2-5 August 2009, Cancun, Mexico.

[32] Ghafar-Zadeh E, Sawan M. Toward Fully Integrated CMOS Based Capacitive Sensor for Lab-on-Chip Applications. Proceeding IEEE International Workshop on Medical Measurements and Applications, MeMeA2008, 9-10 May 2008, Ottawa, Canada.

[33] Barretino D. Design Considerations and Recent Advances in CMOS-Based Microsystems for Point-of-Care Clinical Diagnostics. Proceedings of IEEE International Symposium on Circuits and Systems, ISCAS2006, 21-24 May 2006, Island of Kos, Greece.

[34] Nic, M, Jirat J, Kosata B. Compendium of Chemical Terminology. International Union of Pure and Applied Chemistry; 2006.

[35] Woo LY, Martin LP, Glass RS, Gorte RJ. Impedance Characterization of a Model Au/Yttria-Stabilized Zirconia/Au Electrochemical Cell in Variying Oxygen and NOx Concentrations. Journal of The Electrochemical Society 2007; 154(4) 129-135.

[36] Berggren C, Bjarnason B, Johansson G. Capacitive Biosensors. Journal of Electroanalysis 2001; 13(3) 173-180.

[37] Zhiwei Z, Junhai K, Rust MJ, Jungyoup H, Chong A. Functionalized Nano Interdigitated Electrodes Arrays on Polymer with Integrated Microfluidics for Direct Bio-Affinity Sensing Using Impedimetric Measurement. Journal of Sensors and Actuators A: Physical 2007; 136(2) 518–526.

[38] Aziz JNY, Abdelhalim K, Shulyzki R, Genov R, Bardakjian BL, Derchansky M, Serletis D, Carlen PL. 256-channel Neural Recording and Delta Compression Microsystem With 3D electrodes. IEEE Journal of Solid-State Circuits 2009; 44(3) 995-1005.

[39] Murari K, Thakor N, Stanacevic M, Cauwenberghs G. Wide-Range, Picoampere-Sensitivity Multichannel VLSI Potentiostat for Neurotransmitter Sensing. Proceedings of 26th Annual IEEE Engineering in Medicine and Biological Society, IEMBS2004, 1-5 September 2004, San Francisco, USA.

[40] Kissinger P, Heineman WR. Laboratory Techniques in Electroanalytical Chemistry, Second Edition, Revised and Expanded. CRC press; 1996.

[41] Jichun Z, Trombly N, Mason A. A Low Noise Readout Circuit for Integrated Electrochemical Biosensors Array. Proceedings of 3th IEEE Conference on Sensors, IEEE Sensors 2004, 24-27 October 2004, Vienna, Austria.

[42] Calvo B, Medrano N, Celma S. A Low-Power High-Sensitivity CMOS Voltage-to-Frequency Converter. Proceedings of 52nd IEEE International Midwest Symposium on Circuits and Systems, MWASCAS2009, 2-5 August 2009, Cancun, Mexico.

[43] Ma LY, Khan S, Nordin AN, Alam AZ, Omar J, Al-Khateeb KAS, Islam MR, Naji AW. A Low-Cost First-Order Sigma-Delta Converter Design and Analysis. Proceedings of IEEE Instrumentation and Measurement Technology Conference, I2MTC2011, 10-12 May 2011, Hangzhou, China.

[44] Chun-Yueh H. Design of a Voltammetry Potentiostat for Biochemical Sensors. Analog Integrated Circuits and Signal Processing 2011; 67(3) 375-381.

[45] Scully JR, Silverman DC, Kendig MW. Electrochemical Impedance: Analysis and Interpretation. ASTM Special Technical Publication, 1993.

[46] Barsoukov E, Macdonald JR. Impedance Spectroscopy: Theory, Experiment, and Applications. Wiley, 2005.

[47] Min M, Märtens O, Parve T. Lock-In Measurement of Bio-Impedance Variations. Journal of the International Measurement Confederation, Measurement 2000; 27(1) 21-28.

[48] Gabal M, Medrano N, Calvo B, Celma S, Martinez PA, Azcona C. A Single Supply Analog Phase Sensitive Detection Amplifier for Embedded Applications. Proceedings of 25th Conference on Design of Circuits and Integrated Systems, DCIS2010, 17-19 November 2010, Lanzarote, Spain.

[49] Ferri G, De Laurentis P, D'Amico A, De Natalem C. A Low-Voltage Integrated CMOS Analog Lock-In Amplifier Prototype for LAPS Applications. Journal of Sensors and Actuators A: Physical 2001; 92(1-3) 263-272.

[50] Azzolini C, Magnanini A, Tonelli M, Chiorboli G, Morandi C. A CMOS Vector Lock-In Amplifier for Sensor Applications. Microelectronics Journal 2010; 41(8) 449 –457.

[51] Veeravalli A, Sánchez-Sinencio E, Silva-Martínez J. Transconductance Amplifier Structures with Very Small Transconductances: A Comparative Design Approach. IEEE Journal of Solid-State Circuits 2002; 37(6) 770-775.

[52] Arnaud A, Fiorelli R, Galup-Montoro C. Nanowatt, Sub-nS OTAs, With Sub-10-mV Input Offset, Using Series-Parallel Current Mirrors. IEEE Journal of Solid-State Circuits 2006; 41(9) 2009-2018.

[53] Gaspara J, Chen SF, Gordillo A, Hepp M, Ferreyra P, Marqués C. Digital Lock-In Amplifier: Study, Design and Development with a Digital Signal Processor. Microprocessors and Microsystems: Embedded Hardware Design 2004; 28(4) 157–162.

[54] Hanson S, Seok M, Lin Y, Foo Z, Kim D, Lee Y, Liu N, Sylvester D, Blaauw D. A Low-Voltage Processor for Sensing Applications With Picowatt Standby Mode. IEEE Journal of Solid-State Circuits 2009; 44(4) 1145-1155.

[55] Bo Z, Pant S, Nazhandali L, Hanson S, Olson J, Reeves A, Minuth M, Helfand R, Austin T, Sylvester D, Blaauw D. Energy-Efficient Subthreshold Processor Design. IEEE Transactions on Very Large Scale Integration (VLSI) Systems 2009; 17(8) 1127-1137.

[56] Strong T. D., Martin, S. M., Franklin, R. F., and Brown, R. B. (2006). Integrated Electrochemical Neurosensors. Proceedings of the IEEE International Symposium on Circuits and Systems ISCAS'06, pp.4110-4113, 21-24 May 2006.

[57] Ahmadi, M. M., and Jullien, G. A. (2005). A Very Low Power CMOS Potentiostat for Bioimplantable Applications", Proceedings of the Fifth International Workshop on System-on-Chip for Real-Time Applications, pp.184-189, 20-24 July 2005.

[58] Colomer-Farrarons J, Miribel-Catala P. A CMOS Self-Powered Front-End Architecture for Subcutaneous Event-Detector Devices. Three-Electrodes Amperometric Biosensor Approach. Springer; 2011.

[59] Hogervorst R, Tero JP, Eschauzier RGH, Huijsing JH. A Compact Power-Efficient 3V CMOS Rail-to-Rail Input/Ouput Operational Amplifier for VLSI Cell Libraries. IEEE Journal of Solid-State Circuits 1994; 29(12) 1505-1513.

[60] Baker RJ, Li HW, Boyce DE. CMOS Circuit Design, Layout and Simulation. Wiley – IEEE press; 1997.

[61] Hastings A. The Art of Analog Layout, 2nd Edition. Prentice Hall; 2005.

[62] Sarpeshkar R. Universal Principles for Ultra Low Power and Energy Efficient Design. IEEE Transactions on Circuits and Systems - II: Express Briefs 2012; 59(4) 193-198.

Biosensor Signal Analysis

Nanostructured Biosensors: Influence of Adhesion Layer, Roughness and Size on the LSPR: A Parametric Study

Sameh Kessentini and Dominique Barchiesi

Additional information is available at the end of the chapter

1. Introduction

The development of nanobiosensors dedicated to early disease diagnosis has an utmost societal interest. The biosensors based on gold nanostructures are known to be efficient and tunable [30, 43]. As an illustration, the objective of the European project *Nanonatenna* (F5-2009 241818) under the "Health" research area of the seventh framework program, is the development of a high sensitive and specific nanosensor based on extraordinary optical signal enhancement dedicated to the in vitro proteins detection and disease diagnosis (cancer, cardiovascular or infectious diseases). The diagnosis process is based on the in vitro detection of the presence of small quantities of the target protein. The consortium of 12 partners works on the design of nanoantennas to reach biosensor high sensitivity. For this, gratings of gold nanoparticles are used to enhance locally the optical signal when excited by an adequate illumination.

The underlying physical phenomenon is the localized surface plasmon resonance (LSPR). The surface plasmon resonance is defined by Raether: "The electron charges on a metal boundary can perform coherent fluctuations which are called surface plasma oscillations" [39]. According to [32], "Localized surface plasmons are non-propagating excitations of the conduction electrons of metallic nanostructures coupled to the electromagnetic field... The curved surface of the particles exerts an effective restoring force of the driven electrons, so that a resonance can arise, leading to field amplification both inside and in the near-field zone outside the particle." The position of the LSPR can therefore be tuned by modifying the shape and size of the nanoparticles. It can be adapted to the specific excitation of molecules deposited on the nanoparticle surface, and the detection of a very few quantity is therefore expected.

A periodic arrangement of nanoparticles is used to increase the sensitivity of the biosensor. The prolate and oblate exhibit LSPRs which are related to their asymmetry, leading to a distribution of energy in different LSPR modes [21]. Coupling between transverse and longitudinal LSPR occurs. To prevent this effect which could decrease the efficiency of the biosensor, gratings of gold cylindric nanoparticles have been extensively studied [11, 19, 21–23, 45]. In that papers, the tuning of the LSPR has been proved by varying the diameter of nanocylinders or seldom their height.

Even if the fabrication process of nanodevices has been continuously improved and more control and precision were achieved [25, 46], the process of deposition of metal on the substrate is subject to incertitudes on the size of the nanostructures that are generally rough. Moreover, a thin intermediate layer (chromium, titanium...) is used to stick gold on substrate. This adhesion layer is usually neglected in simulations, excepted in a few studies for Surface Plasmon Resonance based biosensor [3, 6, 7], and in the case of cylinder-based transducer [42]. Nevertheless in the theoretical and numerical studies, the roughness of the gold nanostructures are neglected and the nanometric adhesion layer of gold on substrate is rarely included.

This chapter is dedicated to the parametric study of a specific nanobiosensor, made of a grating of gold nanocylinders deposited on a dielectric substrate and the goal is to investigate the influence of adhesion layer and roughness on the position of the LSPR as a function of the geometrical parameters of cylinders: their diameter D and height h. The first section is devoted to the validation of the model by comparison to previous simulations with Finite Difference Time Domain [42], and to experimental results [20, 21, 23]. The analysis of the parametric study is given in section 3 and a method to deduce heuristic laws for the LSPR is also proposed, before concluding.

2. The model

2.1. The biosensor

Figure 1 shows a schematic of the biosensor. All the parameters related to the experiments can be found in [23]. A grating of gold nanocylinders is deposited on CaF_2 substrate. The distance between cylinders P is supposed to be fixed to 200 nm. Let us note that the periodicity of the grating may influence the detected position of the LSPR [19, 28], in particular in the case of an homogeneous-evanescent switch of a diffracted order [1, 12]. Experimental studies have shown that a variation of the period $P + D = 200 \pm 50$ nm, leads to a LPSR shift lower than 20 nm [20, p. 67], with $D = 50$ nm and a spectral investigation from 575 nm to 680 nm. [19] observed a redshift of 14 nm observed when the interparticle distance P was varied from 200 to 300 nm. Nevertheless this result cannot be generalized as it may depend also on other parameters as the diameter and the height of cylinders. However the influence of the uncertainties on the height and the diameter of nanocylinders on the LSPR has still not been investigated. We focus on the systematic study of the position of LSPR as a function of the diameter D and height h of the gold nanocylinders. The influence of the thickness e of the adhesion layer and the roughness of surface are investigated.

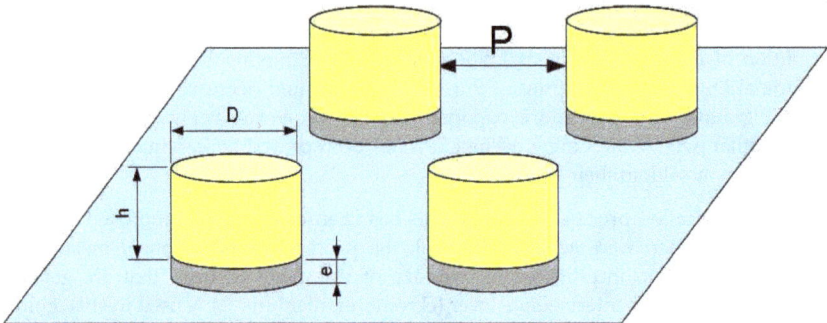

Figure 1. Biosensor: grating of gold cylinders with diameter D, height h and period $P + D$. The gold nanostructures are deposited on a CaF_2 substrate. A layer of Chromium of thickness e is used to improve adhesion of gold on CaF_2.

The nanostructures are deposited on a CaF_2 substrate through electron beam lithography (EBL) and lift-off techniques. To achieve EBL, a 30 kV Hitachi S-3500N scanning electron microscope (SEM) equipped with nanometer pattern generation system (NPGS, by J.C. Nabity) are used [8]. The SEM images [20, 21, 23, 37] reveal the experimental sources of uncertainties on the geometrical parameters in the process of fabrication:

- Height (h): the maximum uncertainty is $\delta_h = \pm 2$ nm. This value is due to both the roughness and the process of metal deposition. The SEM images and AFM (Atomic Force Microscopy) scans reveal a nanometric roughness with RMS (root of mean square) lower than one nanometer [20].

- Thickness (e): the maximum uncertainty is also $\delta_e \pm 2$ nm (same deposition technique), but may depend on the thickness of the intermediate layer.

- Diameter (D): the maximum uncertainty is $\delta_D = \pm 20$ nm. This value is relative to both the fabrication and the resolution of the SEM [41] and to a drift of diameter and shape on the whole grating. This last source of uncertainty is evaluated through statistics on the SEM images and is compatible with that found in literature [29].

- Cylinders separation (P): the maximum uncertainty is the same as that on D, $\delta_P = \pm 20$ nm, with some variations [19].

In the following, a numerical model (DDA) will be used to investigate the propagation of uncertainties on the position of the LSPR.

The illumination comes parallel to the cylinder axis and is linearly polarized. The detection of the LSPR is carried out in transmission, in the same direction (in the specular direction) and in far field. Spectroscopic studies are experimentally performed and the maximum of extinction over the incoming wavelengths λ_0 is supposed to reveal the LSPR position [5]. It is commonly admitted that the nanostructures scatter and absorb a part of the incoming light and that the detected intensity in transmission reveals the extinction of the illumination by the sample [9, 30].

2.2. The numerical model

2.2.1. The Discrete Dipole Approximation method

The measurement of the LSPR is made in far field and is modeled as the maximum of the extinction cross-section $C_{ext}(\lambda_0)$ (nm^2) with respect to the illumination wavelength λ_0 [9, 30]. The extinction cross section is the sum of the scattering and absorption cross sections. It corresponds to the rate of the total amount of incident electromagnetic energy abstracted from the incident beam due to interactions with particles.

The DDA is widely used for absorption, extinction and scattering calculations by nanoparticles [27] C_{ext} [15, 16, 37]. Its accuracy was checked by comparison to analytical solutions for spherical nanoparticles, ellipsoid [31], and infinite cylinder [16]. In what follows, a brief description of this method and of the numerical tool are given.

The method was firstly developed by [13, 14] and [38]. The main idea is to discretize the nanoparticle into a set of N elements or dipoles with polarizabilities α_j, located at r_j. Each dipole has a polarization $P_j = \alpha_j E_j$, where E_j is the electric field at r_j induced by the incident wave and the sum of the dielectric fields induced by interaction with other dipoles. Consequently, a system of complex linear equations must be solved to find polarizations P_j and evaluate the extinction cross section following [15, Eq. (8) p. 1493]:

$$C_{ext} = Q_{ext}\pi\frac{D^2}{4} = \frac{4\pi k_0}{|E_0|^2}\sum_{j=1}^{N}\left\{Im\left[P_j.(\alpha_j^{-1})^*P_j^*\right]\right\},\qquad(1)$$

with Q_{ext}, the extinction efficiency [9], $k_0 = 2\pi/\lambda_0$ the modulus of the wave vector and E_0 the amplitude, of the illumination monochromatic plane wave. Im is the imaginary part of a complex number. C_{ext} is written under the assumption of linearly polarized incident light [15]. The method was extended to periodic structures by [34] and [10, 16].

2.2.2. Numerical parameters

The Fortran code DDSCAT 7.1, developed by Draine and Flatau, is used for calculating extinction of light by irregular particles based on the DDA [17]. DDSCAT offers the possibility of editing new shapes. Therefore, we edit the cylindrical shape to include the adhesion layer and the roughness. To simulate a realistic roughness, a uniform probability law is used to remove or to add a dipole from/to the surface, or to keep it unchanged. Therefore, the root of mean square roughness is RMS=$\sqrt{\frac{(-2)^2+0+2^2}{3}} \approx 1.6$ nm, that is close to the RMS that can be observed in MEB images of samples in abundant literature and websites.

The inter-dipole distance $d = 2$ nm is smaller than 2.6 nm that ensured the validity of the calculations in [19]. The target precision for the inversion of the matrix of coupling between dipoles is 10^{-3}. The magnitude of the electric field at distance r from any dipole decreases as a polynomial function of $1/r$. [16] introduced the factor $\exp\left[-(\gamma k_0 r)^4\right]$ to vanish the coupling between remote dipoles in a periodic lattice (with $k_0 = 2\pi/\lambda_0$ the modulus of the illumination wave vector) and therefore increase the speed and accuracy of computation. The cutoff parameter $\gamma = 0.1$ (which smoothly suppresses the influence of far dipoles in periodic structures) is chosen to achieve both sufficient accuracy in the investigated size range of parameters, and a reasonable computational time over 12400 calculations. Indeed, for each structure with given h, D, e and roughness, the computation of spectrum in the range 550 nm to 850 nm of wavelength requires 31 evaluations of the model, if a precision of 10 nm is assumed sufficient for the LSPR spectral position.

The basis of the DDA being the discretization of materials, and therefore including the thick substrate in the model would be expensive. Consequently, the surrounding medium is modeled by an effective medium [24, 37] including the optical properties of the glass substrate [33] and air:

$$\epsilon_{eff} \approx (\epsilon_{air} + \epsilon_{CaF_2})/2 = 1.5267 \qquad (2)$$

The relative permittivity of CaF_2 ($1.433^2 = 2.0535$) is considered as constant on the whole investigated domain of wavelengths ($\lambda_0 \in [550, 850]$ nm). The effective medium approximation may induce a blue-shift the LSPR [19], but the comparison with experimental data is able to validate directly this approximation and the choice of the optical properties for gold and chromium.

The optical properties of gold and chromium (the adhesion layer of thickness e) are the bulk ones [26, 36]. Chromium is more absorbing than gold and therefore the extinction spectrum is broadened and attenuated for Surface Plasmon resonance [7, 18, 35, 40] and nanostructured biosensors [35, 42].

To perform a parametric study, the optimum of dicretization of the parameter space (D, h, e λ_0) must be determined, by a first evaluation of the propagation of experimental uncertainties through the numerical model. The target is the position of the LSPR.

2.3. Propagation of experimental uncertainties on the position of the LSPR

The experimental uncertainties mentioned above (subsection 2.1) can help to define the step size of D, h, e and λ_0 in order to maintain the computational time of the parametric study in a reasonable range and to obtain significant results. For this, we compute the propagation of experimental uncertainties through the model to check their influence on the shift of the LSPR. Two diameters are considered for this evaluation with the above described model: $D = 100$ nm and $D = 200$ nm, near the boundaries of the investigated domain of the parametric study. The height of cylinders is the reference in experiments $h = 50$ nm [23]. A step of 10 nm is used for the computations of the spectrum $C_{ext}(\lambda_0)$. The cylinders are supposed to have smooth surfaces. The corresponding uncertainty on the position of LSPR ($\lambda_0(LSPR)$) is therefore ± 5 nm in the numerical calculations.

The combined uncertainty $u_B(\lambda_0(LSPR))$ on the LSPR shift can therefore be evaluated from the above uncertainties of type B [44], by considering *a priori* uniform law of probability within the above intervals and no correlation between these parameters. The following results are obtained:

- $D = 100$ nm, $h = 50$ nm, $P = 200$ *nm* [23].
  - For $P \in [180; 220]$ nm, the shift of the LSPR position is ± 5 nm.
  - For $D \in [80; 120]$ nm, the shift of the LSPR position is ± 20 nm.

- For $h \in [48; 52]$ nm, the shift of the LSPR position is ± 30 nm.
- For $e \in [0; 4]$ nm, the shift of the LSPR position is ± 10 nm.

$$u_B(\lambda_0(LSPR), D = 100nm) = \frac{1}{\sqrt{3}}\sqrt{5^2 + 20^2 + 30^2 + 10^2} = 21.8nm \qquad (3)$$

- $D = 200$ nm, $h = 50$ nm, $P = 200$ nm [23].
  - For $P \in [180; 220]$ nm, the shift of the LSPR position is ± 20 nm.
  - For $D \in [180; 220]$ nm, the shift of the LSPR position is ± 30 nm.
  - For $h \in [48; 52]$ nm, the shift of the LSPR position is ± 10 nm.
  - For $e \in [0; 4]$ nm, the shift of the LSPR position is ± 10 nm.

$$u_B(\lambda_0(LSPR), D = 200nm) = \frac{1}{\sqrt{3}}\sqrt{20^2 + 30^2 + 10^2 + 10^2} = 22.4nm \qquad (4)$$

In both cases, uncertainties may produce a shift greater than ± 20 nm within the investigated range of diameters. The results show that h and D are critical parameters for small diameter $D = 100$ nm. The influence of the uncertainty on the period seems to increase for larger diameters. The combined uncertainty is stable, even if the sensitivity to each source of uncertainty is not the same.

Consequently, the influence of the experimental uncertainties on the position of the LSPR can be observed considering a sampling step of 10 nm for λ_0. The corresponding uncertainty on the position of LSPR ($\lambda_0(LSPR)$) is therefore ± 5 nm in the numerical calculations. The evaluation of the experimental uncertainty (Eqs. 3 and 4) is used in the following.

The comparison of the model of smooth cylinders without adhesion layer and that with roughness and $e = 2$ nm, for some diameters reveals that the chromium layer and a RMS equal to 1.6 nm have also significant influence on the shift of the LSPR. For $D = 100$ nm, the shift is -40 nm, for $D = 130, 150$ nm, the shift is $+10$ nm, and for $D = 180, 200$ nm, the shift is $+20$ nm. Therefore, the effect of these parameters on LSPR can be described with the above mentioned sampling for λ_0. Moreover, first trends emerge: the small cylinders exhibits a different behavior of the large ones.

The following subsections are devoted to the validation of the model by comparison with theoretical and experimental data.

2.4. Influence of the effective medium approximation

The comparison with other theoretical results may help to validate the proposed model of effective medium. In this first comparison, the reference is a theoretical study of the influence of a chromium adhesion layer with FDTD (Finite Difference Time Domain) method [42]. In that reference, no roughness was introduced, the surface of cylinders was smooth, $D = 100$ nm and $h = 50$ nm. The period of the grating was $D + P = 300$ nm with $P = 200$ nm (Figure 1). The relative permittivities were found in the same references [26, 36] and the nanostructures were deposited on a glass substrate. The intermediate layer had thickness 0, 1, 5, 30 nm. The location of the LSPR in the absence of Cr occurs at $\lambda_0 = 586$ nm with

FDTD and $\lambda_0 = 590$ nm with DDA. Therefore, we may conclude that the model of effective index is relevant, for the considered sample. [42] showed that if the Chromium adhesion layer is introduced, the spectrum curve is broadened and the maximum is blue-shifted, for $D = 100$ nm. For example, the LSPR was found at 541 nm for $e = 5$ nm and between 570 nm and 600 nm with the DDA (this interval is determined from eight DDA results with combinations of $D = 90$, $D = 110$ and $e = 4$, $e = 6$ nm). The blueshift of LSPR is less with DDA and effective index than with FDTD. Therefore, the comparison with experiments can help to decide on the ability of DDA to describe biosensors.

2.5. Validation of the DDA model with experimental data

The reference experimental data were published in [23, Fig. 2]. In that paper, the thickness of chromium $e = 3$ nm and the height of cylinders is 50 nm, with above mentioned experimental uncertainties. The computation of the position of the LSPR requires 31 computations of the extinction cross-section C_{ext} for $\lambda_0 \in [550; 850]$ nm. The smooth and rough samples are considered for the model. The relevance of the numerical results relies on using the experimental uncertainties to generate the corresponding intervals, for the position of the LSPR.

It is well known that increasing the diameter D or decreasing the height h of cylinders redshift the LSPR [37]. The above mentioned theoretical study [42] and our results show that including the adhesion layer blueshifts the LSPR for $D = 100$ nm. Therefore, the boundaries of the numerical results are chosen to find the minimum and the maximum of the LSPR position, according to the above mentioned sampling steps. For example, for experimental data with $e = 3$ nm, $D = 100$ nm, $h = 50$ nm, the interval of possible numerical position of LSPR is $[\lambda_0(LSPR, e = 4, D = 90, h = 52nm); \lambda_0(LSPR, e = 2, D = 110, h = 48nm)]$. In Figs. 2 and 3, the computed LSPR is between the dashed and the solid lines for smooth (blue) and rough (red) structures.

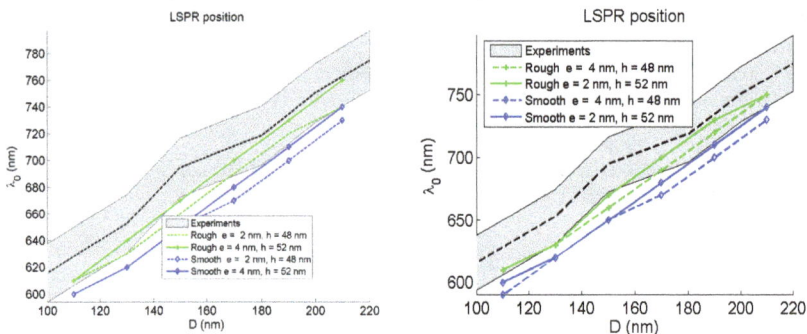

Figure 2. Experimental values (dashed black line) ($h = 50$, $e = 3$ nm) and uncertainties (gray, Eqs. 3 and 4), by courtesy of Nicolas Guillot and Marc Lamy de la Chapelle [23]. Extremal DDA results for smooth and rough structures, obtained by using experimental uncertainties on e and h.

Figure 2 shows a better agreement of simulation with rough surface than with smooth surface over the interval of diameters. The LSPR position for smooth cylinders is clearly blueshifted and does not fall within the interval of experimental uncertainties. For $D = 100$ nm, $h = 50$ nm, $P = 200$ nm (Fig. 1), DDA gives $\lambda_0(LSPR) \in [603; 628]$ which is closer to experiments [23] ($\lambda_0(LSPR) = 615$ nm) than FDTD [42], where the LSPR was found between 541 nm ($e = 5$ nm) and 583 nm ($e = 1$ nm).

To evaluate the sensitivity of the results obtained by DDA with h, reference experimental data are extracted from [20] with $D = 100$ nm. Again, the agreement of experimental data with the numerical results for rough structures is better than for smooth ones. Both studies of LSPR position as a function of D and h show that the roughness produces a redshift, which is necessary to reproduce the experimental data, especially for the largest heights. This effect could also explain the blueshift obtained by [42] with smooth structures.

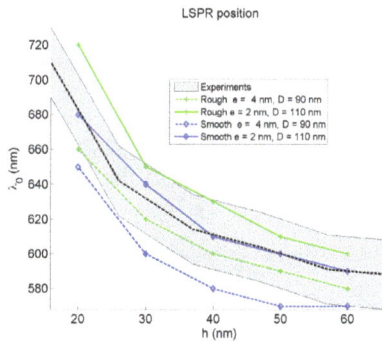

Figure 3. Experimental data from [20] (dashed line) ($D = 100$ nm, $e = 3$ nm) and uncertainty $u_B = \pm 20$ nm. Extremal DDA results for smooth and rough structures, obtained by using experimental uncertainties on e and D.

For each diameter D (Fig. 2) or height h (Fig. 3), the difference of LSPR between the dashed and solid lines can be considered as an evaluation of the sensitivity of the LSPR to the uncertainty on the thickness of chromium and on the size of structures. For all diameters the sensitivity is lower than 20 nm.

The comparisons of numerical results to experimental data show that the rough model with adhesion layer is more efficient to describe the experiments on a wide range of parameters. Moreover, the model of effective index for the external medium and the choice of the relative permittivities of gold and chromium seem to be adequate, as well as the exogenous parameters of the DDA (d, γ). Therefore, the proposed model can be used to provide full parametric study of the biosensor.

3. Parametric study

According to the experimental uncertainties, we have chosen an adapted sampling of each parameter of the model in subsection 2.3. The good agreement with the experimental data shown in the previous section confirms also the adequate choice of the exogenous parameters: the inter-dipole distance d, the cutoff parameter γ, the target precision of matrix inversion 10^{-3}, and the relative permittivities [26, 36]. The effective relative permittivity of the surrounding medium is given in Eq. 2.

Consequently, a systematic study of the LSPR position and its quality is produced for a class of nanostructured biosensors with cylindrical shapes. The quality of LSPR is given by the maximum over the spectral range of the extinction cross-section (C_{ext}): $\max_{\lambda_0} C_{ext}(D, h, e, RMS)$ or its relative dimming. The relative variation $\Delta_r C_{ext}$ of the quality of the plasmon corresponds to the attenuation of the LSPR:

$$\Delta_r C_{ext} = 100 \frac{\max_{\lambda_0}(C_{ext}(e)) - \max_{\lambda_0}(C_{ext}(0))}{\max_{\lambda_0}(C_{ext}(0))} \tag{5}$$

The influence of size parameters (D, h) for various adhesion layer (e) and roughnesses (RMS), on the tuning of the plasmon resonance is proposed.

The LSPR position and its quality are plotted in color levels, as functions of the height h and the diameter D. For the plots of the LSPR position ($\lambda_0(LSPR)$), the colors of rectangles are chosen to be close to the real colors [2] in the visible domain and vary from black to light pink for the near infrared. When the position of the LSPR is out off the considered range of wavelengths [550; 850] nm, a white rectangle is plotted.

The influence of the adhesion layer and of the roughness are successively investigated.

3.1. The adhesion layer

Figures 4 show the position of the LSPR and the maximum value of C_{ext} for an adhesion layer of thickness $e = 2, 4$ and 6 nm. The maximum of C_{ext} is linked to the quality of the LSPR: the sensitivity of the biosensor is expected to be improved with the increase of the extinction cross-section. Increasing the thickness e produces a broadening of the spectrum which results in a decrease of $\max(C_{ext})$ and therefore of the quality of the LSPR as defined above.

For the three thicknesses of chromium the position of the LSPR is about the same but the quality of LSPR is deteriorated if e growth. The position of the LSPR is redshifted if the diameter D of cylinders is increased. This conclusion was underlined in [23]: "Indeed, as it is well known, the position of the LSPR is redshifted for higher diameter". The height h influences also the position of the LSPR: for a given diameter, if h increases, the LSPR is blueshifted. These conclusions are also consistent with previous studies of cylindrical nanorods [37].

The LSPR shift (nm) and $\Delta_r C_{ext}$ are displayed in Fig. 5. Even if the LSPR shift may be considered as negligible with regards to the influence of the uncertainties of fabrication, the relative attenuation of the LSPR can reach more than 45% for the smallest D and h. Therefore, the adhesion layer cannot be neglected if the efficiency of the biosensor is targeted. The chromium (and all adhesion layers) being more absorbing, it deteriorates the efficiency of the biosensor and therefore, its thickness should be reduced as much as possible.

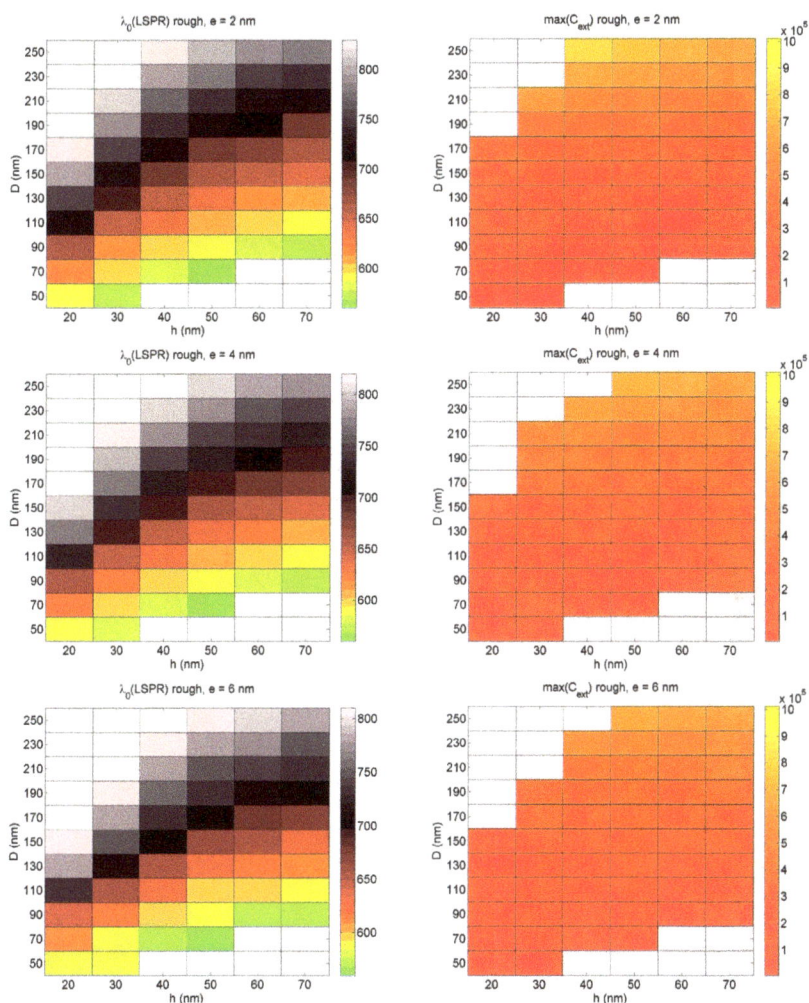

Figure 4. Numerical parametric study of the cylinder based biosensor: position of the LSPR ($\lambda_0(LSPR)$) and its quality illustrated by the maximum of C_{ext} (nm^2) as functions of the diameters D and heights h of the nanocylinders, for different thicknesses of the adhesion layer $e = 2$, $e = 4$ and $e = 6$ nm.

3.2. The roughness

A similar set of simulations could reveal the influence of the roughness (Fig. 6).

The roughness has less influence on the quality of plasmon than e. This influence decreases when D and h increases. The relative influence of the roughness decreases if the thickness e of the adhesion layer increases and becomes lower than 20% if $e > 4$ nm. In the case of pure gold ($e = 0$ nm), the roughness induces a decrease (less than 20%) of max(C_{ext}) for small

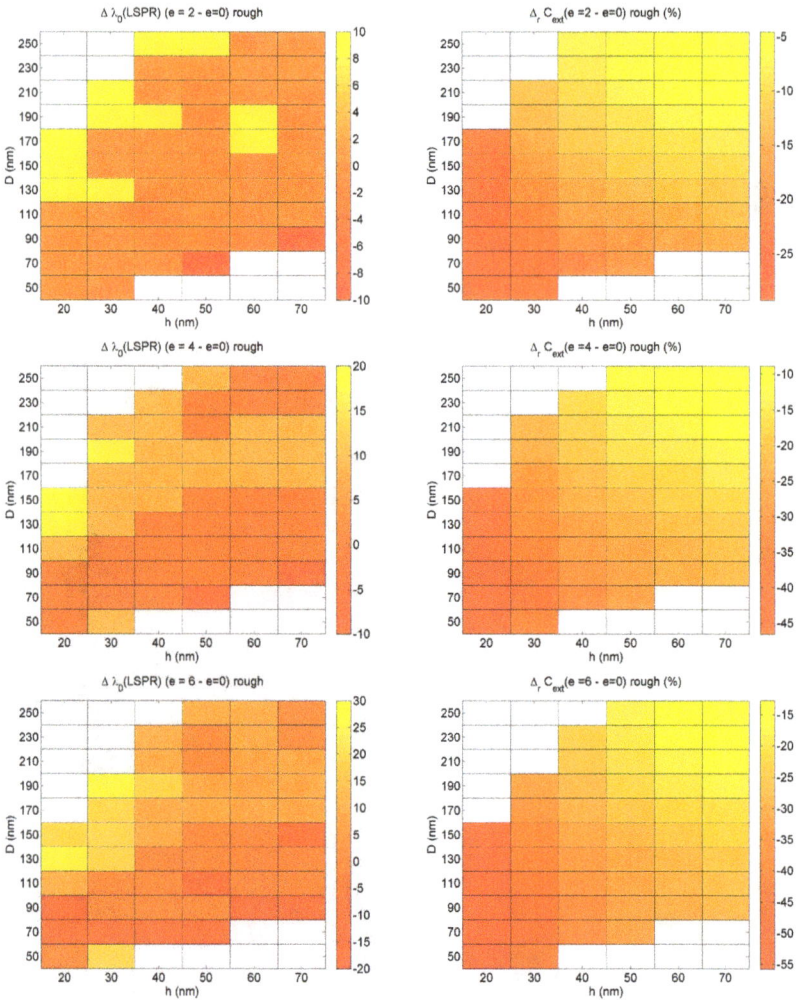

Figure 5. Numerical parametric study of the cylinder based biosensor: shift of the position of the LSPR $\Delta\lambda_0(LSPR)$ (nm) and its quality $\Delta_r C_{ext}$ (Eq. 5) as functions of the diameters D and heights h of the nanocylinders, for thicknesses of the adhesion layer $e = 2$, $e = 4$ and $e = 6$ nm, with reference to $e = 0$.

diameters and small height, and an increase (also less than 20%) for larger values of D and h. The redshift of the LSPR can reach 40 nm for small heights and average diameters. This effect can be explained by the increase of the ratio of the area of rough surface of the cylinder to its height. The effect of the roughness is reduced when the volume of cylinder increases. The shift of the LSPR becomes negligible for the highest and largest cylinders. The LSPR is redshifted in almost all the cases, when roughness is introduced. The redshift is smaller than 40 nm for all the investigated adhesion layers.

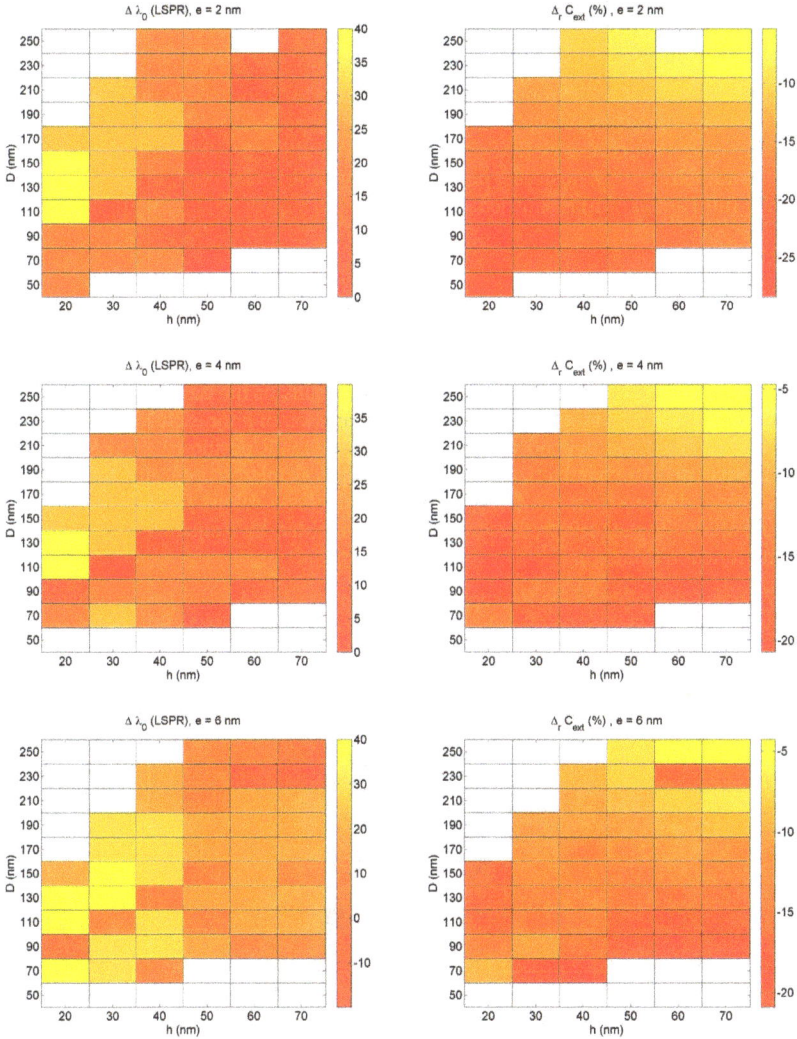

Figure 6. Numerical parametric study of the cylinder based biosensor: shift of the position of the LSPR $\Delta\lambda_0(LSPR)$ (nm) and its relative quality $\Delta_r C_{ext}$ (Eq. 5) as functions of the diameters D and heights h of the nanocylinders, for thicknesses of the adhesion layer $e = 2$, $e = 4$ and $e = 6$ nm, with reference to the smooth structure.

The diameter D and the height h seem to be linearly connected for a given LSPR position as it can be observed in Fig. 4. This linear behavior can lead to heuristic law which can be used to study the propagation of uncertainties.

4. Heuristic law for LSPR

4.1. The LSPR $\lambda_0(LSPR)(D,h)$

For both thicknesses of the chromium adhesion layer $e = 2$ and $e = 6$ nm, the position of LSPR is about the same when both D and h are varied respectively by 20 nm and 10 nm, with $P = 200$ nm, and a roughness of RMS= 1.6 nm. The resulting behavior law can be deduced from the parametric study: a steady position of the LSPR is observed, if the following law between D (nm) and h (nm) is satisfied:

$$D(P = 200, e = 2 - 6) = ah + b \approx (0.0084\lambda_0(LSPR) - 3.86)h + (0.319\lambda_0(LSPR) - 160) \quad (6)$$

This law is deduced from the least square fit of the results of the parametric study, simply by computing the slope and the intercept of D as a function of h for each LSPR position, and then by finding the linear dependence of a and b on the LSPR position $\lambda_0(LSPR)$. A similar approach was used to characterize the near-field optical microscopes and the evanescent near-field around nanostructures [1, 4]. The method uses a robust fit of the results of the parametric study. The algorithm uses iteratively reweighted least squares with the bisquare weighting function (Matlab).

Figure 7 shows the plots of Eq. 6, superimposed on figures 4. The linear behavior of D as a function of h is dependent on the LSPR position. Equation 6 helps to determine the geometrical parameters (D and h) of the biosensor to adjust the LSPR to a given wavelength.

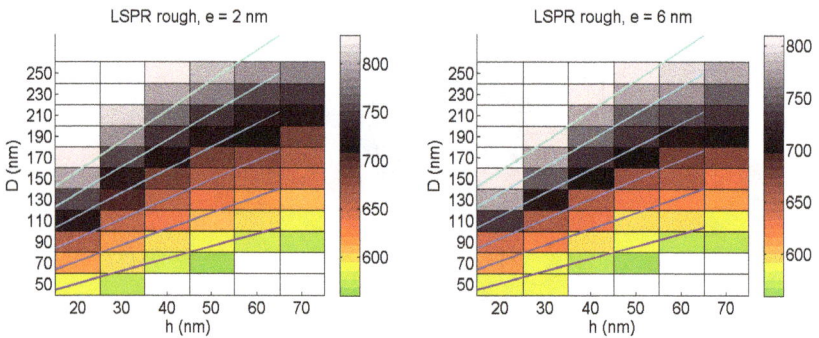

Figure 7. Plots of D as a function of h for $\lambda_0(LSPR) \in \{600; 640; 680; 720; 760; 800\}$ from Eq. 6.

Similarly, the position of the LSPR (nm) can be deduced from Eq. 6 as a function of D (nm) and h (nm):

$$\lambda_0(LSPR) \approx \frac{160 + D + 3.86h}{0.319 + 0.0084h} \quad (7)$$

A comparison of the simple model (Eq. 7) and experimental data in [21] shows a good agreement. Equation 7 seems to be a good approximation of the variation of the LSPR as a function of the geometrical parameters D and h, for a grating of rough gold cylinders with adhesion layer (e = 2 nm) and P = 200 nm (Fig. 1). The specific case of homothetic cylinders with $h = D$, removes a degree of freedom and leads to a simpler formula:

$$\lambda_0(LSPR, h = D) = \frac{160 + 4.86D}{0.319 + 0.0084D} \tag{8}$$

This formula is strictly valid in the interval $h = D$ ranging from 50 to 70 nm, by cons it should still be checked for greater heights.

Asymptotic form of $\lambda_0(LSPR)$ could be of interest especially for specific aspect ratio of the cylinder. For example, for nanodiscs ($h \ll D$), the position of the LSPR can be deduced from the series of Eq. 7.

$$\lambda_0(LSPR, h \ll D) = (501 + 3.1D) - (0.0825D + 1.1)h + o[(h/D)^2], \tag{9}$$

where $o[(h/D)^2]$ represents omitted terms of order higher than 2 in the series.

This equation confirms that the correction induced by the height assuming a small aspect ratio h/D is a blueshift of the LSPR, whatever are the diameters D. These formula can be used to select the best parameters for a given LSPR position. Another interest of these simple laws is the fact that propagation of uncertainties of the fabrication process can be evaluated directly, and therefore, the sensitivity analysis of LSPR position on the geometrical parameters h and D can be deduced, at least within the investigated domain of geometrical parameters and wavelengths. This sensitivity analysis is conducted in what follows.

4.2. Sensitivity of LSPR to uncertainties on size parameters

The propagation of uncertainties and the sensitivity analysis of a process or a physical phenomenon helps to improve the fabrication process of any device, by exhibiting the parameters that should be controlled first. Indeed the identification of critical parameters which uncertainty should be reduced is relevant, for a challenging improvement of technology.

First, the propagation of experimental uncertainties of fabrication can be deduced from the derivative of Eq. 7. The uncertainty on $\lambda_0(LSPR)$ is deduced form the uncertainties on D ($u(D)$) and h ($u(h)$) [44, 5.1.2]:

$$u(\lambda_0(LSPR)) = \sqrt{\left(\frac{\partial \lambda_0(LSPR)}{\partial D}\right)^2 u^2(D) + \left(\frac{\partial \lambda_0(LSPR)}{\partial h}\right)^2 u^2(h)}, \tag{10}$$

where the partial derivatives are called sensitivity coefficients. Using Eq. 7 gives:

$$u(\lambda_0(LSPR)) = \sqrt{\left(\underbrace{\frac{1}{0.319 + 0.0084h}}_{S_D}\right)^2 u^2(D) + \left(\underbrace{\frac{0.11266 + 0.0084D}{(0.319 + 0.0084h)^2}}_{S_h}\right)^2 u^2(h)} \qquad (11)$$

Knowing the experimental uncertainties on h and D, the uncertainty on the LSPR can be deduced. The sensitivity coefficients S_h and S_D can be used to evaluate the effect of the experimental dispersion of values of D and h around a mean value. S_D is almost independent of h and equal to 3. Regarding S_h and within the same approximation $S_h \approx 1.107 + 0.082D$. Therefore, in the investigated domain of D, $S_h \in [5.2; 21.6]$. The uncertainty on h must be 7 times smaller than that on D to balance the contributions of each uncertainty to the overall uncertainty. This condition is becoming increasingly critical as the diameter D of cylinders is increased. Fortunately, the control on the thickness of gold deposition (height of cylinders) is about ± 2 nm and that of diameters is about ± 20 nm. In this case, the uncertainty on the LSPR position remains lower than 30 nm for all diameters if $h > 40$ nm (Fig.8).

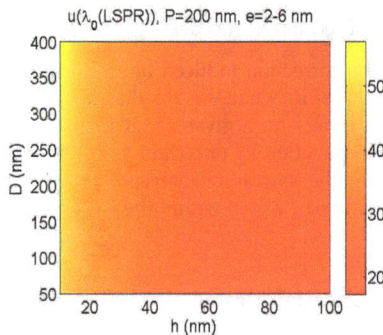

Figure 8. Uncertainty on the position of the LSPR computed from the heuristic law (Eq. 11) deduced from the least square fit of the parametric study results. The parameters of the model are $P = 200$ nm with roughness ($rms = 1.6$ nm) and chromium adhesion layer ($e = 2 - 6$ nm).

Figure 9 gathers the experimental results retrieved from [23] with experimental uncertainties (gray domain and) the heuristic law deduced from Eq. 7 with uncertainties (Eq. 10). The good agreement between experimental results and the heuristic model with associated uncertainties can be observed. The heuristic model fits well the experimental data and the uncertainties deduced from Eq. 11 are coherent with the experimental ones.

In this section, we have obtained an heuristic law to describe the link between the position of the LSPR and the size parameters D and h of the cylinders. This law seems to be valid for an adhesion layer of chromium of thickness $e = 2 - 6$ nm. Moreover, the simplicity of this law helps to determine a first approximation of the size of the cylinder, as well as the effect of the propagation of experimental uncertainties on the position of the LSPR.

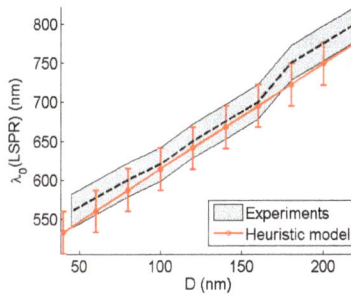

Figure 9. Experimental data retrieved from [23, Fig. 2] ($h = 50$ nm, $e = 3$ nm, $P = 200$ nm) (black) with uncertainties (gray). Heuristic model deduced from the numerical parametric study of gold nanocylinder grating (red curve) with $P = 200$ nm, $h = 50$ nm, RMS 1.6 nm and chromium adhesion layer ($e = 4$ nm).

5. Conclusion

A model based on the Discrete Dipole Approximation (DDA) is used for a parametric study of the biosensors made of a grating of nanocylinders. The Localized Surface Plasmon Resonance (LSPR) is modeled by the position of the maximum over the spectrum of the extinction cross-section. The investigated parameters are the height h, the diameter D, the thickness e of the chromium adhesion layer, and the roughness of the nanocylinders. Thin adhesion layers slightly modify the position of the LSPR but degrade significantly its quality: the sensitivity of the biosensor can be highly decreased. Roughness induces a redshift of the LSPR but less alters the quality of the resonance than the adhesion layer. The relative influence of h and D is more complex. Actually, if D increases, a redshift is always observed. The same effect occurs if h decreases. Consequently the fine tuning of the LSPR can be achieved by varying D or h, assuming a fixed distance between cylinders (P). This property helps to deduce heuristic laws for the LSPR position and the propagation of uncertainties. The basic law, deduced from a least square fit, gives simply an approximation of the position of the LSPR as a function of D and h. The agreement with experimental results [21, 23] is satisfactory, falling within the experimental uncertainties of fabrication. Therefore the heuristic law of behavior of the LSPR position can be used with confidence (for $P = 200$ nm). Varying the period $P + D$ of the grating of nanocylinders could be of interest. The characterization of the LSPR being clarified, the influence of the functionalization layer will be the object of future studies.

Acknowledgment

This work was supported by the "Conseil Régional de Champagne Ardennes", the "Conseil Général de l'Aube" and the *Nanoantenna* European Project (FP7 Health-F5-2009-241818).

Author details

Sameh Kessentini and Dominique Barchiesi

Automatic Mesh Generation and Advanced Methods (GAMMA3 - UTT/INRIA) - Charles Delaunay Institute - University of Technology of Troyes - Troyes, France

References

[1] Barchiesi, D. [1998]. Pseudo modulation transfer function in reflection scanning near-field optical microscopy, *Optics Communications* 154: 167–172.

[2] Barchiesi, D. [2008]. Simulations d'expériences faisant intervenir la couleur : dispersion par un prisme et réflectance, *Bulletin de l'Union des Professeurs de Physique Chimie* 909 (102)(1): 1369–1382.

[3] Barchiesi, D. [2011]. *New Perspectives in biosensors technology and applications*, INTECH Open Access, Rijeka, Croatia, chapter 5, pp. 105–126.
URL: *http://cdn.intechweb.org/pdfs/14870.pdf*

[4] Barchiesi, D., Bergossi, O., Spajer, M. & Pieralli, C. [1997]. Image resolution in reflection scanning near-field optical microscopy (R-SNOM) using shear-force (ShF) feedback: Characterization using spline and Fourier spectrum, *Appl. Opt.* 36(10): 2171–2177.

[5] Barchiesi, D., Kremer, E., Mai, V. & Grosges, T. [2008]. A Poincaré's approach for plasmonics: The plasmon localization, *J. Microscopy* 229: 525–532.

[6] Barchiesi, D., Lidgi-Guigui, N. & Lamy de la Chapelle, M. [2012]. Functionalization layer influence on the sensitivity of surface plasmon resonance (SPR) biosensor, *Optics Communications* 285(6): 1619–1623.

[7] Barchiesi, D., Macías, D., Belmar-Letellier, L., Van Labeke, D., Lamy de la Chapelle, M., Toury, T., Kremer, E., Moreau, L. & Grosges, T. [2008]. Plasmonics: Influence of the intermediate (or stick) layer on the efficiency of sensors, *Appl. Phys. B* 93: 177–181.

[8] Billot, L., M. Lamy de la Chapelle, Grimault, A. S., Vial, A., Barchiesi, D., Bijeon, J.-L., , Adam, P.-M. & Royer, P. [2006]. Surface enhanced Raman scattering on gold nanowire arrays: Evidence of strong multipolar surface plasmon resonance enhancement, *Chem. Phys. Lett.* 422(4-6): 303–307.

[9] Bohren, C. F. & Huffman, D. R. [1998]. *Absorption and Scattering of Light by Small Particles*, John Wiley & Sons, Inc., New York.

[10] Chaumet, P. C., Rahmani, A. & Bryant, G. W. [2003]. Generalization of the coupled dipole method to periodic structures, *Phys. Rev. B* 67(16): 165404(1–5).

[11] Dasgupta, A. & Kumar, G. V. P. [2012]. Palladium bridged gold nanocylinder dimer: plasmonic properties and hydrogen sensitivity, *Appl. Opt.* 51(11): 1688–1693.

[12] Davy, S., Barchiesi, D., Spajer, M. & Courjon, D. [1999]. Spectroscopic study of resonant dielectric structures in near–field, *European Physical Journal (Applied Physics)* 5: 277–281.

[13] Devoe, H. [1964]. Optical properties of molecular aggregates. I. classical model of electronic absorption and refraction, *J. Chem. Phys.* 41: 393–400.

[14] Devoe, H. [1965]. Optical properties of molecular aggregates. II. classical theory of the refraction, absorption, and optical activity of solutions and crystals, *J. Chem. Phys.* 43: 3199–3208.

[15] Draine, B. T. & Flatau, P. J. [1994]. Discrete-dipole approximation for scattering calculations, *J. Opt. Soc. Am. A* 11: 1491–1499.

[16] Draine, B. T. & Flatau, P. J. [2008]. Discrete-dipole approximation for periodic targets: Theory and tests, *J. Opt. Soc. Am. A* 25: 2693–2703.

[17] Draine, B. T. & Flatau, P. J. [2010]. User guide to the discrete dipole approximation code DDSCAT 7.1, http://arXiv.org/abs/1002.1505v1.

[18] Ekgasit, S., Thammacharoen, C., Yu, F. & Knoll, W. [2005]. Influence of the metal film thickness on the sensitivity of surface plasmon resonance biosensors, *Appl. Spectroscopy* 59: 661–667.

[19] Félidj, N., Aubard, J., Lévi, G., Krenn, J. R., Salerno, M., Schider, G., Lamprecht, B., Leitner, A. & Aussenegg, F. R. [2002]. Controlling the optical response of regular arrays of gold particles for surface-enhanced Raman scattering, *Phys. Rev. B* 65: 075419–075427.

[20] Grand, J. [2004]. *Plasmons de surface de nanoparticules : spectroscopie d'extinction en champs proche et lointain, diffusion Raman exaltée*, PhD thesis, Université de technologie de Troyes.

[21] Grand, J., Lamy de la Chapelle, M., Bijeon, J.-L., Adam, P.-M., Vial, A. & Royer, P. [2005]. Role of localized surface plasmons in surface-enhanced raman scattering of shape-controlled metallic particles in regular arrays, *Phys. Rev. B* 72(3): 033407.

[22] Grimault, A.-S., Vial, A. & Lamy de la Chapelle, M. [2006]. Modeling of regular gold nanostructures arrays for SERS applications using a 3D FDTD method, *Appl. Phys. B* 84(1-2): 111–115.

[23] Guillot, N., , Shen, H., Frémaux, B., Péron, O., Rinnert, E., Toury, T. & Lamy de la Chapelle, M. [2010.]. Surface enhanced Raman scattering optimization of gold nanocylinder arrays: Influence of the localized surface plasmon resonance and excitation wavelength, *Appl. Phys. Lett.* 97(2): 023113–023116,.

[24] Haija, A. J., Freeman, W. L. & Roarty, T. [2006]. Effective characteristic matrix of ultrathin multilayer structures, *Optica Applicata* 36: 39–50.

[25] Huang, X., Neretina, S. & El-Sayed, M. A. [2009]. Gold nanorods: From synthesis and properties to biological and biomedical applications, *Advanced Materials* 21: 4880–4910.

[26] Johnson, P. & Christy, R. W. [1972]. Optical constants of the noble metals, *Phys. Rev. B* 6: 4370.

[27] Kessentini, S. & Barchiesi, D. [2012]. Quantitative comparison of optimized nanorods, nanoshells and hollow nanospheres for photothermal therapy, *Biomed. Opt. Express* 3(3): 590–604.

[28] Lamprecht, B., Schider, G., Lechner, R. T., Ditlbacher, H., Krenn, J. R., Leitner, A. & Aussenegg, F. R. [2000]. Metal nanoparticles gratings: influence of dipolar interaction on the plasmon resonance, *Phys. Rev. Lett.* 84: 4721–4723.

[29] Laurent, G., Félidj, N., Aubard, J., Lévi, G., Krenn, J. R., Hohenau, A., Schider, G., Leitner, A. & Aussenegg, F. R. [2005]. Evidence of multipolar excitations in surface enhanced Raman scattering, *Phys. Rev. B* 65: 045430-1–045430-5.

[30] Le Ru, E. C. & Etchegoin, P. G. [2009]. *Principles of Surface-Enhanced Raman Spectroscopy and related plasmonic effects*, 1 edn, Elsevier, Amsterdam.

[31] Lee, K. S. & El-Sayed, M. A. [2005]. Dependence of the enhanced optical scattering efficiency relative to that of absorption of gold metal nanorods on aspect ratio, size, end-cap shape, and medium refractive, *J. Phys. Chem. B* 109: 20331–20338.

[32] Maier, S. A. [2007]. *Plasmonics. Fundamentals and Applications*, Springer, New York, USA.

[33] Malitson, I. H. [1963]. A redetermination of some optical properties of calcium fluoride, *Applied Optics* 2: 1103–1107.

[34] Markel, V. A. [1992]. Scattering of light from two interacting spherical particles, *J. Mod. Opt.* 39(4): 853–861.

[35] Neff, H., Zong, W., Lima, A., Borre, M. & Holzhüter, G. [2006]. Optical properties and instrumental performance of thin gold films near the surface plasmon resonance, *Thin Solid Films* 496: 688–697.

[36] Palik, E. D. [1985]. *Handbook of Optical Constants*, Academic Press Inc., San Diego USA.

[37] Pelton, M., Aizpurua, J. & Bryant, G. W. [2008]. Metal-nanoparticles plasmonics, *Laser & Photon. Rev.* 2(3): 136–159.

[38] Purcell, E. & Pennypacker, C. R. [1973]. Scattering and absorption of light by nonspherical dielectric grains, *Astrophysical Journal* 186: 705–714.

[39] Raether, H. [1988]. *Surface Plasmons on Smooth and Rough Surfaces and on Gratings*, Springer-Verlag, Berlin.

[40] Sexton, B. A., Feltis, B. N. & Davis, T. J. [2008]. Effect of surface roughness on the extinction-based localized surface plasmon resonance biosensor, *Sensors and Actuators A Physical* 141: 471Û475.

[41] Sharma, D., Sharma, R., Dua, S. & Ojha, V. N. [2012]. Pitch measurements of 1D/2D gratings using optical profiler and comparison with SEM / AFM, *AdMet 2012*, Metrology Society of India, ARAI, Pune, India, pp. NM 003, 1–4.

[42] Vial, A. & Laroche, T. [2007]. Description of dispersion properties of metals by means of the critical points model and application to the study of resonant structures using the FDTD method, *J. Phys. D: Appl. Phys.* 40: 7152–7158.

[43] Vidotti, M., Carvalhal, R. F., Mendes, R. K., Ferreira, D. C. M. & Kubota, L. T. [2011]. Biosensors based on gold nanostructures, *J. Braz. Chem. Soc.* 22(1): 3–20.

[44] Working Group 1 [2008]. *Evaluation of measurement data - Guide to the expression of uncertainty in measurement*, 1 edn, Joint Committee for Guides in Metrology, Paris. Corrected version 2010.

[45] Yan, H.-H., Xiao, Y.-Y., Xie, S.-X. & Li, H.-J. [2012]. Tunable plasmon resonance of a touching gold cylinder arrays, *J. At. Mol. Sci.* 3(3): 252–261.

[46] Yanik, A. A., Huang, M., Artar, A., Chang, T.-Y. & Altug, H. [2010]. On-chip nanoplasmonic biosensors with actively controlled nanofluidic surface delivery, *in* M. I. Stockman (ed.), *Plasmonics: Metallic Nanostructures and Their Optical Properties VIII*, Vol. 7757, SPIE, San Diego, California, USA, pp. 775735–1–6.

Chapter 12

Love Wave Biosensors: A Review

María Isabel Rocha Gaso, Yolanda Jiménez,
Laurent A. Francis and Antonio Arnau

Additional information is available at the end of the chapter

1. Introduction

In the fields of analytical and physical chemistry, medical diagnostics and biotechnology there is an increasing demand of highly selective and sensitive analytical techniques which, optimally, allow an in real-time label-free monitoring with easy to use, reliable, miniaturized and low cost devices. Biosensors meet many of the above features which have led them to gain a place in the analytical bench top as alternative or complementary methods for routine classical analysis. Different sensing technologies are being used for biosensors. Categorized by the transducer mechanism, optical and acoustic wave sensing technologies have emerged as very promising biosensors technologies. Optical sensing represents the most often technology currently used in biosensors applications. Among others, Surface Plasmon Resonance (SPR) is probably one of the better known label-free optical techniques, being the main shortcoming of this method its high cost. Acoustic wave devices represent a cost-effective alternative to these advanced optical approaches [1], since they combine their direct detection, simplicity in handling, real-time monitoring, good sensitivity and selectivity capabilities with a more reduced cost. The main challenges of the acoustic techniques remain on the improvement of the sensitivity with the objective to reduce the limit of detection (LOD), multi-analysis and multi-analyte detection (High-Throughput Screening systems-HTS), and integration capabilities.

Acoustic sensing has taken advantage of the progress made in the last decades in piezoelectric resonators for *radio-frequency* (rf) telecommunication technologies. The so-called gravimetric technique [2], which is based on the change in the resonance frequency experimented by the resonator due to a mass attached on the sensor surface, has opened a great deal of applications in bio-chemical sensing in both gas and liquid media.

Traditionally, the most commonly used acoustic wave biosensors were based on QCM devices. This was primarily due to the fact that the QCM has been studied in detail for over 50 years and has become a mature, commercially available, robust and affordable technology [3, 4]. LW acoustic sensors have attracted a great deal of attention in the scientific community during the last two decades, due to its reported high sensitivity in liquid media compared to traditional QCM-based sensors. Nevertheless, there are still some issues to be further understood, clarified and/or improved about this technology; mostly for biosensor applications.

LW devices are able to operate at higher frequencies than traditional QCMs [5]; typical operation frequencies are between 80-300 MHz. Higher frequencies lead, in principle, to higher sensitivity because the acoustic wave penetration depth into the adjacent media is reduced [6]. However, the increase in the operation frequency also results in an increased noise level, thus restricting the LOD. The LOD determines the minimum surface mass that can be detected. In this sense, the optimization of the read out and characterization system for these high frequency devices is a key aspect for improving the LOD [7].

Another important aspect of LW technology is the optimization of the fluidics, specially the flow cell. This is of extreme importance for reducing the noise and increasing the biosensor system stability; aspects that will contribute to improve the LOD.

The analysis and interpretation of the results obtained with LW biosensors must be deeper understood, since the acoustic signal presents a mixed contribution of changes in the mass and the viscoelasticity of the adsorbed layers due to interactions of the biomolecules. A better understanding of the transduction mechanism in LW sensors is a first step to advance in this issue; however its inherent complexity leads, in many cases, to frustration [8].

The fabrication process of the transducer, unlike in traditional QCM sensors, is another aspect under investigation in LW technology, where features such as: substrate materials, sizes, structures and packaging must be still optimized.

This chapter aims to provide an updated insight in the mentioned topics focused on biosensors applications.

2. Basis of LW sensors

The Love wave physical effect was originally discovered by the mathematician Augustus Edward Hough Love. He observed an effect caused by earthquake waves far from the epicenter due to the lower acoustic wave velocity of waves propagating along the stratified geological layers [9]. The LW sensor is a layered structure formed, basically, by a piezoelectric substrate and a guiding layer (see Figure 1a). LW devices belong to the family of *surface acoustic wave* (SAW) devices in which the acoustic wave propagates along a single surface of the substrate. The piezoelectric substrate of a LW device primarily excites a *shear horizontal surface acoustic wave* (SH-SAW) or a *surface skimming bulk wave* (SSBW) depending on the material and excitation mode of the substrate. Both waves have shear horizontal particle displacements (perpendicular to the wave propagation direction and parallel to the waveguide surface). This type of acoustic wave operates efficiently in liquid media, since the radiation of compressional waves into the liquid is minimized.

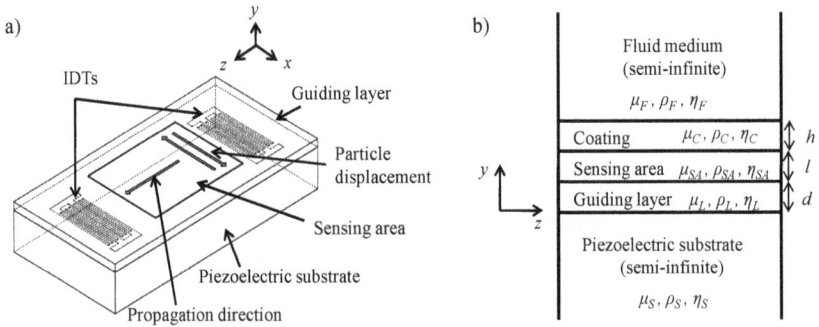

Figure 1. a) Basic structure of a LW sensor. b) Five-layer model of a LW biosensor.

LW sensors consist of a transducing area and a sensing area. The transducing area consists of the *interdigital transducers* (IDTs), which are metal electrodes, sandwiched between the piezoelectric substrate and the guiding layer. The input IDT is excited electrically (applying an rf signal) and launches a mechanical acoustic wave into the piezoelectric material which is guided through the guiding layer up to the output IDT, where it gets transformed back to a measurable electrical signal. The *sensing area* is the area of the sensor surface, located between the input and output IDT, which is exposed to the analyte.

LW sensors can be used for the characterization of processes involving several layers deposited over the sensing area; such is the case of biosensors. A LW biosensor can be described as a layered compound formed by the LW sensor in contact with a finite viscoelastic layer, the so-called coating, contacting a semi-infinite viscoelastic liquid as indicated in Figure 1b. Each layer has its material properties given by: the shear modulus μ, density ρ and viscosity η. Hence, the subscripts S, L, SA, C and F denotes the substrate, guiding layer, sensing area, coating and fluid layers, respectively. Biochemical interactions at the sensing area cause changes in the properties of the propagating acoustic wave which can be detected at the output IDT.

The difference between the mechanical properties of the guiding layer and the substrate creates an entrapment of the acoustic energy in the guiding layer keeping the wave energy near the surface and slowing down the wave propagation velocity. This guiding layer phenomenon makes LW devices very sensitive towards any changes occurring on the sensor surface, such as those related to mass loading, viscosity and conductivity [5]. The higher the confinement of the wave in the guiding layer, the higher the sensitivity [10].

The proper design of a LW device for biosensor applications must consider the advances made on these basic elements. Updated information about each one of these elements is then required and can be found in the following sections.

2.1. Piezoelectric substrate

Thanks to *piezoelectricity* electrical charges can be generated by the imposition of mechanical stress. The phenomenon is reciprocal; applying an appropriate electrical field to a piezoelectric material generates a mechanical stress [11]. In LW sensors an oscillating electric field (rf signal) is applied in the input IDT which, due to the piezoelectric properties of the substrate, launches an acoustic guided wave. The guided wave propagates through the guiding layer up to the output IDT where, again due to the piezoelectric properties of the substrate, is converted back to an electric field for measurement. A remarkable parameter of the piezoelectric substrate is the *electromechanical coupling coefficient* (K^2), which indicates the conversion efficiency from electric energy to mechanical energy; its value depends on the material properties. This is an important design parameter in LW sensors, since higher K^2 lead to low loss LW devices and, therefore, more sensitive LW sensors [12].

When choosing a material for the substrate of LW devices, apart from the desired low losses, other requirements, such as low *temperature coefficient of frequency* (TCF) have to be considered as well. Special crystal cuts of the piezoelectric substrate material[1] can yield an intrinsically temperature-compensated device which minimizes the influence of temperature on the sensor response, thus improving the LOD [13,14].

The shear horizontal polarization required for operation of the LW sensor in liquid media is another aspect to be considered when choosing the substrate material. In this sense, quartz is the only common substrate material that can be used to obtain a *purely* shear polarized wave [13]. The crystal cut and the wave propagation direction, which depends on the IDTs orientation, define the elastic, dielectric and piezoelectric constants of the crystal, and therefore the wave polarization. Possible cuts which generate a purely shear polarized wave are the AT-cut quartz and the ST-cut quartz. AT-cut quartz and ST-cut quartz are both Y-cuts, rotated 35°15′ and 42°45° about the original crystallographic X-axis, respectively.

Initially, LW devices were made in ST-cut quartz [15], however, ST-cut quartz is very sensitive to temperature (its TCF is around 40 ppm/°C) [16]. This was a restrictive factor in terms of sensor LOD and, thus, temperature-compensated systems based on different quartz cuts and different materials for the substrate such as lithium tantalate (LT), $LiTaO_3$, and lithium niobate (LN), $LiNbO_3$, were investigated [17-19]. In particular, AT-cut quartz, 36° YX LT and 36° YX LN were proposed, the last two corresponds to specific cuts of LT and LN materials [10]. Table 1 contains the values of some characteristic parameters of the previously men-

---

1 The substrate crystal cuts (or plates) are obtained by cutting slices of a single-crystal starting material with an arbitrary orientation relative to the three orthogonal crystallographic axes.

tioned substrate materials. In column 2, the *substrate shear velocity* v_S, is defined by the substrate material properties ($v_S = (\mu_S/\rho_S)^{1/2}$).

Substrate	v_s(m/s)	ρ_s(kg/m3)	K^2 (%)	TCF (ppm/°C)
ST-cut Z' propagating quartz	5050	2650	1.9	40
AT-cut Z' propagating quartz	5099	2650	1.4	0-1
36° YX LN	4800	4628	16	-75 to -80*
36° YX LT	4200	7454	5	-30 to -45

Table 1. Most commonly employed crystal cuts for LW devices (modified from [18]).*Approximate value.

LN substrates have higher coupling factor and low propagation loss than LT and quartz substrates. However, these substrates are extremely vulnerable to abrupt thermal shocks.

The low insertion loss, very large electromechanical coupling factor K^2 and low propagation loss which characterize 36° YX LT substrates [20] provide advantages over other substrates such as quartz cuts, where exquisite care in the fluidic packaging is required to prevent excessive wave damping [21]. For this reason, LT seems to be the substrate material of choice in high-loss applications due to its high coupling factor K^2, while in low-loss applications quartz may exhibit better wave characteristic [22]. The main shortcomings of 36° YX LT substrates are: they do not generate a pure shear wave, which increases the damping when is liquid loaded; and they have a poor thermal stability due to their high TFC (-30 to -40 ppm/°C [19]) if compared with AT-quartz.

From the point of view of thermal stability, AT-cut quartz seems to be the most appropriate due to its very low TCF [14]. Although the coupling coefficient of the AT-cut quartz is the lowest compared to other cuts, in our opinion, AT-cut quartz is currently the most suitable substrate for LW biosensing applications among the mentioned substrates, for several reasons: 1) it is capable of generating pure shear waves, diminishing the damping when is liquid loaded; 2) its thermal stability is the highest one, which improves the LOD; 3) the mass sensitivity of quartz substrates is significantly high than that of LT substrates [17,23]; and 4) LT and LN substrates are extremely fragile and must be handled with great care during the device fabrication to prevent them from breaking in pieces.

2.2. Interdigital transducers

Interdigital transducers (IDTs) were firstly reported in 1965 by White and Voltmer [24] for generating SAWs in a piezoelectric substrate. An IDT, in its most simple version, is formed by two identical combs-like metal electrodes whose strips are located in a periodic alternating pattern located on top of the piezoelectric substrate surface. Figure 2a shows the structure of a *single-electrode* IDT which consists of two strips per *period p* and *acoustic aperture W*. The strip width is equal to the space between strips ($p/4$). One comb is connected to ground and the other to the center conductor of a coaxial cable where a rf signal is provided. A pair of two strips with different potential is called a *finger pair*.

The IDT electric equivalent circuit is explained in reference [25]. Figure 2b shows the IDT frequency response, where $A(f)$ is the electrical amplitude of the rf signal. The maximum in $A(f)$ occurs when the *wavelength* λ of the generated acoustic wave is equal to the period p and this arises at the so called *synchronous frequency* f_s. In relation to the *bandwidth* B of an IDT frequency response, this will be narrower when increasing the *number of finger pairs* N. However, there is a limitation in the maximum N recommended, due to the fact that, in practice, when N exceeds 100, the losses associated with mass loading and the scattering from the electrodes increase. This neutralizes any additional advantage associated with the increase of the number of the finger pairs.

Due to symmetry of the IDT in the direction of propagation, the LW energy is emitted in equal amounts in opposite directions, giving an inherent loss of 3 dB. In a two-port device this factor contributes 6 dB to the total insertion loss [25,26].

Aluminum has been widely used as IDTs material and has been extensibility demonstrated in literature as suitable for SAW generation. Aluminum has an ability to resist corrosion and is the third most abundant element on Earth (after oxygen and silicon). It also has a low cost compared to other metals. The metallic layer of the electrodes must be thick enough to present a low electric resistance, but sufficiently thin to avoid an excessive mechanic charge for the acoustic wave (acoustic impedance breaking) [27]. Generally, a thickness between 100 and 200 nm of aluminum is employed.

There are a number of second-order effects, which are often significant in practice, that affect the transducer frequency response. The effect for which the transducer strips reflect surface waves causing mechanical and electrical perturbations of the surface is called *electrode interaction* [30]. Usually, these unwanted reflections cancel each other over wide frequency bands and become negligible. However, in a certain frequency band, the scattered waves are in phase, adding them constructively and causing very strong reflection (*Bragg reflection*) which distorts the transducer frequency response. For a *single-electrode* IDT (see Figure 2a), this situation occurs at the resonance condition $\lambda = p$. Thus, *double-electrode* (or *double finger pair* or *split-electrode*) IDTs are used to avoid this unwanted effect. In double-electrode IDTs there are four strips per period (see Figure 2c) and thus, the Bragg reflection can be suppressed at the LW resonance frequency [28]. One disadvantage of the double-electrode is the increased lithographic resolution required for fabricating the IDTs [29].

Another significant second-order effect is the generation of the *triple-transit signal*. In a device using two IDTs, which is the case of a LW device, the output IDT will in general produce a reflected wave, which is then reflected a second time by the input IDT. Thus, a reflected wave reaches the output IDT after traversing the substrate three times, giving an unwanted output signal known as the *triple-transit signal* [26]. This effect is reduced by making the input and output IDT separation large enough.

Some authors use *reflectors* to improve the frequency response of the LW device. Reflectors are composed of metal gratings placed in the same configuration than IDTs and are located at the ends of the IDTs (see Figure 2d). These components have generally less finger pairs than the IDTs. The space periodicity of the reflectors is equal than in the IDTs [30]. Very narrow low-loss pass band can be realized in a two-port device, when the device is designed so that the reflectors resonate at the IDT resonance frequency, since the transfer admittance becomes very large [28].

2.3. Guiding layer

The difference between the mechanical properties of the piezoelectric substrate and the guiding layer generates a confinement of the acoustic energy in the guiding layer, slowing down the wave propagation velocity, but maintaining the propagation loss [32]. In particular, the condition for the existence of Love wave modes is that the shear velocity of the guiding layer material ($v_L = (\mu_L/\rho_L)^{1/2}$) is less than that of the substrate ($v_S = (\mu_S/\rho_S)^{1/2}$) [31]. When both materials, substrate and guiding layer, have similar density the ratio μ_S/μ_L determine the dispersion of the Love mode; a large value of that ratio (higher μ_S and lower μ_L) leads to a stronger entrapment of the acoustic energy [32] and thus, greater sensitivity. Hence, the benefit of the guiding layer is that an enhanced sensitivity to mass deposition can be obtained [33], but also to viscoelastic interactions.

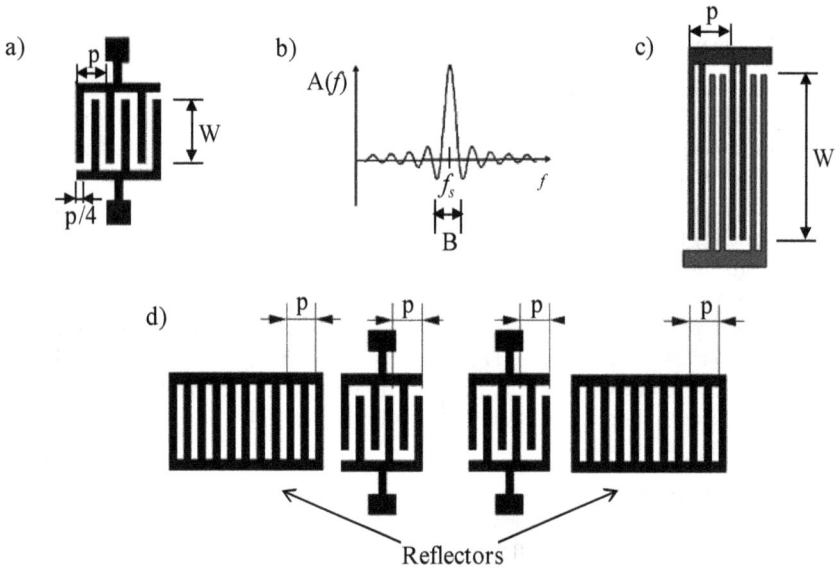

Figure 2. a) Single-electrode Interdigital Transducer (IDT) with period p, electrode width equal to space between electrodes (p/4) and aperture W (modified from [25]). b) Frequency response of an IDT (positive frequencies), where A(f) is the electrical amplitude (modified from [25]). c) Double-electrode IDT with period p, electrode width equal to space between electrodes and aperture W. d) Two grating reflectors are place at both ends of the IDTs (modified from [30]).

The effect of the guiding layer on Love modes influence the substrate coupling factor K^2, increasing it [14]. Also influence the temperature behavior, since modifies the TCF compared to their parent SSBWs device.

In relation to the materials used for the guiding layer, those with a low shear velocity and low insertion loss seem to be the most promising materials for developing sensitive biosensors [22,32,34]. Materials such as polymers [35], silicon dioxide (SiO_2) [17], gold (Au) [36] and zinc oxide (ZnO) [37,38] have been used as guiding layers [21]. In Table 2 some properties of these materials are presented[2]. The use of polymers (like Novolac, polyimide, polydimethylsiloxane (PDMS) and polymethylmethacrylate (PMMA)) is interesting from the point of view of the sensitivity, since they have low shear velocity. Additionally, some polymers, like Novolac photoresist, are very resistant to chemical agents [39,40]. However, polymers have high acoustic damping (losses) [39] and this is a clear disadvantage for biosensing application.

Guiding layer material	μ_L (GPa)	ρ_L (kg/m3)	v_L (m/s)
SiO_2	17.87	2200	2850.04
ZnO	40.17	5720	2650.00
Au	28.50	19300	1215.19
Polyimide	0.87	1420	780.48
PDMS	250×10^{-6}	965	16.09
PMMA	1.70	1180	1200.28

Table 2. Employed materials for guiding layers of LW devices.

Guiding layer/substrate structures made with ZnO as guiding layer have some advantages over those with a different material. This is the case of ZnO/ST-quartz structure, for which significantly high sensitivity, small TCF and high K^2 were reported [38]. ZnO/LT devices were also found to have higher mass sensitivity than SiO_2/LT [23]. However, ZnO has several disadvantages: it is CMOS contaminant, a semiconductor, and thus, it can deteriorate the efficiency of the transducers and make some artifacts. In addition, it gets easily rough when sputtered and it is very reactive with acids, liquids, or moisture, so it will dissolve if exposed to water or humid environment, which is a big problem in biosensors application. Regarding Au guiding layers, they provide very strong wave guiding, since Au has a relatively

---

2 These values are for guidance, since for deposited or grown materials these values depend on the desposition technique and for polymers layers on the cure process.

low shear acoustic velocity and a high density. However, it couples the rf signal from input to output IDT.

Silicon dioxide (SiO_2) -also known as fused silica- is a standard material in semiconductor industry and offers low damping, sufficient low shear velocity and excellent chemical and mechanical resistance [41]. It is the only native oxide of a common semiconductor which is stable in water and at elevated temperatures, an excellent electrical insulator, a mask to common diffusing species, and capable of forming a nearly perfect electrical interface with its substrate. When SiO_2 is needed on materials other than silicon, it is obtained by chemical vapor deposition (CVD), either thermal CVD or Plasma enhanced CVD (PECVD) [42]. The main shortcoming for SiO_2 is that the optimum thickness, at which the maximum sensitivity is reached, is very high (see Section 5), so this complicates the manufacturing process. Nevertheless, at the present, we consider that SiO_2 is the most appropriate material for LW biosensors guiding layer, mainly due to its low damping and excellent chemical and mechanical properties [42].

2.4. Sensing area

The sensing area can be made of different material than the guiding layer. Sensing layers have been reported composed of materials like PMMA [43] and SU-8 [44], but the most commonly employed is gold (Au). Generally, the thickness of this layer varies from 50-100 nm and 2-10 nm of chrome (Cr) or titanium (Ti) is needed to promote adherence to the guiding layer. Au surfaces are very attractive candidates for self-assembly due to their metallic nature, great nobility, and particular affinity for sulphur. This aspect allows functionalization with thiols of various types and adhesion to diverse organic molecules, which are modified to contain a sulphur atom. These coatings, assembled onto the gold surfaces, can serve as biosensors [36]. Immobilization techniques on gold for biosensing are quite common and much utilized in the scientific community. However, immobilization techniques on different materials, like SiO_2, could greatly simplify the LW biosensors fabrication.

3. Measurement techniques

Figure 3a shows a configuration of a two-port LW device where it behaves as a delay line. D is the distance between input and output IDT and L is the center-to-center distance between the IDTs. Thus, the sensor is a transmission line which transmits a mechanical signal (acoustic wave) launched by the input port (input IDT) due to the applied rf electrical signal. After a time delay the traveling mechanical wave is converted back to an electric signal in the output port (output IDT). In general, changes in the coating layer and/or in the semi-infinite fluid medium (see Figure 1b) produce variations in the acoustic wave properties (wave propagation velocity, amplitude or resonant frequency). These variations can be measured comparing the input and output electrical signal, since phasor V_{in} remains unchanged, while phasor V_{out} changes. Thus, from an electric point of view, a LW delay line can be defined by its transfer function $H(f) = V_{out}/V_{in}$, which represents the relationship between input and output electrical signal. $H(f)$ is a complex number which can be defined as $H(f) = Ae^{j\varphi}$, being $A = |V_{out}/V_{in}|$ the amplitude and φ the *phase shift* between V_{out} and V_{in}. In terms of voltage, the *insertion loss* (IL) in dB is given by 20 $\log_{10}(A)$. Figure 3b, presents the frequency response of an AT-cut Z' propagating/SiO$_2$ LW device designed a fabricated by the authors of this chapter.

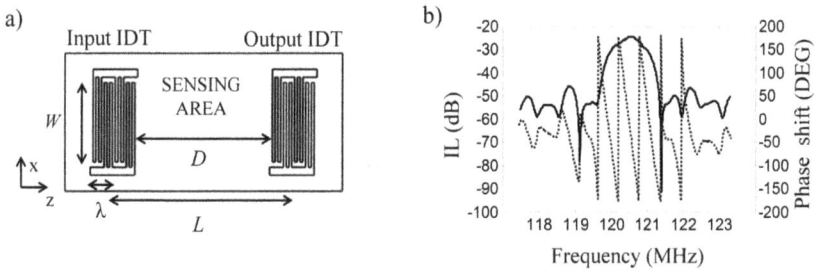

Figure 3. a) Scheme of a LW delay line. It consists of two ports. In the input IDT an rf signal is applied which launches an acoustic propagating wave. The output signal is recorded at the output IDT. *D* is the distance between input and output IDT and *L* is the center-to-center distance between the IDTs. b) Frequency response of a LW device designed a fabricated by the authors of this chapter. The phase shift (dotted line) and IL (solid line) were measured using a network analyzer.

In biosensors, biochemical interactions at the sensing area will modify the thickness and properties of the coating, and therefore variations in the amplitude and phase of the electrical transfer function can be measured. These variations can be monitored in real time, which provides valuable information about the interaction process.

The LW delay line can be used as frequency determining element of an oscillator circuit (*closed loop* configuration). Effectively, in an oscillator circuit the LW device is placed as a delay line in the feedback loop of an rf amplifier in a closed loop configuration [10,45]. Therefore, a change in the wave velocity, due to a sensing effect, produces a time delay in the signal through the LW device which appears as phase-shift; this phase-shift is transferred in terms of frequency-shift in an oscillator configuration. The oscillator is, apparently, the simplest electronic setup: the low cost of their circuitry as well as the integration capability and continuous monitoring are some features which make the oscillators an attractive configuration for the monitoring of the determining parameter of the resonator sensor, which in the case of the LW device is the phase-shift of the signal at resonance [46-49]. However, due to the following drawbacks, in our opinion, the oscillators are not the best option for acoustic wave sensor characterization: 1) they do not provide direct information about signal amplitude; 2) they, eventually, can stop oscillation if insertion losses exceed the amplifier gain during an experiment; and 3) despite of the apparent simple configuration, a very good design is necessary to guarantee that a LW resonator will operate at a specific frequency, and this is not a simple task. In effect, in the same way than in QCM oscillators it is required to assure that the sensor resonates on one defined resonance mode and does not "jump" be-

tween spurious resonances [7], in LW oscillators one must assure that the sensor will operate at one phase ramp in the sensor response band-pass, and does not jump from one to
another which are almost of identical characteristics (see Figure 3b). Moreover, when the
resonator dimensions get smaller and the frequency increases this becomes more difficult to
achieve, since when increasing frequency there is a decrease of the resonator quality factor, a
decrease in frequency stability [50] and in LW the ramps become nearer to each other.

In an *open loop* configuration, the input transducer is excited at a fixed frequency while the
phase shift between V_{out} and V_{in}, φ, is recorded [32]. In this configuration, in the absence of
interferences, phase variations measured experimentally can be related to changes in the
physical properties of the layers deposited over the sensing area.

Network analyzers are the most commonly used instrumentation for characterizing LW delay lines in open loop configurations. Nevertheless, recently, some authors successfully validated a new characterization technique based on the open loop configuration [51]. A read
out circuit based on this technique for high frequency liquid loaded QCM devices has been
developed by the same authors [52], and tested with LW devices with very satisfactory results [53]. The main advantages of this read out circuit are its low cost, high integration,
small size, calibration facility and the possibility of being used as an interface for multi-analysis detection.

4. Modeling methods

As mentioned, there is an open research field regarding the employed materials, and its
physical and geometric properties for achieving more optimized LW devices for biosensors
applications. For instance, the thickness of the Love-wave guiding layer is a crucial parameter that can be varied to achieve a more sensitivity device. The fabrication of LW sensors is
complex and expensive due to their micro sizes [54]; therefore, simulations and modeling of
LW devices as previous steps to their production could be very valuable. Models allow relating changes in some characteristics of the wave, as the velocity, with changes in the physical properties of the layers deposited over the sensing area, and thus, provide information
about the sensing event. Nevertheless, modeling LW devices commonly requires simplified
assumptions or the use of numerical methods [23] due to the complex nature of SAW propagation in anisotropic and piezoelectric materials.

In this section, information regarding the current most popular models used for modeling
LW sensors is provided: the transmission line model, the dispersion equation and the Finite
Element Method.

4.1. Transmission line model

It is well known that the propagation and attenuation of acoustic waves in guiding structures can be obtained by equivalent transmission line models [8,55]. The theory of sound
wave propagation is very similar mathematically to that of electromagnetic waves, so tech-

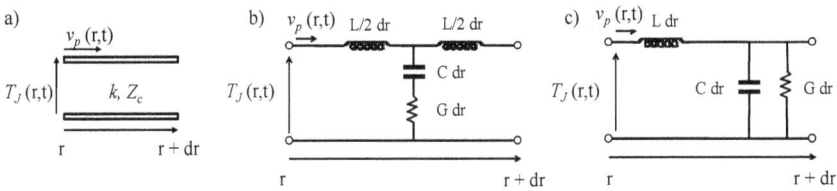

Figure 4. a) Pictorial representation of a transmission line. b) Transmission line equivalent series model for acoustic propagation in a viscoelastic layer. c) Transmission line equivalent parallel model for acoustic propagation in a viscoelastic layer.

niques from transmission line theory are also used to build structures to conduct acoustic waves; and these are also called transmission lines. The *transmission line model* (TLM) for acoustic waves take advantage of the concepts and techniques of proven value in electromagnetic microwaves to corresponding problems in elastic guided waves [8].

A transmission line is characterized by its *secondary parameters* which are the propagation *wavevector k* (when scalar *wavenumber*) and the *characteristic impedance* Z_c (see Figure 4a). It is important to mention that these parameters do not depend on the transmission line length. In each plane of an electric transmission line it is possible to define a magnitude for a voltage and other for the current in the line. For acoustics transmission lines current and voltage in electromagnetic are replaced by the *particle velocity* v_p and the *stress* $-T_J$, respectively, where J indicates de stress direction ($J = 1, 2,..., 6$) [55]. In an acoustic transmission line Z_c represents the relation between the stress $-T_J$ and the particle velocity v_p of the material, and k quantifies how the wave energy will be propagated along the transmission line. To quantify the variations of $-T_J$ and v_p when the wave propagates through the transmission line, the lumped elements models presented in Figure 4b and 4c are introduced. Figure 4b corresponds to the series model, and Figure 4c to the parallel model. The lumped elements of these models are called the transmission line *primary parameters*, which are dependent on the line length. Analyzing the parallel model of Figure 4c, the following coupled differential equations are obtained:

$$\frac{d^2T_J}{dr^2} = Z \cdot Y \cdot v_p$$
$$\frac{d^2v_p}{dr^2} = Z \cdot Y \cdot T_J$$

(1)

where $Z=j\omega L$, $Y=G+j\omega C$ and $\omega=2\pi f$. Being Z, L, Y, C, G, f, the impedance, inductance, admittance, capacitance and conductance per unit of length, respectively and f the frequency of T_J and v_p.

The solutions for these equations are given by:

$$T_j(\vec{r}) = T_j^+ e^{-\gamma \cdot \vec{r}} + T_j^- e^{\gamma \cdot \vec{r}}$$

$$v_p(\vec{r}) = \frac{T_j^+}{Z_c} e^{-\gamma \cdot \vec{r}} - \frac{T_j^-}{Z_c} e^{\gamma \cdot \vec{r}} \qquad (2)$$

where T^+ and T^- are arbitrary values for the intensity of the incident an reflected waves, respectively. The *linear propagation exponent* or *complex propagation factor* γ is directly related to the wavevector ($\gamma=jk$) and is given by $\gamma = (ZY)^{1/2} = \alpha + j\beta$. The real part of γ, denoted as *attenuation coefficient* α, represents the attenuation suffered by the wave when propagating through the transmission line. The imaginary part of γ, denoted as *phase coefficient* β, when multiplied by a distance, quantifies the phase shift that the wave suffers when traveling that distance. The characteristic impedance of the line Z_c is given by $Z_c=(Z/Y)^{1/2}$. First row of Table 3 shows the relationship between primary and secondary parameters of a transmission line for the series and parallel model.

The relation between the secondary (or primary) parameters and the properties of the transmission line materials is achieved by comparing Eqs. (1) with the motion equations of the LW assuming isotropic layers [55]. These relations are given in the second row of Table 3 for the series and parallel model.

Series model	Parallel model
$k = -j\gamma = \omega \sqrt{\dfrac{LC}{1 + j\omega C / G}}$	$k = -j\gamma = -j\sqrt{j\omega L\ (G + j\omega C)}$
$Z_c = \sqrt{\dfrac{L}{C} + j\omega \dfrac{L}{G}}$	$Z_c = \sqrt{\dfrac{j\omega L}{G + j\omega C}}$
$C = \dfrac{1}{\mu}$	$C = \dfrac{\mu}{\omega^2 \eta^2 + \mu^2}$
$G = \dfrac{1}{\eta}$	$G = \dfrac{\eta \omega^2}{\omega^2 \eta^2 + \mu^2}$
$L = \rho$	$L = \rho$

Table 3. Equivalent transmission line model parameters in terms of the layer properties.

Figure 5a shows a simplify description in which a LW propagates in a waveguide structure. Two assumptions were made: 1) the materials in the figure are isotropic (for piezoelectric substrates this assumption is valid because of a low anisotropy) and 2) the main wave propagating in the z direction results from the interaction of two partial waves with the same component in z direction and opposites components in y direction [12].

If all the properties of the layers which are involved in the LW transmission line are known, it is possible to obtain the *phase velocity of the Love mode* v_φ. The LW propagates in each layer i of the device in two directions z and y. In the case of a typical biosensor the device consists of 5 layers with the subscripts i equal to: S for the substrate, L for the guiding layer, SA for the sensing area, C for the coating and F for the fluid. Direction z is known as the *longitudinal direction* and direction y as the *transverse direction*. The wave does not find

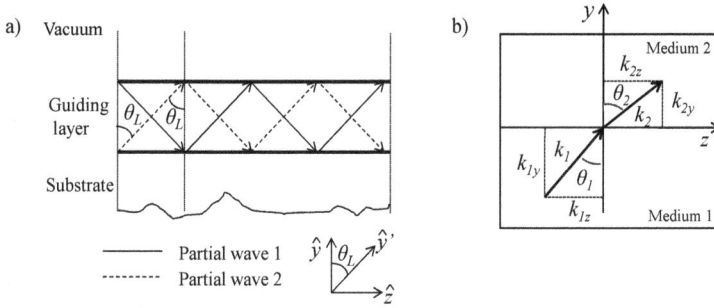

Figure 5. a) Simplified description of a LW traveling in a guided structure. b) Schematic representation of the wave-vectors k_1 and k_2 in two different media.

any material properties change in the longitudinal direction; hence it propagates in this direction acquiring a phase shift and attenuation (in case of material with losses). However, in the transverse direction a stationary wave exists when the resonance condition is met. Thus, each layer counts with two transmission lines, one in the transverse direction and the other in the longitudinal direction. When a wave propagates in y' direction (see Figure 5a), as it happens with the partial waves of a layer i, k_i has that same direction (see Figure 5b), and therefore, it has components in z and y directions. In this way, the secondary parameters of each transmission line are determined from the projection of the parameter in the proper direction. The relation between the wavevector and the wave velocity is $k_i = \omega/v_i$. Therefore, k_i and v_i have the same directions, so the wave velocity in y' direction also has components in z and y.

Equations (3) and (4) give the expression of the secondary parameters and wave velocity in the transverse and longitudinal directions, respectively:

$$Z_{ciy} = Z_{ci}\cos\theta_i, \quad k_{iy} = |k_i|\cos\theta_i, \quad v_{iy} = |v_i|\cos\theta_i \tag{3}$$

$$Z_{ciz} = Z_{ci}\sin\theta_i, \quad k_{iz} = |k_i|\sin\theta_i, \quad v_{iz} = |v_i|\sin\theta_i \tag{4}$$

The angles θ_i are called *complex coupling angles* and depend on the material properties of each layer, but also on the material properties of the adjacent layers, since in each interface the Snell's laws have to be satisfied. In isotropic solids, the incident and reflected waves must all have the same component of k_i in the longitudinal direction [55]; therefore:

$$k_{Sz} = k_S\sin\theta_S = k_L\sin\theta_L = k_{Lz} = k_{SA}\sin\theta_{SA} = k_{SAz} = k_C\sin\theta_C = k_{Cz} = k_F\sin\theta_F = k_{Fz} \tag{5}$$

These conditions, together with the *transverse resonance relation* [8] obtained from the transmission line models in the direction of resonance (transverse direction) (see Figure 6) provides the phase velocity of the Love wave v_φ.

The transverse resonance relation establishes that:

$$\overleftarrow{Z}(P) + \overrightarrow{Z}(P) = 0 \tag{6}$$

where $\overrightarrow{Z}$ represents the acoustic impedance seen by the wave at the right of the plane P and $\overleftarrow{Z}$ the acoustic impedance seen by the wave at the left of the plane. This relation states that the acoustic impedances looking both ways from some reference plane P must sum to zero. The location of P is arbitrary. The solution of applying this condition provides an angle θ_i for one layer, depending on where P was located. From the material properties of each layer and applying the Snell's laws the angles for the rest of the layers can be found. Once these angles are known the z and y components of k_i and v_i can be obtained, and thus the phase velocity v_φ. Assuming that almost all the energy of the wave is confined in the waveguide, the phase velocity can be defined as the wavefront velocity of the acoustic wave propagating in the guiding layer, which in this case propagates in the z direction. In lossless materials k_{iz} and k_{iy} are real numbers and therefore v_φ is given by:

$$v_\varphi = \frac{\omega}{k_{Lz}} = \frac{\omega}{|k_L|\sin\theta_L} \tag{7}$$

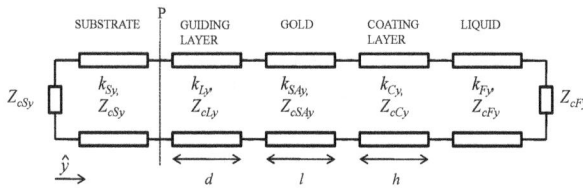

Figure 6. Equivalent transmission line model of the LW layered structure in the direction of resonance y. The lines are connected in series to satisfy the boundary conditions and the two semi-infinite layers are loaded with its characteristic impedance.

When the material has losses, k_{iz} and k_{iy} are complex numbers with real and imaginary parts, and then the attenuation coefficients appear:

$$k_i = k_{iz}\hat{z} + k_{iy}\hat{y} = \left(\beta_{iz} - j\alpha_{iz}\right)\hat{z} + \left(\beta_{iy} - j\alpha_{iy}\right)\hat{y} \tag{8}$$

In this case, the phase velocity is given by:

$$v_\varphi = \frac{\omega}{\beta_{Lz}} = \frac{\omega}{\Re\{k_{Lz}\}} = \frac{\omega}{\Re\{|k_L|\sin\theta_L\}} = \frac{1}{\Re\left\{\frac{\sin\theta_L}{|v_L|}\right\}} \tag{9}$$

On the other hand, the *attenuation of the Love Wave* α_{LW} is considered to happen mostly in the propagation direction z, since in the resonance direction, y, a stationary wave takes place. Therefore:

$$\alpha_{LW} = \alpha_{Lz} = -\Im\{k_{Lz}\} = -\Im\{|k_L|\sin\theta_L\} \tag{10}$$

Thus, following this procedure it is possible to obtain the phase velocity and attenuation of a LW propagating in a layer. Nevertheless, to complete this, it is necessary to know the material properties of all the layers which integrate the LW device. However, when the device is used as a sensor, and in particular in biosensor applications, the coating layer properties are unknown parameters. Thus, quantifying the variations suffered by the mechanical and geometric layer properties over the sensing layer when measuring the electrical parameters is what is really interesting.

Variations in amplitude and phase of the transfer function $H(f) = V_{out}/V_{in}$ (due to perturbations in the acoustic wave) can be monitored in real-time. These perturbations can occur due to variations of the mechanical and geometrical properties of the layers deposited over the sensing area. Such physical changes affect the propagation factor of the wave, and thus, the attenuation and phase velocity of the Love Wave. Next, the relations between LW electrical parameters defined in Section 3 (φ and IL) and the complex propagation factor are explained.

The relation between the output and input voltage in a delay line (DL) of length z can be modeled by its transfer function $H_{DL}(f)$ in the following way:

$$H_{DL}(f) = e^{-\gamma z} \tag{11}$$

where γ corresponds to the propagation factor of the wave in the line, which in this case corresponds to that of the guiding layer in the z direction γ_{Lz}. Assuming that the transfer function of the input and output IDTs is equal to unit [26], the relation between the electrical signal in the output and input IDTs $H(f)$ is the same than the one for the delay line:

$$H(f) = \frac{V_{out}}{V_{in}} = \frac{|V_{out}|}{|V_{in}|}e^{j(\varphi_{out}-\varphi_{in})} = \frac{|V_{out}|}{|V_{in}|}e^{j\varphi} = e^{-\gamma_{Lz}z} = e^{-(\alpha_{Lz}+j\beta_{Lz})z} \tag{12}$$

Thus, taking into account the relations defined in Section 3, the normalized phase shift and IL are given by:

$$\frac{IL}{z} = -\alpha_{Lz} 20 \log e$$
$$\frac{\varphi}{z} = -\beta_{Lz}$$

(13)

The increment in IL/z and φ/z from a non perturbed state γ_{Lz0} to a perturbed state γ_{Lz1} is the following:

$$\Delta\frac{IL}{z} = \left(\alpha_{Lz1} - \alpha_{Lz0}\right) \cdot 20 \log e$$
$$\Delta\frac{\varphi}{z} = \beta_{Lz1} - \beta_{Lz0}$$

(14)

The last set of equations provides a relation between the experimental data and the physical parameters of the layers. The extraction of the layers physical parameters is a major problem. Assuming that the physical properties of the substrate, guiding layer, gold and fluid medium are known and that these properties do not change during the sensing process, which can be the case in biosensing, still the parameters of the coating layer are not known. In comparison, the wave propagation direction in QCM coincides with the resonant direction. Therefore, for low frequency QCM applications it is possible to assume that the biochemical interaction is translated to simple mass changes, since it is reasonable to assume that the thickness of the coating layer is acoustically thin. This simplifies enormously the parameters extraction. However, in LW sensors, this assumption is not valid, and then the only two experimental data obtained in LW devices (Eq. (14)) are not enough to extract the unknown parameters of the coating. This, together with the complex equations which relate the measured data with the material properties, results in a complex issue that, to our knowledge, is not yet solved. Hence, there is an open research field of great interest related to this issue.

4.2. Dispersion equation

The *dispersion equation* provides the wave phase velocity as a function of the guiding layer thickness. The procedure for obtaining this equation for a two-layer system (guiding layer and substrate) is detailed in reference [56]. Broadly, this equation is reached after imposing the boundary conditions to determine the constants appearing in the particle displacement expressions of the waveguide and the substrate. These displacements are the solution of the equation of motion in an isotropic and non-piezoelectric material. After extensive algebraic manipulation [56], the dispersion equation for a two-layer system is found, resulting [57]:

$$\tan\left(k_{Ly}d\right) = \frac{\mu_S}{\mu_L}\sqrt{\frac{1-\left(v_\varphi^2/v_S^2\right)}{\left(v_\varphi^2/v_L^2\right)-1}}$$

(15)

where k_{Ly} is the guiding layer transverse wavenumber in y direction, given by:

$$k_{Ly} = \sqrt{\frac{\omega^2}{v_L^2}-\frac{\omega^2}{v_\varphi^2}}$$

(16)

Taking into account the relation between the frequency and wave wavelength, $f = v_\varphi/\lambda$, the argument of the tangent in Eq.(15) can be written as:

$$k_{Ly}d = 2\pi v_\varphi\frac{d}{\lambda}\sqrt{\frac{1}{v_L^2}-\frac{1}{v_\varphi^2}}$$

(17)

where the ratio d/λ is the normalized guiding layer thickness.

From the dispersion equation, the phase velocity can be solved numerically. On the other hand, the *group velocity*, v_g, as a function of the normalized guiding layer thickness can also be determined from the phase velocity by the formula [34]:

$$v_g = v_\varphi\left(1+\frac{d/\lambda}{v_\varphi}\frac{dv_\varphi}{d(d/\lambda)}\right)$$

(18)

The phase velocity and group velocity of an AT-cut quartz/SiO$_2$ Z' propagating layered structure were calculated using the dispersion equation (Eq. (15)) solved numerically through the bisection method. The data used to solve the equation for this LW structure were: v_S = 5099 m/s, ρ_S = 2650 kg/m³, v_L = 2850 m/s, ρ_L = 2200 kg/m³ and λ = 40 μm. The respective shear modulus were obtained through $\mu_i = v_i^2\,\rho_i$. Results are depicted in Figure 7. This figure provides information about the changes in the wave phase velocity due to changes in the guiding layer thickness. Assuming that the perturbing mass layer deposited over the sensing area is of the same material than the guiding layer, this equation will provide information over the sensing event. Nevertheless, this assumption is far from the biosensors reality, where the consideration of a five–layer model is required (see Figure 1b).

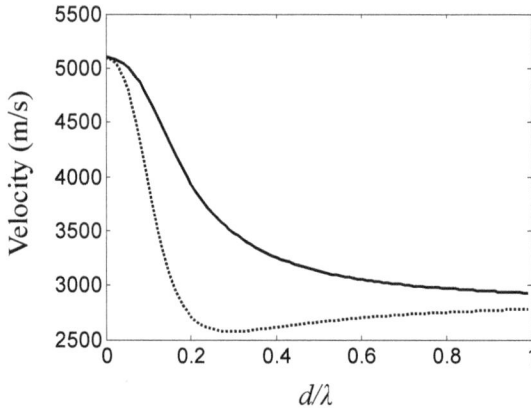

Figure 7. a) AT-cut Z' propagating quartz/SiO$_2$ phase velocity (continuous line) and group velocity (dashed line) for λ = 40 µm.

Mc Hale et al. developed the dispersion equation of three and four-layer systems neglecting piezoelectricity of the substrate [33]. When the substrate piezoelectricity is not considered, the dispersion equation is simplified. This can be the case of quartz substrates, which piezoelectricity is low. However, as the piezoelectricity of a substrate increases (like in the case of LT and LN), neglecting piezoelectricity, or assuming it is accounted for a stiffening effect in the phase velocity, may be less valid [58]. Liu et al. provided a theoretical model for analyzing the LW in a multilayered structure over a piezoelectric substrate [58]. Nevertheless, in our opinion, for those applications with a high number of layers, as in the case of biosensors, the use of the TLM is more convenient, since it is a very structured and intuitive model where the addition of an extra layer does not make the procedure more complex. From the programming point of view, this is an enormous advantage.

The phase velocity provided by the dispersion equation can be used to determine the optimal guiding layer thickness, which provides a maximum sensitivity. This issue will be addressed in Section 5.

4.3. LW sensor 3D FEM simulations

The models mentioned before, applied simplifying assumptions like considering the device substrate as isotropic and neglecting the substrate piezoelectricity. This makes the models far from the LW device reality. For a more accurate calculation of piezoelectric devices operating in the sonic and ultrasonic range, numerical methods such as finite element method (FEM) or boundary element methods (BEM) are the preferred choices [59]. The combination of the FEM and BEM (FEM/BEM) has been used by many authors for the simulation of SAW devices [60]. Periodic Green's functions [61] are the basis of this model. However, the FEM/BEM only applies to infinite periodic IDTs and it is a 2 dimensional (2D) analysis. Thus, this method is only an approximation of a real finite SAW device.

Simulations of piezoelectric media require the complete set of fundamental equations relating mechanical and electrical phenomena in 3 dimensions (3D). The 3D-FEM can handle these types of differential equations. The FEM formulation for piezoelectric SAW devices is well explained in [59]. In general, the procedure for simulating LW devices using the 3D-FEM is the following: 1) the 3D model of the device is build using a computer-aided design (CAD) software; 2) the 3D model is imported to a commercial finite element software, which allows piezoelectric analysis; 3) the material properties of the involved materials are introduced in the software; 4) the convenient piezoelectric finite element is selected; 5) the model is meshed with the selected finite element; and 6) the simulation is run in the software. As result the software obtains the particle displacements and voltage at every node of the model.

Although 3D-FEM simulations are extremely useful tools for studying LW device electroacoustic interactions, LW simulations in real size are still a challenge. Delay lines in practice are of many wavelengths and simulate them would require having too many finite elements. Thus, further efforts are required in order to achieve simulations able to reproduce real cases, which do not consume excessive computational resources. Nevertheless, some authors have simulated scaled LW sensors using this method [62-65].

5. Sensitivity and limit of detection

A key parameter when designing a LW biosensor is the device sensitivity [13]. In general terms, the sensitivity is defined as the derivative of the response (R) with respect to the physical quantity to be measured (M):

$$S = \lim_{\Delta M \to 0} \frac{\Delta R}{\Delta M} = \frac{dR}{dM} \tag{19}$$

It is possible to have different units of sensitivity depending on the used sensor response. E.g. for frequency output sensors, frequency ($R = f$), relative frequency ($R = f / f_0$), frequency shift ($R = f - f_0$) and relative frequency shift ($R = (f - f_0)/ f_0$) can be found, being f_0 the non perturbed starting frequency. Hence, four different possibilities of sensitivity formats are possible, and therefore it is extremely important to mention which case is been used in each application.

The sensitivity of LW sensors gives the correlation between measured electric signals delivered by the sensor and a perturbing event which takes place on the sensing area of the sensor. A high sensitivity relates a strong signal variation with a small perturbation [32]. Depending on the used electronic configuration, the electrical signal measured in LW devices can be: operation frequency, amplitude (or Insertion Loss), and phase. From this signal, and using the theoretical models, the phase velocity and group velocity can be obtained (see Sections 4.1 and 4.2).

When the sensor response R is the phase velocity, and the perturbing event which takes place on the sensing area is a variation of the *surface mass density* of the coating σ ($\sigma = h \rho_C$), the *velocity mass sensitivity* S^v_σ of a LW device at a constant frequency is given by [32,66]:

$$S^v_\sigma = \frac{1}{v_{\varphi 0}} \frac{\partial v_\varphi}{\partial \sigma}\bigg|_f \tag{20}$$

where $v_{\varphi 0}$ is the unperturbed phase velocity and v_φ is the phase velocity after a surface mass change.

Hence, S^v_σ has units of m²/kg. In Eq. (20) partial derivatives have been considered as the phase velocity depends on several variables. The surface mass sensitivity reported for LW devices in literature are between 150-500 cm²/g [45,67].

In sensor applications the phase velocity shift must be obtained from the experimental values of phase or frequency shifts. For the closed loop configuration, where the experimentally measured quantity is the frequency, the *frequency mass sensitivity* S^f_σ is defined as:

$$S^f_\sigma = \frac{1}{f_0} \frac{df}{d\sigma} \tag{21}$$

Jakoby and Vellekoop [66] noticed that the sensitivity measured by frequency changes in an oscillator (Eq. (21)) differs from the estimated velocity mass sensitivity (Eq. (20)) by a factor v_g/v_φ, since Love modes are dispersive. Thus, a typical 10% difference can be noted between S^f_σ and S^v_σ.

For the open loop configuration, where the experimentally measured quantity is the phase, the *phase sensitivity* (also called *gravimetric sensitivity*) S^φ_σ in absence of interference is defined as:

$$S^\varphi_\sigma = \frac{1}{k_{Lz}D} \frac{d\varphi}{d\sigma} = S^v_\sigma \tag{22}$$

where D is the distance between input and output IDT (see Figure 3) and k_{Lz} is the wavenumber of the Love mode, therefore $k_{Lz}D$ is the unperturbed phase φ_0.

The theoretical mass sensitivity of LW devices, derived from the perturbation theory[55], can be determined from the phase velocity according to [57]:

$$S_\sigma^v = \frac{-1}{\rho_L d}\left(1 + \frac{\sin(k_{Ly}d)\cos(k_{Ly}d)}{k_{Ly}d} + \frac{\rho_S}{\rho_L}\frac{\cos^2(k_{Ly}d)}{k_{Sy}d}\right)^{-1} \tag{23}$$

where k_{Ly} is the wavenumber in the guiding layer (Eq. (16)) and k_{Sy} is the wavenumber in the substrate (Eq. (24)). The minus sing in Eq. (23) indicates that the phase velocity of the perturbed event due to an increment in the surface mass density is less than the unperturbed phase velocity ($\Delta v_\varphi = v_\varphi - v_{\varphi 0}$).

$$k_{Sy} = \sqrt{\frac{\omega^2}{v_\varphi^2} - \frac{\omega^2}{v_S^2}} \tag{24}$$

The phase velocity can be obtained with the dispersion equation (see Figure 7) or transmission line model. Once the phase velocity is known, the mass sensitivity curve can be obtained using Eq. (23). In Figure 8, the dependence of the sensitivity with the guiding layer thickness for an AT-cut quartz/SiO$_2$ Z' propagating structure is depicted. A maximum mass sensitivity S_σ^v of -39.44 m^2/kg is observed at d/λ of 0.171 which corresponds to a d = 6.84 μm. The phase velocity at d/λ = 0.171 is v_φ = 4140 m/s (see Figure 7) which leads to a synchronous frequency $f_s = v_\varphi/\lambda$ = 103.5 MHz.

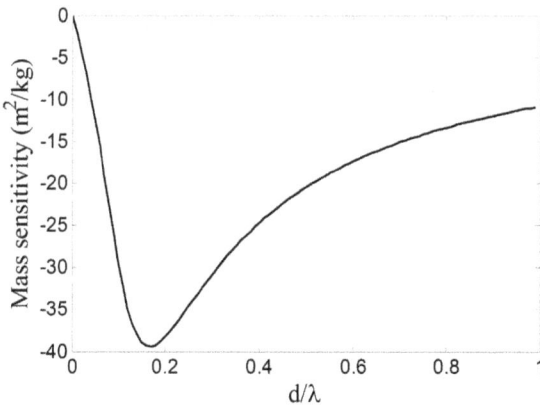

Figure 8. a) AT-cut quartz Z' propagating /SiO$_2$ mass sensitivity considering λ = 40 μm.

Thus, as mentioned previously, for very small thicknesses compared to the wavelength, the acoustic field is not confined to the surface and deeply penetrates into the piezoelectric substrate, resulting in low mass sensitivity [68]. If the thickness is increased the sensitivity rises, as the acoustic energy is more efficiently trapped in the guiding layer. However, an excessive increase in the guiding layer thickness leads to a reduction of wave energy density and the mass sensitivity decreases [69]. Therefore, there is an optimal guiding layer thickness at which a maximum mass sensitivity is achieved for a specific wavelength.

In those applications where a coating layer in contact with a liquid is deposited over the sensing area, such is the case for biosensors, both changes in the surface mass density $\Delta\sigma$ and in the mass viscosity $\Delta(\rho_C\eta_C)^{1/2}$ of the coating occur, due to the biochemical interaction between the coating and the liquid medium. In this case, the sensitivity can be modeled by the four components matrix shown in Eq.(25) [15]. These components relate shifts in surface mass density and in mass viscosity to the measured electrical signals: *phase shift φ* and *Insertion Loss (IL)*. The matrix components relate shifts of surface density $\Delta\sigma$ and mass viscosity $\Delta(\rho\eta)^{1/2}$ to shifts of electrical phase $\Delta\varphi$ and signal attenuation ΔIL. Notice that $S_{\varphi,\sigma}$ is not the same than S^φ_σ in Eq.(22).

$$\begin{bmatrix} \Delta\varphi \\ \Delta IL \end{bmatrix} = \begin{pmatrix} S_{\varphi,\sigma} & S_{\varphi,\sqrt{\rho_C\eta_C}} \\ S_{IL,\sigma} & S_{IL,\sqrt{\rho_C\eta_C}} \end{pmatrix} \begin{bmatrix} \Delta\sigma \\ \Delta\sqrt{\rho_C\eta_C} \end{bmatrix} \tag{25}$$

Generally, the high sensitivity of microacoustic sensors is closely related to the fact that they show a high temperature stability (low TCF) and a large signal-to-noise ratio, which, in turn yields low detection limits and a high resolution of the sensor [13]. The *limit of detection (LOD)* is a very important characteristic of acoustic biosensors, since it gives the minimum surface mass that can be detected by the device. It can be directly derived from the ratio between the noise in the measured electrical signal N_f and the sensitivity of the device. For instance, in a closed loop configuration, this noise N_f is the RMS value of the frequency measured over a given period of time in stable and constant conditions [32]. It is usually recommended to measure a signal variation higher than 3 times the noise level in order to conclude from an effective variation [70]. From this recommendation, it comes out that the LOD is given by [32,71]:

$$LOD = \sigma_{min} = \frac{3 \cdot N_f}{S^f_\sigma \cdot f} \tag{26}$$

where f is the operation frequency.

The LOD is improved by minimizing the influence of temperature on the sensor response [13]. The stability with respect to temperature can be achieved by implementing temperature control in the biosensor system and choosing materials of low TCF, as seen in Section 2.1.

6. LW packaging and flow cells

Flow cells are usually constructed with an inlet at one side, an outlet at the other, and the channels facing the sensor (see Figure 9). In most cases, a rubber seal is used for sealing, and in LW flow-cells additional absorbers made with rubber materials at the ends of the IDTs are recommended to improve the signal response (see Figure 9b). Factors such as the flow cell design, flow patterns and flow-through system influence the binding efficiency and the course of binding kinetics, resulting in possible variations of the true results [72].

The permittivity and the dielectric losses of the liquid medium, necessary in LW biosensors applications, influence the propagation of Love modes, since this medium acts as an additional layer. When the liquid medium is deposited over the device surface, the presence of this medium over the IDTs modifies the electrodes transfer function [73]. Permittivity and dielectric losses of the liquid lead to a corresponding change of the IDT input admittance, which influences the amplitude of the signal. A flow–through cell is crucial to eliminate this electric influence of the liquid. Such flow-cell isolates the IDTs from the liquid, confining the liquid in the region between the IDTs (sensing area). LW device packaging or flow-cells generally use walls to accomplish this purpose. These walls, when pressed onto the device surface, disturb the acoustic wave, resulting in an increase in overall loss and distortion of the sensor response. Walls must be designed to minimize the contact area in the acoustic path in order to obtain the minimum acoustic attenuation (see Figure 9c). It is known that the acoustic wave is significantly affected when increasing the walls width [27] and that materials used for these walls play an important role. Hence, great care must be taken to ensure that the designed LW flow cells do not greatly perturb the acoustic signal. Recently, some researchers have explored different possibilities to achieve the packaging of LW sensors for fluidic applications [27,74] and other authors have being exploring different LW flow-cell approaches [53,75,76].

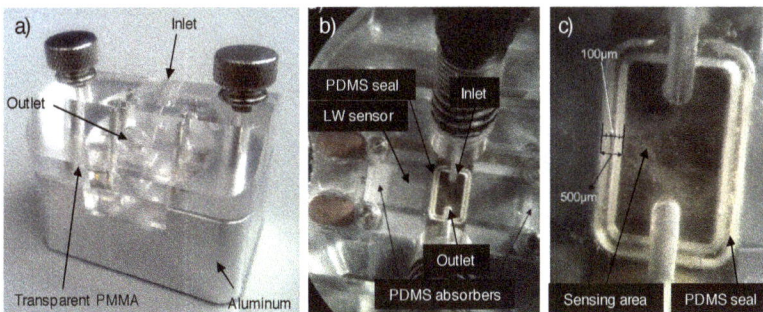

Figure 9. A developed LW flow cell for immunosensors application designed and fabricated by the authors of this chapter [53]. a) Overview of the flow cell b) Microscope view through the PMMA of the LW sensor and flow cell elements. c) PDMS square seal with pick end to minimize the contact area.

Further improvements for LW biosensors flow cells can be achieved, since it is an entirely new field and the development trends moves towards smaller flow cells which allows the use of less analyte. For instance, investigation on the materials used for creating flow cells, packaging, and flow patterns in microfluidic channels [72] would enhance LW performance and move them faster to the lab-on-a-chip arena.

7. LW biosensors state-of-the-art

The first approaches employing LW for biochemical sensing were reported in 1992 by Kovacs et al. [77] and by Gizeli et al. [78], who first demonstrated the use of such devices as mass sensing biosensors in liquids. However, it was not until 1997 that LW acoustic devices were used to detect real-time antigen-antibody interactions in liquid media [46].

In 1999, a contactless LW device was built in order to protect electrodes from the conductive and chemically aggressive liquids used in biosensing [79]. The advantage of this technique is that no bonding wires are required.

In 2000, a dual channel LW device was used as a biosensor to simultaneously detect *Legionella* and *E. coli* by Howe and Harding [80]. In this approach a novel protocol for coating bacteria on the sensor surface prior to addition of the antibody was introduced. Quantitative results were obtained for both species down to 10^6 cells/mL, within 3 h.

In 2003, a LW immunosensor was designed as a model for virus or bacteria detection in liquids (drinking or bathing water, food, etc.) by Tamarin et al. [47]. They grafted a monoclonal antibody (AM13 MAb) against M13 bacteriophage on the device surface (SiO_2) and sensed the M13 bacteriophage/AM13 immunoreaction. The authors suggested the potentialities of such acoustic biosensors for biological detection. The same year, it was shown that mass sensitivity of LW devices with ZnO layer was larger than that of sensors with SiO_2 guiding layers [48]. The authors of this work monitored adsorption of rat immunoglobulin G, obtaining mass sensitivities as high as 950 cm^2/g. They pointed out that such a device was a promising candidate for immunosensing applications.

An aptamer-based LW sensor which allowed the detection of small molecules was developed by Schlensog et al. in 2004 [81]. This biosensor offers an advantage over immunosensors, since it does not require the production of antibodies against toxic substances. A LW biosensor for the detection of pathogenic spores at or below inhalational infectious levels was reported by Branch et al. in 2004 [20]. A monoclonal antibody with a high degree of selectivity for anthrax spores was used to capture the non-pathogenic simulant *Bacillus thuringiensis* B8 spores in aqueous conditions. The authors stated that acoustic LW biosensors will have widespread application for whole-cell pathogen detection.

Moll et al. developed an innovative method for the detection of *E. coli* employing an LW device in 2007 [49]; it consisted of grafting goat anti-mouse antibodies (GAM) onto the sensor surface and introducing *E. coli* bacteria mixed with anti-*E. coli* MAb in a second step. The sensor response time was shorter when working at 37°C, providing results in less than 1 hour with a detection threshold of 10^6 bacteria/mL. More recently, the same group described a multipurpose LW immunosensor for the detection of bacteria, virus and proteins [82]. They successfully detected bacteriophages and proteins down to 4 ng/mm^2 and *E.coli* bacteria up to 5.0×10^5 cells in a 500 µL chamber, with good specificity and reproducibility. The authors stated that whole bacteria can be detected in less than one hour.

Andrä et al. used a LW sensor to investigate the mode of action and the lipid specificity of human antimicrobial peptides [83]. They analyzed the interaction of those peptides with model membranes. These membranes, when attached to the sensor surface, mimic the cytoplasmic and the outer bacterial membrane. A LW immunosensor was used in 2008 by Bisoffi et al. [84] to detect Coxsackie virus B4 and Sin Nombre Virus (SNV), a member of the hantavirus family. They described a robust biosensor that combines the sensitivity of SAW at a frequency of 325 MHz with the specificity provided by monoclonal and recombinant antibodies for the detection of viral agents. Rapid detection (within seconds) for increasing virus concentrations was reported. The biosensor was able to detect SNV at doses lower than the load of virus typically found in a human patient suffering from hantavirus cardiopulmonary syndrome.

In 2009, it was shown the possibility to graft streptavidin-gold molecules onto a LW sensor surface in a controlled way and was demonstrated the capability of the sensor to detect nano-particles in aqueous media by Fissi et al. [85]. In 2010, a complementary metal–oxide semiconductor CMOS-LW biosensor for breast cancer biomarker detection was presented by Tigli et al. [36]. This biosensor was fabricated using CMOS technology and used gold as guiding layer and as interface material between the biological sensing medium and the transducer.

LW devices were used as sensors for okadaic acid immono-detection through immobilized specific antibodies by Fournel et al. [76]. They obtained three times higher frequency shifts with the okadaic acid than with an irrelevant peptide control line. A LW based bacterial biosensor for the detection of heavy metal in liquid media was reported in 2011 by Gammoudi et al. [86]. Whole bacteria (*E. coli*) were fixed as bioreceptors onto the acoustic path of the sensor coated with a polyelectrolyte multilayer using a layer by layer electrostatic self-assembly procedure. Changes of bacteria viscoelastic equivalent parameters in presence of toxic heavy metals were monitorized.

A LW-based wireless biosensor for the simultaneous detection of Anti- Dinitrophenyl-KLH (anti-DNP) immunoglobulin G (IgG) was presented by Song et al. in 2011 [87]. They used poly(methyl-methacrylate) (PMMA) guiding layer and two sensitive films (Cr/Au). A LW sensor whose phase shifts as a function of the immobilized antibody quantity, combined with an active acoustic mixing device, was proposed by Kardous et al. [88] in 2011. They assessed that mixing at the droplet level increases antibodies transfer to a sensing area surface and increases the reaction kinetics by removing the dependency with the protein diffusion coefficient in a liquid, while inducing minimum disturbance to the sensing capability of the Love mode. LW sensors have been also used to study the properties of protein layers [40], DNA [89,90] and detect the adsorption and desorption of a lipid layer [91].

Currently, the only commercial LW biosensor system available in the market is commercialized by the German company *SAW instruments GmbH*. The sensor system can achieve a limit of detection (LOF) of 0.05 ng/cm^2 with a sample volume of 40-80 μL. *Senseor* company (Mougins, France) has a commercially available microbalance development kit (SAW-MDK1) which consists of a two-channel LW delay lines.

8. Trends and future challenges of LW sensors

LW biosensors have not been very well recognized by the scientific community [72] nor by the market yet. This might be due to the technological hindrances found for applying this device as biosensor, since it is sensitive to changes in the viscoelastic properties of the coating, which complicates the results interpretation. Reports about applications where mass alterations are separated from viscoelastic effects can enhance the acceptance of LW sensors [72]. Hence, it is necessary to investigate different alternatives for carrying out the procedure of parameters extraction with LW sensors. Nowadays, the trend is the placement of multiple, small, versatile sensors into a network configured for a specific location [10]. LW devices will move into the lab-on-a chip arena during the next years. Nevertheless, LW devices still have some hurdles to clear. LW biosensors packaging needs further development and cost reduction. In addition, much research and efforts are still required addressing the fluidic technology issue. Integration and automation with electronics and flow cells reduce costs of the system and increase the throughput. For a better performance of LW sensors, the combination with other detection methods such as optical [32] or chromatographic [92] are being considered.

Mathematical modeling and simulations of these devices are also essential for the development of new sensors, especially with respect to the study of new materials and wave propagation [72]. Numerical calculations and FEM analysis of LW sensors could help for further understanding of these devices.

Author details

María Isabel Rocha Gaso[1,2], Yolanda Jiménez[1], Laurent A. Francis[2] and Antonio Arnau[1]

*Address all correspondence to: marocga@doctor.upv.es

1 Wave Phenomena Group, Department of Electroic Engineering, Universitat Politècnica de València, Spain

2 Sensors, Microsystems and Actuators Laboratory of Louvain (SMALL), ICTEAM Institute, Université Catholique de Louvain, Belgium

References

[1]   March C, Manclús JJ, Jiménez Y, Arnau A, Montoya A. A piezoelectric immunosensor for the determination of pesticide residues and metabolites in fruit juices. Talanta 2009;78(3) 827-833.

[2]   Lucklum R, Soares D, Kanazawa K. Models for resonant sensors. In: Arnau A. (ed.) Piezoelectric transducers and applications. Springer; 2008. p63-96.

[3]   Janshoff A, Galla HJ, Steinem C. Piezoelectric mass-sensing devices as biosensors- an alternative to optical biosensors? Angew. Chem. Int. Ed. 2000;39 4005-4032.

[4]   Andle JC and Vetelino JF. Acoustic wave biosensors. Sens. Actuators, A 1994;44 167-176.

[5]   Länge K, Rapp BE, Rapp M. Surface acoustic wave biosensors: a review. Anal. Bioanal. Chem. 2008;391(5) 1509-1519.

[6]   Smith JP and Hinson-Smith V. Commercial SAW sensors move beyond military and security applications. Anal. Chem. 2006; 3505-3507.

[7]   Weber J, Link M, Primig R, Pitzer D, Wersing W, Schreiter M. Investigation of the scaling rules determining the performance of film bulk acoustic resonators operating as mass sensors. IEEE Trans. Ultrason. Ferroelectr. Freq. Cont. 2007;54(2) 405-412.

[8]   Oliner AA. Microwave network methods for guided elastic waves. IEEE Trans. Microwave Theory Tech. 1969;17(11) 812-826.

[9]   Voinova MV. On Mass Loading and Dissipation Measured with Acoustic Wave Sensors: A Review. Journal of Sensors 2009; 1-13.

[10] Rocha-Gaso M-I, March-Iborra C, Montoya-Baides A, Arnau-Vives A. Surface generated acoustic wave biosensors for the detection of pathogens: a review. Sensors 2009;9 5740-5769.

[11] Ballato A. Piezoelectricity: old effect, new thrusts. IEEE Trans. Ultrason. Ferroelectr. Freq. Cont. 1995;42(5) 916-926.

[12] Francis LA. Investigation of Love waves sensors. Optimisation for biosensing applications. Master Thesis. Université Catholique de Louvain; 2001.

[13] Jakoby B, Bastemeijer J, Vellekoop MJ. Temperature-compensated Love-wave sensors on quartz substrates. Sens. Actuators, A 2000;82((1-3)) 83-88.

[14] Herrmann F and Büttgenbach S. Temperature-compensated Shear Horizontal surface acoustic wave in layered quartz/SiO_2- structures. Physica status solidi 1998;170 R3-R4.

[15] Du J and Harding GL. A multilayer structure for Love-mode acoustic sensors. Sens. Actuators, A 1998;65 152-159.

[16] Tamarin O, Déjous C, Rebiere D, Pistré J, Comeau S, Moynet D, Bezian J. Study of acoustic Love wave devices for real time bacteriophage detection. Sens. Actuators, B 2003;91 275-284.

[17] Herrmann F, Weihnacht M, Buttgenbach S. Properties of sensors based on shear-horizontal surface acoustic waves in LiTaO3/SiO2 and quartz/SiO2 structures. IEEE Trans. Ultrason. Ferroelectr. Freq. Cont. Jan.2001;48(1) 268-273.

[18] Kalantar-Zadeh K, Powell DA, Sadek AZ, Wlodarski W, Yang QB, Li YX. Comparison of ZnO/64° LiNbO3 and ZnO/36° LiTaO3 surface acoustic wave devices for sensing applications. Sens. Lett. 2006;4(2) 135-138.

[19] Hickernell FS, Knuth HD, Dablemont RC, Hickernell TS. The surface acoustic wave propagation characteristics of 64° YX LiNbO$_3$ and 36° YX LiTaO$_3$ substrates with thin-film SiO$_2$. In: proceedings of the IEEE Ultrason.Symp., Seattle, USA. 1995.

[20] Branch DW and Brozik SM. Low-level detection of a Bacillus anthracis simulant using Love-wave biosensors on 36° YX LiTaO$_3$. Biosens. Bioelectron. 2004;19 849-859.

[21] Branch DW and Thayne LE. 4D- 4 Love wave acoustic array biosensor platform for autonomous detection. In: proceedings of the IEEE Ultrason.Symp., 2007.

[22] Gizeli E, Bender F, Rasmusson A, Saha K, Josse F, Cernosek R. Sensitivity of the acoustic waveguide biosensor to protein binding as a funcion of the waveguide properties. Biosens. Bioelectron. 2003;18 1399-1406.

[23] Powell DA, Kalantar-Zadeh K, Wlodarski W. Numerical calculation of SAW sensitivity: application to ZnO/LiTaO$_3$ transducers. Sens. Actuators, A 2004;115 456-461.

[24] White RM and Voltmer FW. Direct piezoelectric coupling to surface elastic waves. Appl. Phys. Lett. 1965;7(12) 314-316.

[25] Nieuwenhuizen MS and Venema A. Surface acoustic wave chemical sensors. Sens. Mater. 1989;5 261-300.

[26] Morgan DP. Surface-Wave Devices for Signal Processing. Elsevier Science Publishers B.V.; 1991.

[27] El Fissi L. Détection et mesure de nanoparticules pour les applications de capteurs en milieu liquide. PhD Thesis. Université de Franche-Comté; 2009.

[28] Hashimoto K-Y. Surface Acoustic Wave Devices in Telecommunications. Springer-Verlag; 2000.

[29] Campbell C. Surface acoustic wave devices and their signal processing applications. Academic Press; 1989.

[30] Mazein P. Étude de dispositifs à ondes de Love par modélisation numérique de la propagation d'ondes acoustiques. Application à l'optimisation de structures et à la

caractérisation de matériaux en vue de la réalisation de capteurs chimiques. PhD Thesis. L'Université Bordeaux I; 2005.

[31] Du J, Harding GL, Ogilvy JA, Dencher PR, Lake M. A study of Love-wave acoustic sensors. Sens. Actuators, A 1996;56 211-219.

[32] Francis LA. PhD Thesis. Thin film acoustic waveguides and resonators for gravimetric sensing applications in liquid. PhD Thesis. Université Catholique de Louvain; 2006.

[33] McHale G, Newton MI, Martin F. Theoretical mass,liquid,and polymer sensitivity of acoustic wave sensorswith viscoelastic guiding layers. Appl. Phys. Lett. 2003;93(1) 675-690.

[34] Barié N, Wessa T, Bruns M, Rapp M. Love waves in SiO_2 layers on STW-resonators based on $LiTaO_3$. Talanta 2004;62 71-79.

[35] Gizeli E, Stevenson AC, Goddard NJ, Lowe CR. A novel Love-plate acoustic sensor utilizing polymer overlayers. IEEE Trans. Ultrason. Ferroelectr. Freq. Cont. 1992;39(5) 657-659.

[36] Tigli O, Binova L, Berg P, Zaghloul M. Fabrication and characterization of a Surface-Acoustic-Wave biosensor in CMOS Technology for cancer biomarker detection. IEEE Trans. Biomedical. Circuit. Systems. 2010;4(1) 62-73.

[37] Powell DA, Kalantar-Zadeh K, Ippolito S, Wlodarski W. 3E- 2 A layered SAW device based on $ZnO/LiTaO_3$ for liquid media sensing applications, 1 ed 2002, pp. 493-496.

[38] Kalantar-Zadeh K, Trinchi A, Wlodarski W, Holland A. A novel Love-mode device based on a ZnO/ST-cut quartz crystal structure for sensing applications. Sens. Actuators, A 2002;100 135-143.

[39] Matatagui D, Fontecha J, Fernández MJ, Aleixandre M, Gracia I, Cané C, Horrillo MC. Array of Love-wave sensors based on quartz/Novolac to detect CWA simulants. Talanta 2011;85 1442-1447.

[40] Saha K, Bender F, Rasmusson A, Gizeli E. Probing the viscoelasticity and mass of a surface-bound protein layer with an Acoustic Waveguide Device. Langmuir 2003;19 1304-1311.

[41] Herrmann F, Hahn D, Büttgenbach S. Separate determination of liquid density and viscosity with sagitally corrugated Love mode sensors. Sens. Actuators, A 1999;78 99-107.

[42] Franssila S. Introduction to Microbabrication. Wiley; 2004.

[43] Gizeli E, Liley M, Lowe CL. Design considerations for the acoustic waveguide biosensor. Smart Mater. Struct. 1997;6 700-706.

[44] El Fissi L, Friedt J-M, Ballandras S, Robert L, Chérioux F. Acoustic characterization of thin polymer layers for Love mode surface acoustic waveguide. In: proceedings of the IEEE Int.Freq.Control Symp., 2008.

[45] Grate JW, Stephen JM, Richard MW. Acoustic wave microsensors. Anal. Chem. 1993;65(21) 940A-848A.

[46] Harding GL, Du J, Dencher PR, Barnett D, Howe E. Love wave acoustic immunosensor operation in liquid. Sens. Actuators, A 1997;61(1-3) 279-286.

[47] Tamarin O, Comeau S, Déjous C, Moynet D, Rebière D, Bezian J, Pistré J. Real time device for biosensing: design of a bacteriophage model using love acoustic wave. Biosens. Bioelectron. 2003;18 755-763.

[48] Kalantar-Zadeh K, Wlodarski W, Chen YY, Fry BN, Galatsis K. Novel Love mode surface acoustic wave based immunosensors. Sens. Actuators, B 2003;91 143-147.

[49] Moll N, Pascal E, Dinh DH, Pillot JP, Bennetau B, Rebiere D, Moynet D, Mas Y, Mossalayi D, Pistre J, Dejous C. A Love wave immunosensor for whole E. coli bacteria detection using an innovative two-step immobilisation approach. Biosens. Bioelectron. Apr.2007;22(9-10) 2145-2150.

[50] Zimmermann B, Lucklum R, Hauptmann P, Rabe J, Büttgenbach S. Electrical characterisation of high-frequency thickness-shear-mode resonators by impedance analysis. Sens. Actuators, B 2001;76 47-57.

[51] Montagut Y, Garcia JV, Jimenez Y, March C, Montoya A, Arnau A. Validation of a phase-mass characterization concept and interface for acoustic biosensors. Sensors (Basel) 2011;11(5) 4702-4720.

[52] Montagut YJ, Garcia JV, Jimenez Y, March C, Montoya A, Arnau A. Frequency-shift vs phase-shift characterization of in-liquid quartz crystal microbalance applications. Rev. Sci. Instrum. 2011;82(6) 064702.

[53] Rocha-Gaso MI, March C, García J, El Fissi L, Francis LA, Jiménez Y, Montoya A, Arnau A. User-friendly love wave flow cell for biosensors. In: proceedings of the Biosensors 2012, Cancun, Mexico. 2012.

[54] Abdollahi A, Jiang A, Arabshahi SA. Evaluation on mass sensitivity of SAW sensors for different piezoelectric materials using finite-element analyisis. IEEE Trans. Ultrason. Ferroelectr. Freq. Contr. 2007;54(12) 2446-2455.

[55] Auld BA. Acoustic fields and waves in solids. Krieger; 1990.

[56] McHale G, Newton MI, Matin F. Layer guided shear horizontally polarized acoustic plate modes. Appl. Phys. Lett. 2002;91 5735.

[57] Wang Z, Cheeke JDN, Jen CK. Sensitivity analysis for Love mode acoustic gravimetric sensors. Appl. Phys. Lett. 1994;64(22) 2940-2942.

[58] Liu J and He S. Theoretical analysis on Love waves in layered structure with a piezo-electric substrate and multiple elastic layers. J. Appl. Phys. 2010;107 073511.

[59] Sankaranarayanan SKRS, Bhethanabotla VR, Joseph B. Modeling of Surface Acoustic Wave Sensor Response. In: Ram MK, Bhethanabotla VR. (ed.) Sensors for Chemical and Biological Applications. CRC Press; 2010. p97-134.

[60] Laude V, Reinhardt A, Ballandras S, Khelif A. Fast FEM/BEM computation of SAW harmonic admittance and slowness curves. In: proceedings of the IEEE Ultra-son.Symp., 2004.

[61] Plessky VP and Thorvaldsson T. Rayleigh waves and leaky SAW's in periodic systems of electrodes: Periodic Green functions analysis. Electronics Letters 1992;28 1317-1319.

[62] Atashbar MZ, Bazuin BJ, Simpeh M, Krishnamurthy S. 3D FE simulation of H_2 SAW gas sensor. Sens. Actuators, B 2005;111-112 213-218.

[63] Xu G. Finite element analysis of second order effects on the frequency response of a SAW device. In: proceedings of the IEEE Ultrason.Symp., 2000.

[64] Ippolito SJ, Kalantar-Zadeh K, Powell DA, Wlodarski W. A 3-dimensional finite element approach for simulating acoustic wave propagation in layered SAW devices. In: proceedings of the IEEE Ultrason.Symp., 2003.

[65] Rocha-Gaso M-I, Fernandez-Díaz R, March-Iborra C, Arnau-Vives A. Mass sensitivity evaluation of a Love wave sensor using the 3D Finite Element Method. In: proceedings of the IEEE Int.Freq.Control Symp., 2010.

[66] Jakoby B and Vellekoop MJ. Properties of Love waves: applications in sensors. Smart Mater. Struct. 1997;6 668-679.

[67] Ferrari V and Lucklum R. Overview of Acoustic-Wave Microsensors. In: Arnau A. (ed.) Piezoelectric transducers and applications. Springer; 2008. p39-59.

[68] Lee HJ, Namkoong K, Cho EC, Ko C, Park JC, Lee SS. Surface acoustic wave immunosensor for real-time detection of hepatitis B surface antibodies in whole blood samples. Biosens. Bioelectron. June2009;24(10) 3120-3125.

[69] Kovacs G and Venema A. Theoretical comparison of sensitivities of acoustic shear wave modes for (bio) chemical sensing in liquids. Appl. Phys. Lett. 1992;61(6) 639-641.

[70] MacDougall D, Amore FJ, Cox GV, Crosby DG, Estes FL, et.al. Guidelines for data acquisition and data quality evaluation in environmental chemistry. Anal. Chem. 1980;52(14) 2242-2249.

[71] Du J, Harding DR, Collings AF, Dencher PR. An experimental study of Love-wave acoustic sensors operating in liquids. Sens. Actuators, A 1997;60 54-61.

[72] Gronewold TM. Surface acoustic wave sensors in the bioanalytical field: recent trends and challenges. Anal. Chim. Acta Nov.2007;603(2) 119-128.

[73] Jakoby B and Vellekoop MJ. Viscosity sensing using a Love-wave device. Sens. Actuators, A 1998;68 275-281.

[74] Francis LA, Friedt J-M, Bartic C, Campitelli A. A SU-8 liquid cell for surface acoustic wave biosensors. In: proceedings of the SPIE - The International Society for Optical Engineering, 2004.

[75] Tarbague H, Lachaud L, Vellutini L, Pillot JP, Bennetaur B, Moynet D, Rebière D. PDMS microfluidic chips combined to saw biosensors for ultra-fast biodetection of antibodies and E. coli bacteria. In: proceedings of the Biosensors 2012, Cancun, Mexico. 2012.

[76] Fournel F, Baco E, Mamani-Matsuda M, Degueil M, Bennetau B, Moynet D, Mossalayi D, Vellutini L, Pillot J-P, Dejous C, Rebiere D. Love wave biosensor for real-time detection of okadaic acid as DSP phycotoxin. In: proceedings of the Eurosensors XXIV, Linz, Austria. 2010.

[77] Kovacs G, Lubking GW, Vellekoop MJ, Venema A. Love waves for (bio)chemical sensing in liquids. In: proceedings of the IEEE Ultrason.Symp., Tucson, USA. 1992.

[78] Gizeli E, Goddard NJ, Lowe CR, Stevenson AC. A Love plate biosensor utilising a polymer layer. Sens. Actuators, B 1992;6 131-137.

[79] Freudenberg J, Schelle S, Beck K, von Schickfus M., Hunklinger S. A contactless surface acoustic wave biosensor. Biosens. Bioelectron. 1999;14 423-425.

[80] Howe E and Harding G. A comparison of protocols for the optimisation of detection of bacteria using a surface acoustic wave (SAW) biosensor. Biosens. Bioelectron. 2000;15(11-12) 641-649.

[81] Schlensog MD, Thomas MA, Gronewold TM, Tewes M, Famulok M, Quandt E. A Love-wave biosensor using nucleic acids as ligands. Sens. Actuators, B 2004;101 308-315.

[82] Moll N, Pascal E, Dinh DH, Lachaud J-L, Vellutini L, Pillot J-P, Rebière D, Moynet D, Pistré J, Mossalayi D, Mas Y, Bennetau B, Déjous C. Multipurpose Love acoustic wave immunosensor for bacteria, virus or proteins detection. ITBM-RBM 2008;29 155-161.

[83] Andrä J, Böhling A, Gronewold TMA, Schlecht U, Perpeet M, Gutsmann T. Surface acoustic wave biosensor as a tool to study the interactions of antimicrobial peptides with phospholipid and lipopolysaccharide model membranes. Langmuir 2008;24(16) 9148-9153.

[84] Bisoffi M, Hjelle B, Brown DC, Branch DW, Edwards TL, Brozik SM, Bondu-Hawkins VS, Larson RS. Detection of viral bioagents using a shear horizontal surface acoustic wave biosensor. Biosens. Bioelectron. Apr.2008;23(9) 1397-1403.

[85] El Fissi L, Friedt J-M, Luzet V, Chérioux F, Martin G, Ballandras S. A Love-wave sensor for direct detection of biofunctionalized nanoparticles. IEEE 2009; 861-865.

[86] Gammoudi I, Tarbague H, Lachaud JL, Destor S, Othmane A, Moyenet D, Kalfat R, Rebiere D, Dejous C. Love Wave Bacterial Biosensors and Microfluidic Network for Detection of Heavy Metal Toxicity. Sensor Letters 2011;9(2) 816-819.

[87] Song T, Song SY, Yoon HC, Lee K. Development of a wireless Love wave biosensor platform for multi-functional detection. Japanese J. of Appl. Phys. 2011;50 1-6.

[88] Kardous F, El Fissi L, Friedt J-M, Bastien F, Boireau W, Yahiaoui R, Manceau JF, Ballandras S. Integrated active mixing and biosensing using low frequency vibration mixer and Love-wave sensor for real time detection of antibody binding event. J. Appl. Phys. 2011;109 1-8.

[89] Tsortos A, Papadakis G, Mitsakakis K, Melzak KA, Gizeli E. Quantitative determination of size and shape of surface-bound DNA using an acoustic wave sensor. Biophys. J. 2008;94 2706-2715.

[90] Papadakis G, Tsortos A, Gizeli E. Triple-helix DNA structural studies using a Love wave acoustic biosensor. Biosens. Bioelectron. 2009;25 702-707.

[91] Gizeli E, Lowe CL, Liley M, Vogel H. Detection of supported lipid layers with the acoustic Love waveguidedevice: application to biosensors. Sens. Actuators, B 1996;34 295-300.

[92] Marth M, Maier D, Stahl U, Rapp M, Wessa T, Honerkamp J. Optimization of surface acoustic wave sensor arrays and application to high performance liquid chromatography. Sens. Actuators, B 1999;61 191-198.

Chapter 13

Calibrating Biosensors in Flow-Through Set-Ups: Studies with Glucose Optrodes

K. Kivirand, M. Kagan and T. Rinken

Additional information is available at the end of the chapter

1. Introduction

The research and development of selective analytical systems, not requiring pre-treatment of samples and rapidly providing information in real time, is receiving increasing attention in various areas of human life (Chen *et al*, 2011;Palmisano *et al*, 2000;Maestre *et al*, 2005;Jia *et al*, 2004;Surareungchai *et al*, 1999;Tsai & Doong, 2005;Gülce *et al*, 2002;Akin *et al*, 2011;Mishra *et al*, 2012). For example, as environmental pollution poses a serious threat to the human health, there is an urgent demand to monitor pollutants and initiate appropriate environmental pollution treatment in the real time course. In the field of food quality control, the product quality and healthiness are the main factors, influencing customer satisfaction. The number of food processing and manufacturing mistakes can be minimized with risk assessment and continuous checking of the production process, e.g. in dairy farms, it is necessary to control the quality of raw milk on site in order to detect the presence of the residues of different antibiotics and other potentially harmful compounds before loading milk into the dairy production process.

On-site monitoring requires enhanced sensitivity, selectivity, rapidity, and ease of operation of the analytical equipment, which should provide reliable continuous information in real-time and demonstrate sufficient stability of action. Biosensors fulfil all the above-mentioned requirements and have already been applied in clinical diagnostics, food quality control, forensic chemistry, environmental monitoring and other areas (Castillo *et al*, 2004;Reder-Christ & Bendas, 2011;Kivirand & Rinken, 2011).

According to the IUPAC definition, a biosensor is a self-contained, integrated receptor-transducer device, which is capable of providing selective quantitative or semi-quantitative analytical information and which uses a biological recognition element (bio-receptor) and a transducer in direct special contact (Thevenot *et al*, 2001). A biosensor consists of three parts:

(1) the sensitive biological element (such as tissues, microorganisms, cell receptors, enzymes, antibodies, nucleic acids, etc.); (2) the transducer or the detector element (physiochemical, optical, piezoelectric, etc.) that transforms the signal, resulting from the interaction between the analyte and the biological element into another signal that can be measured and quantified; and (3) associated electronics or signal processors that are primarily responsible for the display of the results in a user-friendly way. All these three parts are associated with an information management system. The principle of biosensors is shown in figure 1.

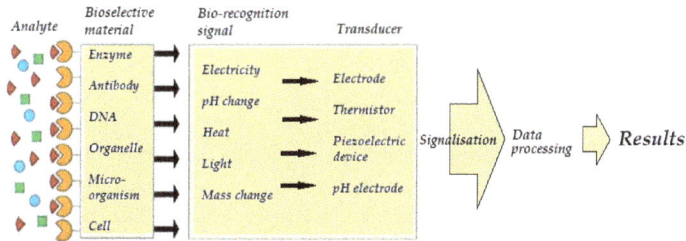

Figure 1. The principle of biosensor systems.

Enzyme-based biosensors technology relies upon the natural specificity of a given enzymatic protein to act selectively on a target analyte or group of analytes. Enzymes are catalysts bearing some excellent properties that may permit to perform the most complex chemical processes under the most benign experimental and environmental conditions. Enzyme-based biosensors have emerged as a valuable technique for qualitative and quantitative analysis of a variety of target analytes. Although biosensors based on other biorecognition elements are rapidly progressing, enzyme biosensors are still the ones most frequently used for practical applications and as model systems in scientific studies. There are several advantages of enzyme-based biosensors: a known reaction mechanism, a stable bio-renewable source of material and possibilities to modify the catalytic properties or substrate specificity by means of genetic engineering or to use catalytic amplification by the modulation of enzyme activity with respect to the target analyte (Castillo *et al*, 2004;Hu *et al*, 2011).

For the implementation at industrial scale, the properties of enzymes have to be improved further, as soluble enzymes should be immobilized for their multiple utilizations. The immobilization of enzymes and the choice of an insoluble carrier are important features in designing the biorecognition part of enzyme-based biosensors. Various immobilization strategies can be envisioned: adsorption, entrapment, covalent cross-linking or affinity (Sassolas *et al*, 2012;Cao, 2005;Mateo *et al*, 2007;Gibson, 1999). In some cases, enzyme immobilization protocols are also based on the combination of several immobilization methods: for example, an enzyme can be pre-immobilized on a carrier by adsorption, affinity or covalent bonding before further entrapment into a porous polymer. Biosensors based on immobilized enzymes have good operational and storage stability, high sensitivity and selectivity, good reproducibility and additionally, as enzyme immobilization reduces the time of enzymatic response, these biosensors can be easily used in continuous-flow and flow-through systems (Sassolas et al, 2012;Yang et al, 2010;Castillo et al, 2004;Cao, 2005).

The other moiety of a biosensor is the signal transduction system, which can be based on the measurement of electrochemical, magnetic, piezoelectric, thermometric or optical signals (Mehrvar & Abdi, 2004;Mello & Kubota, 2002;Sarma et al, 2009;Castillo et al, 2004). Among the above-mentioned systems, fibre-optic sensors are gradually achieving popularity. In comparison with electrochemical transducers, they do not consume any analytes and are insensible to electrical or magnetic interference (in fact, the oxygen detection capability has been demonstrated on single luminescent molecules) (D.R.Walt, 2006;Leung et al, 2007). Fluorescence measurements can be used whenever a fluorescent analyte is detected. Naturally fluorescent compounds are not common in biosensor development and the technique is usually applied in combination with artificially labelled compounds. Fluorescence is applied, for instance in biosensors with oxidase-type enzymes, which catalyze the consumption of oxygen, resulting the decrease in the luminescent signal of a fluorescent dye attached to the surface of an optical fibre. Besides a direct detection of the analyte of interest, the optical biosensor format may also involve indirect detection through optically labelled probes. Optical transducers may detect changes of absorbance, luminescence, polarization or refractive index and can be adapted for the assembly of different enzyme-substrate systems. For example, using a fibre-optic sensor assay which senses pH changes, Viveros et al. have demonstrated a rapid detection of organophosphates (insecticides and potent neurotoxins) (Viveros et al, 2006) and Bidmanova et al. developed an enzyme-based fibre-optic biosensor by co-immobilization of purified enzyme and a fluorescent pH indicator (Bidmanova et al, 2010). Polster et al. have immobilized enzymes onto an array of optical fibers for use in the simultaneous detection of penicillin and ampicillin. These biosensors employ an interferometric technique based on following the shifts in the reflectance spectrum, caused by the pH changes of the solution during the penicillinase - catalyzed hydrolysis of the analytes, penicillin and ampicillin (Polster et al, 1995).

The incorporation of an optical fibre into a biochemical sensor results in several advantages: (1) numerous optically based methods are available for chemical analysis, as almost every chemical analyte can be determined by measuring its spectroscopic properties; (2) fibres can be used to transmit light over long distances; (3) fibres have a multiplex capability (because they can guide light of different wavelengths at the same time and in different directions, multiple- or single-analyte monitoring in single locations can be performed with a single central unit); (4) fibres can be used in harsh environments and are immune to electric or magnetic interferences, and so can be safer than electrochemical biosensors; (5) fibres can be easily miniaturized at low cost (6) fibres can be made biocompatible and thus used for in-vivo measurements; (7) a light guide can carry more information than electric wire; and (8) the temperature-dependence of the fibre is lower than that of common electrodes (Marazuela & Moreno-Bondi, 2002).

Some drawbacks of optical-fibre sensors can limit their applicability: (1) interference of ambient light, although this can be avoided by use of suitable light isolation or modulated light sources; (2) background absorbance or fluorescence of the fibre itself; (3) long response times if mass transfer to the reagent phase is needed; and (4) limited availability of optimized commercial accessories for use with optical fibres (Marazuela & Moreno-Bondi, 2002).

Fibre-optic sensors are a perspective replacement of wide-spread Clark-type amperometric sensors for the detection of oxygen, although the application of oxygen measuring is hindered due to their sensitivity to oxygen fluctuations in natural samples (Li & Walt, 1995). To overcome the problem of the variability of oxygen concentrations, a two-sensor (or dual-sensor) approach with a reference sensor included in the system can be used. Pasic et al. have used a microdialysis-based glucose-sensing system based on a fibre-optic hybrid sensor (Pasic *et al*, 2006; Pasic *et al*, 2007). They used a reference oxygen optrode to detect and compensate response changes caused by side events, like bacterial growth, temperature fluctuations or failure of the peristaltic pump. The constructed sensor was evaluated in vitro using a 3-day continuous testing. As a result, they found that all glucose readings were clinically accurate and acceptable. With the purpose of analyzing complex biological samples without the need for any sample pre-treatment, Chen et al. developed a thermal based flow injection biosensor system, where a reference column was used to detect the non-specific thermal response (Chen *et al*, 2011). This sensor system was used to detect urea and lactate in non-standard milk products (such as lactose free milk). They found that when using that kind of biosensing systems, it was not necessary to remove the interfering compounds during milk analysis. The sensitivity and accuracy of the analysis were in the ranges required by the dairy industry.

One of the key problems of real-time measurements is the calibration of the measuring system and the management of data acquisition to obtain the results as quickly as possible. It is common in biosensor studies that the only information used, is the steady-state output (or the presumed 95% of it, the value of T_{95}) (Baker & Gough, 1996). Unfortunately, the determination of the value of T_{95} is often imprecise because of the difficulties in estimating the attainment of steady state – thus there is an urgent need for exact modelling of processes taking place in different set-ups of experimental measurements (Baker & Gough, 1996;Li & Walt, 1995; Lammertyn *et al*, 2006;Baronas *et al*, 2011). Measurements in continuous-flow systems require additional consideration of the flowing effects, both laminar and turbulent, in the biosensor output signal. The nature of the flow profile of the plug of solution introduced into the flowing stream of carrier solution is normally affected mainly by laminar flow. The resistance between solution and the wall of the tubing causes the solution to travel slower near the wall of the tubing. However, some reports demonstrate that laminar flow in the small tubing does not have much effect on distortion of the zone to the point that affects the accuracy of kinetic studies (Konermann, 1999;Zhou *et al*, 2003;Hartwell & Grudpan, 2012).

The utilization of biosensors in continuous-flow manifolds allows samples to be manipulated or modified as required for the execution of various operations such as separation, automatic dilution or pre-concentration, or some kind chemical or biochemical reaction prior to the final detection step (Hansen, 1996). It is also necessary to distinguish clearly between continuous-flow biosensing (referred to as "sensor system") systems and flow-through biosensors. The primary difference between these two systems lies in whether or not the detection of the analyte of interest is performed simultaneously with other analytical steps (chemical reaction, separation, or both) in the continuous system. Thus in a continuous-flow sensor system, the biochemical reaction takes place before the sample has reached the flow cell for detection, while the flow-through biosensors involve the development of either the overall process or only the last step of the biochemical reaction in the flow cell (Hansen, 1996).

The properties of a continuous-flow biosensor must meet the requirement of being able to follow the maximal anticipated concentration fluctuations within a specific acceptable error (Baker & Gough, 1996). In the flow systems the biosensor contacts the substrate for a short time only. When the analyte disappears, a buffer solution swills the enzyme surface, reducing the substrate concentration at this surface to zero. Because of (analyte) remaining in the enzyme membrane substrate, the mass diffusion as well as the reaction still continues for some time even after the disconnection of the biosensor and substrate (Baronas *et al*, 2002). Compared to a batch system, the flow system present the advantages of the reduction in analysis time allowing a high sample throughput and the possibility to work with small volumes of the substrate (Baronas *et al*, 2002;Baronas *et al*, 2011). The flow arrangement also presents a wide response range and high sensitivity. While modelling a biosensor in a flowing system, it is of crucial importance to take into consideration the external diffusion limitations, because of the mass transport outside the enzyme region (Baronas *et al*, 2011;Ivanauskas & Baronas, 2008;Baronas *et al*, 2002).

Recently, the interest of constructing the fibre-optic flow biosensing systems is growing and these are mostly made for glucose monitoring (Pasic *et al*, 2007; Pasic *et al*, 2006;Zhu *et al*, 2002;Akin *et al*, 2011). Usually glucose oxidase is selected as a model enzyme in biosensing systems because of its low cost, stability and high solubility in different medium. Glucose oxidase is an enzyme which catalyzes the oxidation of β-D-glucose by molecular oxygen to δ-gluconolactone, which subsequently hydrolyzes to gluconic acid and hydrogen peroxide (Bankar *et al*, 2009). The enzyme is also of considerable commercial importance, as it is used in the removal either of glucose or oxygen from food products and in the production of gluconic acid. The most important application of glucose oxidase is as a molecular diagnostic tool. The enzyme is used in biosensors for the quantitative determination of D-glucose in various samples of natural origin, such as body fluids, foodstuff, beverages, and fermentation products (Bankar *et al*, 2009).

The aim of the present research was to study the modelling of a biosensor response and to propose optional biosensor calibration parameters in a flowing medium. We used a glucose optrode, which was a dual-sensor system, enabling to eliminate the fluctuations in the initial dissolved oxygen concentration, temperature and fluidic flow. The system consisted of two oxygen optrodes, one covered with glucose oxidase-containing nylon thread and the other with a similar, but "blank" thread, which were placed into isolated parallel flow channels. Glucose biosensor was selected for these studies as the bioactive compound of this biosensor; enzyme glucose oxidase is a well-characterized robust enzyme of high stability and selectivity towards glucose. The effect of the speed of the flow on different calibration parameters, obtained from the transient phase of the biosensor signal, was studied.

2. Materials and experimental procedures

Glucose oxidase (GOD, EC 1.1.3.4. from *Aspergillus niger*, 17 300 U/g protein) was obtained from Sigma. All other reagents used in the study were of analytical grade. Glucose stock solutions were prepared in phosphate buffer (PB) (pH 6.50, $I = 0.1$ M) and allowed to mutarotate overnight at 37⁰C; glucose working solutions were prepared immediately before use.

2.1. Enzyme immobilization

Glucose oxidase was immobilized onto nylon-6,6 threads, used as a carrier, according to previously published protocol with some minor modifications. Nylon is a perfect carrier for enzyme immobilization, because it is inert, hydrophilic and mechanically strong. However, its inertness prevents enzyme binding without a specific treatment (Isgrove *et al*, 2001;Segura-Ceniceros *et al*, 2006;Sassolas *et al*, 2012;Nan *et al*, 2009). Hence, activation of nylon is essential in immobilizing of an enzyme. One of the possibilities to activate nylon surface is by *O*-alkylation with dimethyl sulfate (DMS). Thus, pieces of nylon thread with a length of 100 cm were immersed into 98% (w/w) dimethyl sulfate at 50⁰C for 10 min; washed thoroughly at first with ice-cold methanol and after that with 0.1 M PB (pH 6.50). Thereafter, the threads were immersed into 12.5% glutaraldehyde solution (in 0.1 M PB, pH 6.50) for 1 h at room temperature. Glutaraldehyde (GA) was used as a linker between the activated carrier and enzyme (Betancor *et al*, 2006;Pahujani *et al*, 2008). The threads were washed with 0.1 M PB (pH 6.50) and incubated overnight at 4⁰C in GOD solution (100 U/ml) (Scheme 1) (Kivirand & Rinken, 2009). Finally, the threads were thoroughly washed and stored in a 0.1 M PB (pH 6.50) at 4⁰C until further use.

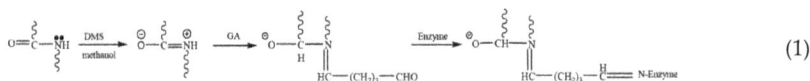

$$O=\overset{}{\underset{}{C}}-\overset{\bullet\bullet}{N}H \xrightarrow[\text{methanol}]{\text{DMS}} \overset{\ominus}{O}-\overset{}{\underset{}{C}}\overset{\oplus}{=}NH \xrightarrow{\text{GA}} \overset{\ominus}{O}-\overset{}{\underset{\underset{\underset{HC-(CH_2)_3-CHO}{H}}{\parallel}}{C}}-N \xrightarrow{\text{Enzyme}} \overset{\ominus}{O}-\overset{}{\underset{\underset{HC-(CH_2)_3-\overset{H}{\underset{}{C}}=N\text{-Enzyme}}{\parallel}}{CH}}-N \tag{1}$$

The GOD-containing threads kept at least 80% of their initial enzymatic activity for 35 days at 37⁰C. The thread's activity was controlled before each series of measurements and the sensor output corrected according to the actual activity of the enzyme.

2.2. Biosensor system

Glucose oxidase catalyzes the oxidation of β-D-glucose by dissolved oxygen causing a decrease of dissolved oxygen concentration in the reaction medium:

$$\beta-D-\text{glucose} + O_2 \xrightarrow{GOD} D-\text{gluconic acid} + H_2O_2 \tag{2}$$

The applied oxygen optrode was constructed in the Institute of Physics at the University of Tartu. This sensor was based on measuring the oxygen-induced phosphorescence quenching of Pd-tetraphenylporphyrin molecules, encapsulated into thermally aged polymethyl methacrylate (PMMA) film, covering the cylindrical surface of a 30 mm long PMMA optical fibre with the diameter of 1 mm (Õige *et al*, 2005;Jaaniso *et al*, 2005). The dissolved oxygen concentration was calculated automatically with the help of Stern-Volmer relationship, using original software Oxysens 2.0.

For the preparation of a glucose optrode, 18 cm of GOD-containing thread was cut and coiled around the oxygen-sensitive surface of the oxygen optrode (Fig.2) (Kivirand *et al*, 2011). If the activity of the thread dropped below 80% of its initial activity, the GOD-containing thread was replaced. The reference optrode was covered with 18 cm of „blank" thread.

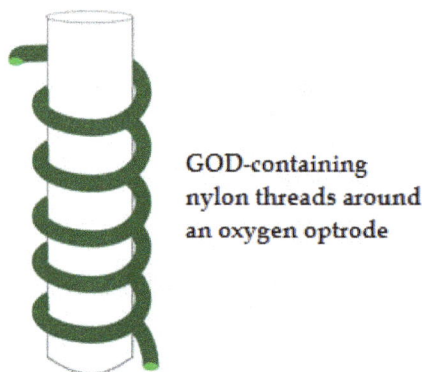

GOD-containing
nylon threads around
an oxygen optrode

Figure 2. Construction of the biosensor: glucose oxidase-containing thread coiled spirally around an oxygen optrode.

The glucose and the reference optrodes were placed into identical and parallel isolated channels of the measuring cell. A schematic cross-section of the cell system used in the studies is presented in Fig.3. The flow channels were 50 mm long and with the diameter of 3 mm. All measurements were carried out at flow speed varying from 0 to 5.1 cm/sec at 37 ± 0.02^0C. A peristaltic pump was used to deliver samples through the system and to wash the system; the temperature was stabilized by a specially constructed oven. To minimize temperature fluctuations, the flow tubing was going through the oven for 10 times before entering the measuring cell, making the temperature very stable, although slightly increasing the analysis time. The stabilization of the temperature was really important, as the analytical performance of the biosensor was anticipated to be greatly dependent on temperature (Pasic *et al*, 2007;Peedel & Rinken, 2012). At temperature 37^0C no enzyme denaturation could be detected and this temperature was used to gain the highest biosensor sensitivity. All measurements at different flow rates and different substrate concentrations were carried out in 0.1 M PB at pH 6.50 (the oxygen saturation concentration at 37^0C is 6.7 mg/ml).

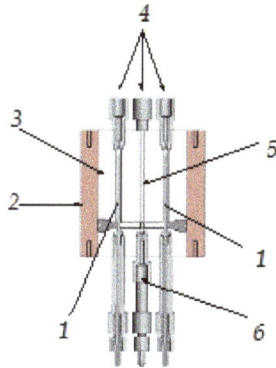

Figure 3. Schematic cross–section of the measuring cell 1 – glucose and oxygen optrodes, covered with nylon thread; 2 – cylindrical messing oven for the stabilization of temperature (± 0.02°C); 3 – measuring cell with flow channels; 4 – outflows; 5 – temperature sensor; 6 – inflow.

In case the biosensor signal parameters were studied in a standing liquid, the glucose assays were injected at the speed of 1.1 cm/sec. After each measurement the system was washed with 0.1 M PB (pH 6.50) until the sensor signals reached their initial values. The sensor output signal was recorded with the interval of 1 sec.

2.3. Data processing

The change of oxygen concentration was found as the difference between the signals of glucose and reference optrodes and normalized to bring the data from different sensors onto a common scale. From the reaction transient phase data, we calculated the total signal change parameter (at $t \rightarrow \infty$) using the earlier – proposed biosensor dynamic model, taking into account the ping-pong mechanism of enzyme kinetics, diffusion phenomena and the inertia of the signal transduction system (Rinken & Tenno, 2001). According to this model, the normalized oxygen concentration $c_{O_2}(t)/c_{O_2}(0)$ during the bio-recognition process in a biosensor is expressed as a 3-parameter function of time t:

$$\frac{c_{O_2}(t)}{c_{O_2}(0)} = A\exp\left(-Bt\right) + (1-A) - 2A\sum_{n=1}^{\infty}(-1)^n \frac{\tau_s}{n^2/B - \tau_s}\left[\exp\left(-Bt\right) - \exp\left(-n^2\frac{t}{\tau_s}\right)\right] \qquad (3)$$

where $c_{O_2}(t)$ is the biosensor output current at time moment t; $c_{O_2}(0)$ is the output current at the start of the reaction; t is time.

The parameter A is a complex coefficient, corresponding to the total possible biosensor signal change at the steady–state and parameter B is the initial maximal slope of process curve; both parameters A and B depend hyperbolically on substrate concentration; τ_s is the time constant of the transducer's response (Rinken & Tenno, 2001). Parameters A, B and τ_s are all independent on each other. According to the applied model, the total signal change parameter A is expressed as

$$ A = \frac{k^*_{cat}\left[E\right]_{total}c^{bulk}_s}{k^{O_2}_{diff}K_{O_2}K_s + (k^*_{cat}\left[E\right]_{total} + k^{O_2}_{diff}K_{O_2})c_{bulk}} \tag{4} $$

where k^*_{cat} is the apparent catalytic constant of the reaction; $[E]_{total}$ is the overall concentration of the immobilized enzyme; $k^{O_2}_{diff}$ is the apparent diffusion constant of the oxygen; K_{O_2} is the dissociation constant of the enzyme-oxygen complex; K_s the dissociation constant for the enzyme-substrate complex; and c_{bulk} is the substrate concentration in solution.

Additionally, from the transient phase data, collected between 20 to 60 seconds from the start of the reaction, the apparent maximal speed parameter of the reaction was calculated and used for the biosensor calibration. The starting moment of the reaction was determined experimentally for the particular measuring system: due to the length of the tubing the probe reached the optrodes after a time interval dependent on the flow speed. The values of all points on biosensor calibration curves are the results of at least 3 parallel measurements.

3. Results

3.1. Output of the biosensor system

The oxygen optrodes acting as oxygen transducers were employed to measure the rate of oxygen consumption in the enzymatic oxidation reactions. At fixed oxygen concentration, the response of the reference oxygen sensor is virtually constant with increasing glucose concentration, while the response of the glucose sensor decreases due to consumption of oxygen during glucose oxidation. The difference between the reference and the glucose sensor responses corresponds to the glucose concentration. Some examples of the signal curves of glucose and reference optrodes are shown in Fig. 4 (A).

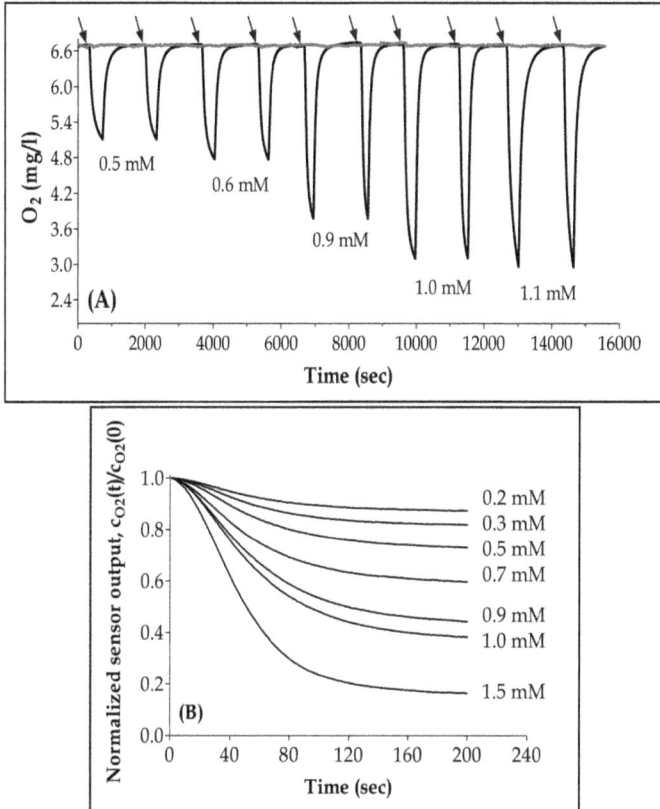

Figure 4. (A): Oxygen concentration responses obtained with the glucose oxidase optrode for different concentrations of glucose solution: — glucose biosensor response (black line); — reference optrode response (grey line) (at 37°C at flow rate 1.1 cm/sec in a 0.1 M phosphate buffer of pH 6.50). Arrows indicate the substrate adding time. **(B):** Normalized sensor outputs at different glucose concentrations (at 37°C at flow rate 1.1 cm/sec in 0.1 M phosphate buffer of pH 6.50).

For the detection of the starting moment of the bio-recognition reaction, the dependence of the time gap between the injection of the probe into the tubing and the probe front reaching the optrodes on the speed of the flow was determined. For the particular measuring cell, it was linearly dependent on the speed of the flow as expected:

$$t = 103.30 - (36.05 \pm 0.95)v \qquad (5)$$

where t is the lag period and v is the speed of the flow. Based on the value of this lag period for every measured flow speed, the biosensor data collected during this lag period was extracted from the databank, used for the calculation of the signal parameters at different glucose concentrations.

The response signal of the reference oxygen sensor was stable and its fluctuations did not exceed 1% of the working range of the sensor at any measured glucose concentration and flow speed. Still, to eliminate all potential experimental noise, the difference between the signals of the reference and the glucose sensor response was used to determine the normalized output of the system. An example of the normalized biosensor output curves at $v = 1.1$ cm/sec at different glucose concentrations are shown on Fig. 4 (B). These curves were used for the determination of the calibration parameters of the biosensor.

3.2. System regeneration

To use the bio-sensing system for real-time analysis, it is necessary to regenerate the system as quickly as possible. Regeneration involves passing a background flow of fluid without reactive components through the flowing system. The speed of cleaning of the flow system depends on the flow rate and slightly on the substrate concentration analyzed, as expected (Fig. 5).

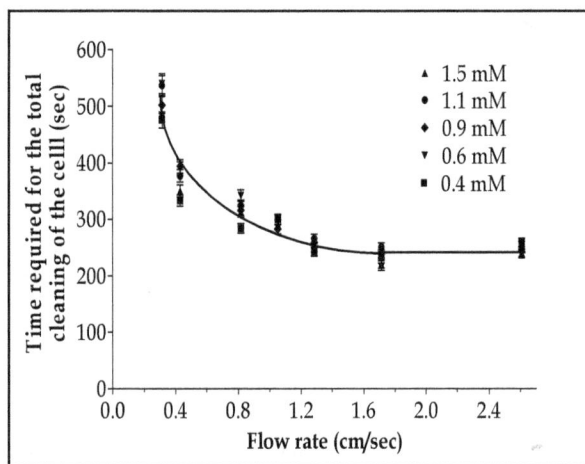

Figure 5. The speed of the cleaning of the biosensing system at different flow rates. Measurements were performed at 37°C in a 0.1 M phosphate buffer (pH 6.50). The values of all points are the results of at least 3 parallel measurements.

The flow rate was varied between 0.3 to 5.1 cm/sec. At lower flow rates (0.3 to 1.3 cm/sec) the system regeneration time increased by increasing of the flow rate. With further increase of the flow rate, the regeneration time became independent on the flow rate, which could be explained with the limits substrate diffusion in the GOD-containing threads. From Fig. 5, it can be seen that the regeneration time was also dependent on the substrate concentration. At lower flow rates (0.3 to 1.1 cm/sec) the dependence was clearly seen – increasing the substrate concentration the regeneration time also increased. At higher flow rates the regeneration time did not depend on the substrate concentration any more. From these results it could be concluded that the minimum required flow rate for system regeneration was at least 1.1 cm/sec. At this flow rate, the time for cleaning the system was 4.5 to 5 min. Studies to minimize the regeneration time required, are ongoing.

3.3. Calibration of the biosensor at different flow rates

The flow rate in the system affected the values of the reaction parameters and thus the sample throughput, biosensor sensitivity and detection limit. The choice of optimal flow rate is the presumption of obtaining accurate and reliable results in flow-through biosensor set-ups. At low flow rates, the apparent speed of the enzyme - catalyzed reaction, registered with a biosensor, is smaller than at high flow speeds, but the steady state signal can be calculated more accurately. For practical biosensor applications, it is important that the time required for the acquirement of the results and for biosensor regeneration is as short as possible.

The flow rate in the system was varied between 0 (stopping the flow for the measurements) and 5.1 cm/sec; at flow rates over 5.1 cm/sec the waste of reagents became unreasonable. In case the flow rates were below 0.3 cm/sec, the experimental noise was very big due to the air bubbles, gathering on the surfaces of the sensors and walls of the flow channels and the value of the signal to noise ratio was below 3.

3.3.1. Sensitivity of the biosensor system based on different calibration parameters

As described earlier, two different calibration parameters were used: the maximum signal change parameter A and the apparent maximal speed parameter v_{app}, determined as described in chapter 2.3. The biosensor calibration curves were made by plotting these parameters versus glucose concentration, as presented in Fig. 6 (A and B).

The glucose assay had a linear range up to 1.2 mM; at higher glucose concentrations the dependence became nonlinear. In case the measurements were carried out with the stopped flow, the biosensor showed linearity up to 0.8 mM. The linear part of these calibration curves and the values of slopes characterize the sensitivity of the biosensing system. Due to the different nature of the used calibration parameters, the dependences of the value of their slopes on flow rate are different (Fig. 7).

The maximum signal change parameter did not substantially depend on the flow rate in the range of the studied glucose concentrations (0.2 to 1.5 mM). Actually, this reaction parameter is also indifferent towards the determination of time, at which the analyte front reaches

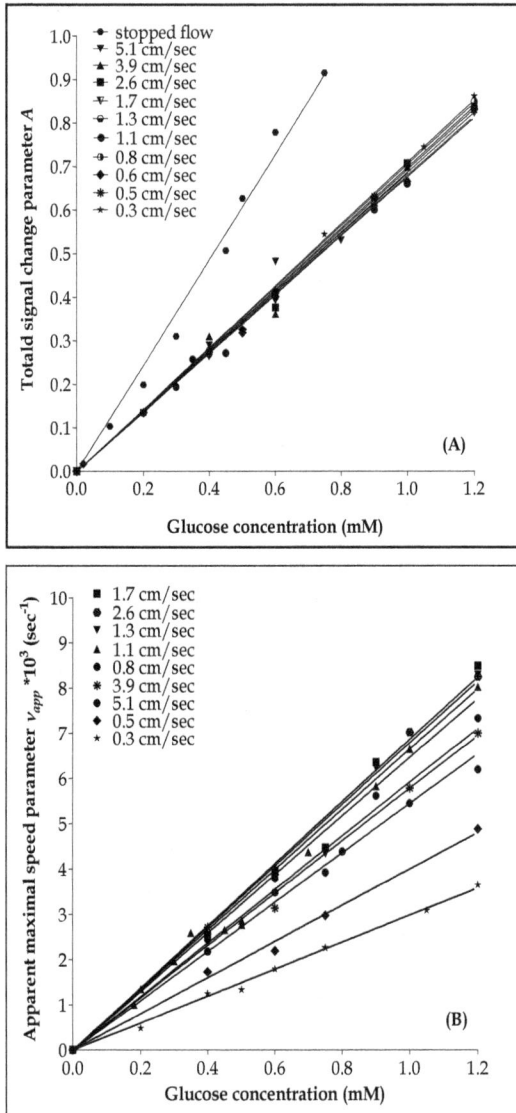

Figure 6. (A). Glucose calibration curves based on maximum signal change parameter A at different flow rates. Measurements were performed at 37°C in a 0.1 M phosphate buffer (pH 6.50). The values of all points are the results of at least 3 parallel measurements. **(B).** Glucose calibration curves based on the apparent maximal speed parameter v_{app} at different flow rates. Measurements were performed at 37°C in a 0.1 M phosphate buffer (pH 6.50). The values of all points are the results of at least 3 parallel measurements.

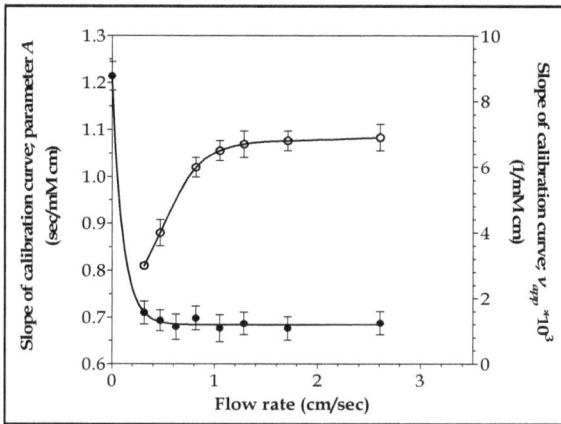

Figure 7. Dependence of the slopes of the calibration curves on different flow rates. The slopes were calculated from the calibration curves. On the left side (●) is the biosensor system response parameter A and on the right side (○) the apparent maximal speed parameter v_{app}.

the biosensor, as it is defined as a biosensor maximum signal change in steady-state conditions ($t \to \infty$). In case this parameter for the glucose oxidation reaction was measured in the standing medium (the flow was stopped), the values of this parameter at different glucose concentrations were significantly higher (Fig. 6A) and the slope of the calibration curve was about 1.7 times higher than it should be, if the same signal rising mechanism had been considered. Actually this increase has only a qualitative value, as the hydraulic stroke of halting the flow influences the parameter values. Actually, in the standing medium the diffusion layer of oxygen and glucose at the surface of the sensors is much thicker and the impact of the reaction kinetics of the measured signal is much bigger than in the flowing mediums. Due to the accumulation of air bubbles in the flow system at small flow rates, it was not possible to carry out experimental measurements at flow rates under 0.8 cm/sec and it is not clear, at which flow rates the signal rising mechanism changes.

The slope of biosensor calibration curve constructed with the apparent maximum speed parameter v_{app}, increases along with the increase of the flow rate until 1.1 cm/sec; at higher flow rates it reaches its maximum value and glucose calibration curves are similar. Thus, applying this parameter, the sensitivity of the biosensor can be modified according to the aim of analysis. As already pointed out, it was not possible to conduct measurements at flow rates under 0.8 cm/sec.

The flow rate also influences the biosensor response time. In standing solutions, 8 minutes were the minimal time of acquiring results with acceptable precision. So the flow rate of 1.3 cm/sec was chosen for the studies of the system repeatability, as it offers acceptable response time and sufficient sensitivity.

The repeatability of the experimental measurements was studied at glucose concentration of 0.5 mM (15 experiments per day and four days in a row). The repeatability of the measurements was very good considering that the standard deviation of the vertical distances of the points from the line $S_{y.x}$ was 0.0051 and the coefficient of determination R^2 was 98% (Fig.8). The results indicated the biosensor to exhibit a fairly analytical feature of repeatability.

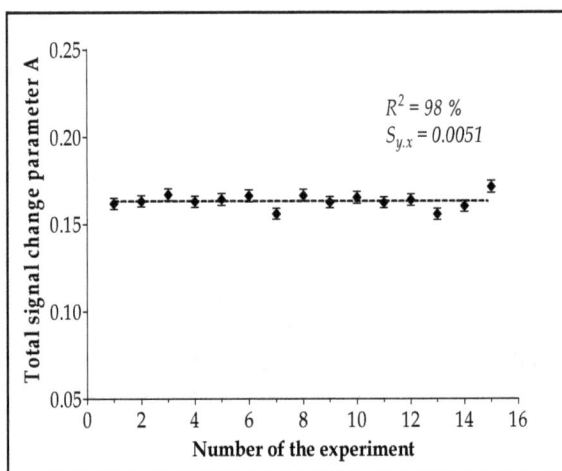

Figure 8. Repeatability of the measurements with the glucose biosensor. Measurements were carried out at 37°C in 0.5 mM glucose solutions in 0.1 M phosphate buffer (pH 6.50) at flow rate 1.3 cm/sec.

3.4. Operational stability of the biosensor

The loss of sensitivity under operational conditions is one of the most serious limits of the practical utility of biosensors. Besides possible leaching of the bio-selective material, the biosensors are ascribed to the inactivation and denaturation of their bioactive compounds. The operational stability of the present biosensor system was assessed by a continuous long-term experiment, in which we repeatedly analysed 0.5 mM glucose solutions. The biosensor system was in everyday exploitation – used for about a 15-measurement-serie per day – after which it was washed with 0.1 M PB (pH 6.50) and left overnight at 37°C. The initial activity of the sensor dropped for about 20% during the first 3 days; after that the biosensor response remained constant for over 35 days operation period with no significant loss of activity (Fig. 9).

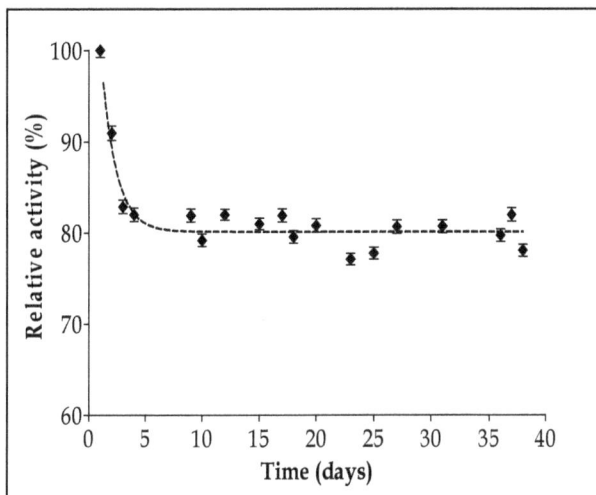

Figure 9. Stability of the biosensor with a GOD-containing nylon thread held at 37°C. Measurements were carried out in 0.5 mM glucose solutions in 0.1 M phosphate buffer (pH 6.50) at flow rate 1.3 cm/sec. The values of all points are the results of at least 3 parallel measurements.

4. Conclusions

A differential optrode based biosensor system for real-time monitoring of glucose in flow-through set-up has been studied and the selection of different calibration parameters analyzed. The influences of the flow rate and oxygen fluctuations on the sensor response have been studied. It was found that even at quite low flow rates the rising of the biosensor signal was controlled by diffusion and only in standing solutions the kinetics of the bio-recognition reaction had a substantial impact on the measurable output. The biosensor steady-state signal, calculated from the transient response was not dependent on the flow rate, if the latter exceeded 0.8 cm/sec.

The applied enzyme immobilizing procedure ensured a good operational stability of the system. Thus, an interference and cross-talk free device for the real-time monitoring of glucose concentration was successfully established. Used sensing system can be generalized for the other biologically important compounds catalyzed by oxidase-class enzymes and for the construction of biosensor arrays for different applications.

List of symbols

$c_{O_2}(t)$ Biosensor output current at time moment t

$c_{O_2}(0)$ Output current at the start of the reaction

t Time

A Complex coefficient, corresponding to the total possible biosensor signal change at the steady-state

B Initial maximal slope of process curve

τ_s Time constant of the transducer's response

v_{app} Apparent maximal speed parameter

$S_{y.x}$ Standard deviation of the vertical distances of the points from the line

Acknowledgements

This work was supported by Estonian Science Foundation grant No. 9061. Special thanks to Dr. R. Jaaniso and A. Floren for providing the oxygen optrodes and constructing the flow cell.

Author details

K. Kivirand, M. Kagan and T. Rinken

Institute of Chemistry, University of Tartu, Estonia

References

[1] Akin,M., Prediger,A., Yuksel,M., Höpfner,T., Demirkol,D.O., Beutel,S., Timur,S., & Scheper,T (2011) A new set up for multi-analyte sensing: At-line bio-process monitoring *Biosensors and Bioelectronics*, 26, 4532-4537.

[2] Baker,D.A & Gough,D.A (1996) Dynamic Delay and Maximal Dynamic Error in Continuous Biosensors *Analytical Chemistry*, 68, 1292-1297.

[3] Bankar,S.B., Bule,M.V., Singhal,R.S., & Ananthanarayan,L (2009) Glucose oxidase - An overview *Biotechnology Advances*, 27, 489-501.

[4] Baronas,D., Ivanauskas,F., & Baronas,R (2011) Mechanisms controlling the sensitivity of amperometric biosensors in flow injection analysis systems *Journal of Mathematical Chemistry*, 49, 1521-1534.

[5] Baronas,R., Ivanauskas,F., & Kulys,J (2002) Modelling dynamics of amperometric biosensors in batch and flow injection analysis *Journal of Mathematical Chemistry*, 32, 225-237.

[6] Betancor,L., Lopez-Gallego,F., Hidalgo,A., Alonso-Morales,N., Mateo,G.D.O.C., Fernandez-Lafuente,R., & Guisan,J.M (2006) Different mechanisms of protein immobilization on glutaraldehyde activated supports: Effect of support activation and immobilization conditions *Enzyme and Microbial Technology*, 39, 877-882.

[7] Bidmanova,S., Chaloupkova,R., Damborsky,J., & Prokop,Z (2010) Development of an enzymatic fiber-optic biosensor for detection of halogenated hydrocarbons *Analytical and Bioanalytical Chemistry*, 398, 1891-1898.

[8] Cao,L (2005) Immobilised enzymes: science or art? *Current Opinion in Chemical Biology*, 9, 217-226.

[9] Castillo,J., Gaspar,S., Leth,S., Niculescu,M., Mortari,A., Bontidean,I., Soukharev,V., Dorneanu,S.A., Ryabov,A.D., & Csöregi,E (2004) Biosensors for life quality - Design, development and applications *Sensors and Actuators, B: Chemical*, 102, 179-194.

[10] Chen,Y., Andersson,A., Mecklenburg,M., Xie,B., & Zhou,Y (2011) Dual-signal analysis eliminates requirement for milk sample pretreatment *Biosensors and Bioelectronics*, 29, 115-118.

[11] D.R.Walt (2006) Fiber Optic Array Biosensors *BioTechniques*, 41, 529-535.

[12] Gibson,T.D (1999) Biosensors: The stability problem *Analusis*, 27, 630-638.

[13] Gülce,H., Ataman,I., Gülce,A., & Yildiz,A (2002) A new amperometric enzyme electrode for galactose determination *Enzyme and Microbial Technology*, 30, 41-44.

[14] Hansen,E.H (1996) Principles and Applications of Flow Injection Analysis in Biosensors *Journal of Molecular Recognition*, 9, 316-325.

[15] Hartwell,S.K & Grudpan,K (2012) Flow-Based Systems for Rapid and High-Precision Enzyme Kinetics Studies *Journal of Analytical Methods in Chemistry*, 2012.

[16] Hu,N., Zhao,M.H., Chuang,J.L., & Li,L (2011) The application study of biosensors in environmental monitoring *Cross Strait Quad-Regional Radio Science and Wireless Technology Conference*, 2, 976-979.

[17] Isgrove,F.H., Williams,R.J.H., Niven,G.W., & Andrews,A.T (2001) Enzyme immobilization on nylon-optimization and the steps used to prevent enzyme leakage from the support *Enzyme and Microbial Technology*, 28, 225-232.

[18] Ivanauskas,F & Baronas,R (2008) Modelling an amperometric biosensor acting in a flowing liquid *International Journal for Numerical Methods in Fluids*, 56, 1313-1319.

[19] Jaaniso,R., Avarmaa,T., Suisalu,A., Floren,A., Ruudi,A., & Õige,K (2005) Stability of luminescence decay parameters in oxygen sensitive polymer films doped with Pd-porphyrins *Optical Materials and Applications, Proceedings of SPIE*, 5946, 1-10.

[20] Jia,N.Q., Zhang,Z.R., Zhu,J.Z., & Zhang,G.X (2004) Multianalyte biosensors for the simultaneous determination of glucose and galactose based on thin film electrodes *Chinese Chemical Letters*, 15, 322-325.

[21] Kivirand,K., Rebane,R., & Rinken,T (2011) A simple biosensor for biogenic diamines, comprising amine oxidase - Containing threads and oxygen sensor *Sensor Letters*, 9, 1794-1800.

[22] Kivirand,K & Rinken,T (2009) Preparation and Characterization of Cadaverine Sensitive Nylon Thread *Sensor Letters*, 7, 580-585.

[23] Kivirand,K & Rinken,T (2011) Biosensors for Biogenic Amines: The Present State of Art Mini-Review *Analytical Letters*, 44, 2821-2833.

[24] Konermann,L (1999) Monitoring Reaction Kinetics in Solution by Continuous-Flow Methods: The Effects of Convection and Molecular Diffusion under Laminar Flow Conditions *Journal of Physical Chemistry A*, 103, 7210-7216.

[25] Lammertyn,J., Verboven,P., Veraverbeke,E.A., Vermeir,S., Irudayaraj,J., & Nicolai,B.M (2006) Analysis of fluid flow and reaction kinetics in a flow injection analysis biosensor *Sensors and Actuators, B: Chemical*, 114, 728-736.

[26] Leung,A., Shankar,P.M., & Mutharasan,R (2007) A review of fiber-optic biosensors *Sensors and Actuators B: Chemical*, 125, 688-703.

[27] Li,L & Walt,D.R (1995) Dual-analyte fiber-optic sensor for the simultaneous and continuous measurement of glucose and oxygen *Analytical Chemistry*, 67, 3746-3752.

[28] Maestre,E., Katakis,I., Narvez,A., & Dominguez,E (2005) A multianalyte flow electrochemical cell: Application to the simultaneous determination of carbohydrates based on bioelectrocatalytic detection *Biosensors and Bioelectronics*, 21, 774-781.

[29] Marazuela,M.D & Moreno-Bondi,M.C (2002) Fiber-optic biosensors - An overview *Analytical and Bioanalytical Chemistry*, 372, 664-682.

[30] Mateo,C., Palomo,J.M., Fernandez-Lorente,G., Guisan,J.M., & Fernandez-Lafuente,R (2007) Improvement of enzyme activity, stability and selectivity via immobilization techniques *Enzyme and Microbial Technology*, 40, 1451-1463.

[31] Mehrvar,M & Abdi,M (2004) Recent developments, characteristics, and potential applications of electrochemical biosensors *Analytical Science*, 20, 1113-1126.

[32] Mello,L.D & Kubota,L.T (2002) Review of the use of biosensors as analytical tools in the food and drink industries *Food Chemistry*, 77, 237-256.

[33] Mishra,R.K., Dominguez,R.B., Bhand,S., Munoz,R., & Marty,J.L (2012) A novel automated flow-based biosensor for the determination of organophosphate pesticides in milk *Biosensors and Bioelectronics*, 32, 56-61.

[34] Nan,C., Zhang,Y., Zhang,G., Dong,C., Shuang,S., & Choi,M.M.F (2009) Activation of nylon net and its application to a biosensor for determination of glucose in human serum *Enzyme and Microbial Technology*, 44, 249-253.

[35] Õige,K., Avarmaa,T., Suisalu,A., & Jaaniso,R (2005) Effect of long-term aging on oxygen sensitivity of luminescent Pd-tetraphenylporphyrin/PMMA films *Sensors and Actuators, B: Chemical*, 106, 424-430.

[36] Pahujani,S., Kanwar,S.S., Chauhan,G., & Gupta,R (2008) Glutaraldehyde activation of polymer Nylon-6 for lipase immobilization: Enzyme characteristics and stability *Bioresource Technology*, 99, 2566-2570.

[37] Palmisano,F., Rizzi,R., Centonze,D., & Zambonin,P.G (2000) Simultaneous monitoring of glucose and lactate by an interference and cross-talk free dual electrode amperometric biosensor based on electropolymerized thin films *Biosensors and Bioelectronics*, 15, 531-539.

[38] Pasic,A., Koehler,H., Klimant,I., & Schaupp,L (2007) Miniaturized fiber-optic hybrid sensor for continuous glucose monitoring in subcutaneous tissue *Sensors and Actuators, B: Chemical*, 122, 60-68.

[39] Pasic,A., Koehler,H., Schaupp,L., Pieber,T.R., & Klimant,I (2006) Fiber-optic flow-through sensor for online monitoring of glucose *Analytical and Bioanalytical Chemistry*, 386, 1293-1302.

[40] Peedel,D & Rinken,T (2012) Effect of Temperature on the Catalytic Properties of Enzymes, Used in Lactose Cascade Biosensors and the Sensitivity of Lactose Biosensing System *Proceedings of the Estonian Academy of Sciences* Article in press.

[41] Polster,J., Prestel,G., Wollenweber,M., Kraus,G., & Gauglitz,G (1995) Simultaneous determination of penicillin and ampicillin by spectral fibre-optical enzyme optodes and multivariate data analysis based on transient signals obtained by flow injection analysis *Talanta*, 42, 2065-2072.

[42] Reder-Christ,K & Bendas,G (2011) Biosensor applications in the field of antibiotic research-a review of recent developments *Sensors*, 11, 9450-9466.

[43] Rinken,T & Tenno,T (2001) The dynamic signal lag of amperometric biosensors Characterisation of glucose biosensor output *Biosensors and Bioelectronics*, 16, 53-59.

[44] Sarma,A.K., Vatsyayan,P., Goswami,P., & Minteer,S.D (2009) Recent advances in material science for developing enzyme electrodes *Biosensors and Bioelectronics*, 24, 2313-2322.

[45] Sassolas,A., Blum,L.J., & Leca-Bouvier,B.D (2012) Immobilization strategies to develop enzymatic biosensors *Biotechnology Advances*, 30, 489-511.

[46] Segura-Ceniceros,E.P., Dabek,K.R., & Ilyina,A.D (2006) Invertase immobilization on nylon-6 activated by hydrochloric acid in the presence of glutaraldehyde as cross-linker *Vestnik Moskovskogo Universiteta, Seriya 2: Khimii*, 47, 143-148

[47] Surareungchai,W., Worasing,S., Sritongkum,P., Tanticharoen,M., & Kirtikara,K (1999) Dual electrode signal-subtracted biosensor for simultaneous flow injection determination of sucrose and glucose *Analytica Chimica Acta*, 380, 7-15.

[48] Thevenot,D.R., Toth,K., Durst,R.A., & Wilson,G.S (2001) Electrochemical biosensors: Recommended definitions and classification *Biosensors and Bioelectronics*, 16, 121-131.

[49] Tsai,H.c & Doong,R (2005) Simultaneous determination of pH, urea, acetylcholine and heavy metals using array-based enzymatic optical biosensor *Biosensors and Bioelectronics*, 20, 1796-1804.

[50] Viveros,L., Paliwal,S., McCrae,D., Wild,J., & Simonian,A (2006) A fluorescence-based biosensor for the detection of organophosphate pesticides and chemical warfare agents *Sensors and Actuators B: Chemical*, 115, 150-157.

[51] Yang,C., Zhang,Z., Shi,Z., Xue,P., Chang,P., & Yan,R (2010) Application of a novel co-enzyme reactor in chemiluminescence flow-through biosensor for determination of lactose *Talanta*, 82, 319-324.

[52] Zhou,X., Medhekar,R., & Toney,M.D (2003) A continuous-flow system for high-precision kinetics using small volumes *Analytical Chemistry*, 75, 3681-3687.

[53] Zhu,L., Li,Y., & Zhu,G (2002) A novel flow through optical fiber biosensor for glucose based on luminol electrochemiluminescence *Sensors and Actuators B: Chemical*, 86, 209-214.

Permissions

The contributors of this book come from diverse backgrounds, making this book a truly international effort. This book will bring forth new frontiers with its revolutionizing research information and detailed analysis of the nascent developments around the world.

We would like to thank Dr. Toonika Rinken, for lending his expertise to make the book truly unique. He has played a crucial role in the development of this book. Without his invaluable contribution this book wouldn't have been possible. He has made vital efforts to compile up to date information on the varied aspects of this subject to make this book a valuable addition to the collection of many professionals and students.

This book was conceptualized with the vision of imparting up-to-date information and advanced data in this field. To ensure the same, a matchless editorial board was set up. Every individual on the board went through rigorous rounds of assessment to prove their worth. After which they invested a large part of their time researching and compiling the most relevant data for our readers. Conferences and sessions were held from time to time between the editorial board and the contributing authors to present the data in the most comprehensible form. The editorial team has worked tirelessly to provide valuable and valid information to help people across the globe.

Every chapter published in this book has been scrutinized by our experts. Their significance has been extensively debated. The topics covered herein carry significant findings which will fuel the growth of the discipline. They may even be implemented as practical applications or may be referred to as a beginning point for another development. Chapters in this book were first published by InTech; hereby published with permission under the Creative Commons Attribution License or equivalent.

The editorial board has been involved in producing this book since its inception. They have spent rigorous hours researching and exploring the diverse topics which have resulted in the successful publishing of this book. They have passed on their knowledge of decades through this book. To expedite this challenging task, the publisher supported the team at every step. A small team of assistant editors was also appointed to further simplify the editing procedure and attain best results for the readers.

Our editorial team has been hand-picked from every corner of the world. Their multi-ethnicity adds dynamic inputs to the discussions which result in innovative outcomes. These outcomes are then further discussed with the researchers and contributors who give their valuable feedback and opinion regarding the same. The feedback is then collaborated with the researches and they are edited in a comprehensive manner to aid the understanding of the subject.

Apart from the editorial board, the designing team has also invested a significant amount of their time in understanding the subject and creating the most relevant covers. They scrutinized every image to scout for the most suitable representation of the subject and create an appropriate cover for the book.

The publishing team has been involved in this book since its early stages. They were actively engaged in every process, be it collecting the data, connecting with the contributors or procuring relevant information. The team has been an ardent support to the editorial, designing and production team. Their endless efforts to recruit the best for this project, has resulted in the accomplishment of this book. They are a veteran in the field of academics and their pool of knowledge is as vast as their experience in printing. Their expertise and guidance has proved useful at every step. Their uncompromising quality standards have made this book an exceptional effort. Their encouragement from time to time has been an inspiration for everyone.

The publisher and the editorial board hope that this book will prove to be a valuable piece of knowledge for researchers, students, practitioners and scholars across the globe.

List of Contributors

Donna E. Crone, Yao-Ming Huang, Derek J. Pitman, Christian Schenkelberg, Keith Fraser, Stephen Macari and Christopher Bystroff
Department of Biology, Rensselaer Polytechnic Institute, Troy, New York, USA

Xiuyun Wang
School of Chemistry, Dalian University of Technology, Dalian, China

Shunichi Uchiyama
Saitama Institute of Technology Fukaya, Saitama, Japan

Joanna Cabaj and Jadwiga Sołoducho
Faculty of Chemistry, Wrocław University of Technology, Wrocław, Poland

M. B. de la Mora
Instituto de Física., Universidad Nacional Autónoma de México. Circuito de la Investigación Científica Ciudad Universitaria, México

M. Ocampo, R. Doti, J. E. Lugo and J. Faubert
Visual Psychophysics and Perception Laboratory, School of Optometry, University of Montreal, Canada

Shengbo Sang, Wendong Zhang and Yuan Zhao
MicroNano System Research Center, Taiyuan University of Technology, Shanxi, China
Key Lab of Advanced Transducers and Intelligent Control System of the Ministry of Education, Taiyuan University of Technology, Shanxi, China

Tatiana Duque Martins, Antonio Carlos Chaves Ribeiro, Henrique Santiago de Camargo, Paulo Alves da Costa Filho, Hannah Paula Mesquita Cavalcante and Diogo Lopes Dias
Chemistry Institute, Campus II, Federal University of Goias, Goiania, Brazil

Kazuo Nakazato
Department of Electrical Engineering and Computer Science, Graduate of Engineering, Nagoya University, Nagoya, Japan

Tran Quang Hung
Department of Materials Science and Engineering, Chungnam National National University, Daejeon,
Korea
Laboratoire Charles Coulomb, CNRS-University Montpellier 2, Montpellier, France

Dong Young Kim
Department of Physics, Andong National University, Andong, Korea

B. Parvatheeswara Rao
Department of Physics, Andhra University, Visakhapatnam, India

CheolGi Kim
Department of Materials Science and Engineering, Chungnam National National University, Daejeon, Korea

Hilmiye Deniz ErtuğruL
Ondokuz Mayıs University, Turkey

Zihni Onur Uygun
Çanakkale Onsekiz Mart University, Turkey

Jaime Punter Villagrasa, Jordi Colomer-Farrarons and Pere Ll. Miribel
Department of Electronics, Bioelectronics and Nanobioengineering Research Group (SIC-BIO), University of Barcelona, Spain

Sameh Kessentini and Dominique Barchiesi
Automatic Mesh Generation and Advanced Methods (GAMMA3 - UTT/INRIA) – Charles Delaunay Institute - University of Technology of Troyes - Troyes, France

María Isabel Rocha Gaso
Wave Phenomena Group, Department of Electroic Engineering, Universitat Politècnica de València, Spain
Sensors, Microsystems and Actuators Laboratory of Louvain (SMALL), ICTEAM Institute, Université Catholique de Louvain, Belgium

Antonio Arnau and Yolanda Jiménez
Wave Phenomena Group, Department of Electroic Engineering, Universitat Politècnica de València, Spain

Laurent A. Francis
Sensors, Microsystems and Actuators Laboratory of Louvain (SMALL), ICTEAM Institute, Université Catholique de Louvain, Belgium

K. Kivirand, M. Kagan and T. Rinken
Institute of Chemistry, University of Tartu, Estonia

www.ingramcontent.com/pod-product-compliance
Lightning Source LLC
Chambersburg PA
CBHW070718190326
41458CB00004B/1017